Transactions on Computational Science and Computational Intelligence

Series Editor
Hamid R. Arabnia, Department of Computer Science
The University of Georgia
Athens, GA, USA

Computational Science (CS) and Computational Intelligence (CI) both share the same objective: finding solutions to difficult problems. However, the methods to the solutions are different. The main objective of this book series, "Transactions on Computational Science and Computational Intelligence", is to facilitate increased opportunities for cross-fertilization across CS and CI. This book series publishes monographs, professional books, contributed volumes, and textbooks in Computational Science and Computational Intelligence. Book proposals are solicited for consideration in all topics in CS and CI including, but not limited to, Pattern recognition applications; Machine vision; Brain-machine interface; Embodied robotics; Biometrics; Computational biology; Bioinformatics; Image and signal processing; Information mining and forecasting; Sensor networks; Information processing; Internet and multimedia; DNA computing; Machine learning applications; Multi-agent systems applications; Telecommunications; Transportation systems; Intrusion detection and fault diagnosis; Game technologies; Material sciences; Space, weather, climate systems, and global changes; Computational ocean and earth sciences; Combustion system simulation; Computational chemistry and biochemistry; Computational physics; Medical applications; Transportation systems and simulations; Structural engineering; Computational electro-magnetic; Computer graphics and multimedia; Face recognition; Semiconductor technology, electronic circuits, and system design; Dynamic systems; Computational finance; Information mining and applications; Biometric modeling; Computational journalism; Geographical Information Systems (GIS) and remote sensing; Ubiquitous computing; Virtual reality; Agent-based modeling; Computational psychometrics; Affective computing; Computational economics; Computational statistics; and Emerging applications.

For further information, please contact Mary James, Senior Editor, Springer, mary.james@springer.com.

Manuel Mora • Fen Wang
Jorge Marx Gomez • Hector Duran-Limon
Editors

Development Methodologies for Big Data Analytics Systems

Plan-driven, Agile, Hybrid, Lightweight Approaches

Springer

Editors
Manuel Mora
Information Systems
Autonomous University of Aguascalientes
Aguascalientes, Mexico

Fen Wang
Information Technology and Administrative Management
Central Washington University
Ellensburg, WA, USA

Jorge Marx Gomez
Informatics
University of Oldenburg
Oldenburg, Germany

Hector Duran-Limon
Information Systems
University of Guadalajara
Zapopan, Mexico

ISSN 2569-7072 ISSN 2569-7080 (electronic)
Transactions on Computational Science and Computational Intelligence
ISBN 978-3-031-40955-4 ISBN 978-3-031-40956-1 (eBook)
https://doi.org/10.1007/978-3-031-40956-1

This Springer imprint is published by the registered company Springer Nature Switzerland AG
The registered company address is: Gewerbestrasse 11, 6330 Cham, Switzerland

Paper in this product is recyclable.

Editorial Preface

Big Data Analytics (BDA) systems are software systems developed to provide valuable insights to decision-makers exploiting Big Data sources [1, 2]. Successful BDA systems have been reported in the literature [3] in diverse domains such as Healthcare, Logistics, Finance, Marketing, Retail, and Education in the last decade [4].

BDA systems are the main outcomes of the new Data Science discipline [5–7] that emerged as a result of the convergence of Statistics, Computer Science, and Business Intelligent Analytics with the practical aim to provide concepts, models, methods, and tools required for exploiting the wide variety, volume, and velocity of available business internal and external data – i.e. Big Data – to lately provide decision-making value to decision-makers [8]. "Through Data Science, one can identify relevant issues, collect data from various data sources, integrate the data, conclude solutions, and communicate the results to improve and enhance organizations' decisions and deliver value to users and organizations" [7; pp. v].

BDA systems have been mainly developed and used for large business organizations due to the nature of the implicated human, technological, organizational, and data resources required for such developments [9, 10]. Additionally, it has been recently identified that the systematic development of BDA systems has not been usually pursued by organizations, and despite the adaptation of a few comprehensive development methodologies for Data Analytics systems [11] such as CRISP-DM, SEMMA, and KDD, many failed BDA system development projects are frequent [12]. From a Systems and Software Engineering perspective, the utilization of software processes and development methodologies – plan-driven, agile, hybrid, and lightweight types – are necessary to fit the expected "Iron Triangle" metrics of schedule, budget, and quality [13–15]. Hence, initial top research has realized the need to incorporate software and systems engineering development methods for complying the business expectations of BDA systems [16, 17].

This research-oriented co-edited book entitled *Development Methodologies for Big Data Analytics Systems – Plan-Driven, Agile, Hybrid, Lightweight Approaches* contributes to advance on this relevant current research problem through the study of development methodologies for BDA systems based on plan-driven, agile, hybrid, and lightweight approaches [16–21].

For this aim, we asked to the international research communities both on Software Engineering and Data Science disciplines to submit high-quality chapters on the following themes:

- *Foundations on Big Data Analytics Systems.* Topics: Big Data Analytics foundations; Big Data Science foundations; Big Data Analytics Systems Frameworks; Big Data Analytics Systems Architectures; Big Data Analytics Tools and Platforms; Big Data Analytics Computational Techniques.
- *Development Methodologies for Big Data Analytics Systems.* Topics: Review of specific plan-driven methodologies such as CRISP-DM, SEMMA, KDD, and generic ones used for Big Data Analytics Systems as RUP, MBASE, and MSF; review of specific agile, hybrid, and lightweight methodologies based on Scrum, XP, ISO/IEC 29110, and Microsoft TDSP and combinations from them.
- *Applications, Challenges, and Future Directions of Big Data Analytics Systems.* Topics: Real-world applications in diverse domains such as Healthcare, Marketing, Financial, Education, Sports, Retail, Logistics, Manufacturing, among others; review of challenges, current problems and limitations, trends, and future directions.

After a six-month double-cycle blind-mode peer-based review process, 11 high-quality chapters were accepted for final publishing in this book from international researchers located in the USA, Germany, and Mexico. The first three chapters provide research-oriented content on foundations for understanding the technological bases of Big Data Analytics systems through a selective review of open sources tools, a review of machine learning processing mechanisms, and a data value ontology. The next five chapters correspond to the core content of this book regarding studies on methods, methodologies, and frameworks for developing and implementing successfully Big Data Analytics systems. These chapters provide updated research on software requirement management tools, comparison of lead traditional, lightweight, and agile development methodologies, as well as implementation frameworks. Finally, the last three chapters address relevant illustrative application cases in the medical, supply chains, and other relevant business domains.

Hence, we consider that these 11 high-quality chapters are useful mainly for researchers, academics, and practitioners interested in the systematic development of Big Data Analytics systems, as well as for Ph.D. involved in these relevant topics.

Aguascalientes, Mexico — Manuel Mora
Ellensburg, WA, USA — Fen Wang
Oldenburg, Germany — Jorge Marx Gómez
Zapopan, Mexico — Hector Duran-Limon

References

1. Laney, D.: 3-D Data Management: Controlling Data Volume, Velocity and Variety. META Group Research File 949 (2001)
2. Davoudian, A., Liu, M.: Big Data systems: A software engineering perspective. ACM Comput. Surv. **53**(5), 1–39 (2020)
3. Davenport, T.H.: Competing on analytics. Harv. Bus. Rev. **84**(1), 98–107 (2006)
4. Watson, H.J.: Tutorial: Big Data analytics: Concepts, technologies, and applications. Commun. Assoc. Inf. Syst. **34**(1), 1247–1268 (2014)
5. Cao, L.: Data science: Challenges and directions. Commun. ACM. **60**(8), 59–68 (2017)
6. Weihs, C., Ickstadt, K.: Data science: The impact of statistics. Int. J. Data Sci. Anal. **6**(3), 189–194 (2018)
7. Arabnia, H.R., Daimi, K., Stahlbock, R., Soviany, C., Heilig, L., Brüssau, K. (eds.): Principles of Data Science. Springer (2020)
8. Mikalef, P., Pappas, I.O., Krogstie, J., Giannakos, M.: Big data analytics capabilities: A systematic literature review and research agenda. Inf. Syst. e-Bus. Manag. **16**(3), 547–578 (2018)
9. Maroufkhani, P., Ismail, W.K.W., Ghobakhloo, M.: Big Data analytics adoption model for small and medium enterprises. J. Sci. Technol. Policy Manag. **11**(4), 483–513 (2020)
10. Davenport, T., Bean, R.: The Quest to achieve data-driven leadership: a progress report on the state of corporate data initiatives – foreword. Special report, New advantage partners (2022)
11. Martinez, I., Viles, E., Olaizola, I.G.: Data science methodologies: current challenges and future approaches. Big Data Res. **24**, 100183 (2021)
12. Davenport, T., Malone, K.: Deployment as a critical business data science discipline. Harvard Data Sci. Rev. (2021). https://doi.org/10.1162/99608f92.90814c32
13. Humphrey, W.S.: The software process: Global goals. In: Software Process Workshop, pp. 35–42. Springer, Berlin/Heidelberg (2005)
14. Agarwal, N., Rathod, U.: Defining 'success' for software projects: an exploratory revelation. Int. J. Proj. Manag. **24**(4), 358–370 (2006)
15. Humphrey, W.S., Konrad, M.D., Over, J.W., Peterson, W.C.: Future directions in process improvement. Crosstalk–J. Def. Softw. Eng. **20**(2), 17–22 (2007)
16. Martínez-Plumed, F., Contreras-Ochando, L., Ferri, C., Orallo, J.H., Kull, M., Lachiche, N., et al.: CRISP-DM twenty years later: From data mining processes to data science trajectories. IEEE Trans. Knowl. Data Eng. **33**(8), 3048–3061 (2019)
17. Haakman, M., Cruz, L., Huijgens, H., van Deursen, A.: AI lifecycle models need to be revised. Emp. Softw. Eng. **26**(5), 1–29 (2021)
18. Beck, K.: Embracing change with extreme programming. Computer. **32**(10), 70–77 (1999)

19. Boehm, B., Turner, R.: Using risk to balance agile and plan-driven methods. Computer. **36**(6), 57–66 (2003)
20. Sutherland, J.: Jeff Sutherland's Scrum Handbook. Scrum Training Institute, Boston (2010)
21. ISO/IEC: ISO/IEC TR 29110–5–1-2:2011 Software Engineering - Lifecycle Profiles for Very Small Entities (VSES) - Part 5-1-2: Management and Engineering Guide: Generic Profile Group: Basic Profile. ISO - International Organization for Standardization (2011)

Acknowledgments

We thank our following colleagues for their academic effort spent in the blind-mode double-cycle review process:

- Dr. Ahmad Alnafoosi, DePaul University, USA
- Dr. Alena Buchalcevova, Prague University of Economics and Business, Czech Republic
- Dr. Angel Muñoz-Zavala, Autonomous University of Aguascalientes, Mexico
- Dr. Christoph Wunck, Emden/Leer University of Applied Sciences, Germany
- Dr. Dora Gonzalez-Bañales, Durango Institute of Technology, Mexico
- Dr. Efosa Idemudia, Arkansas Tech University, USA
- Dr. Fen Wang, Central Washington University, USA
- Dr. Francisco Alvarez-Rodriguez, Autonomous University of Aguascalientes, Mexico
- Dr. Gabriela C. Lopez-Torres, Autonomous University of Aguascalientes, Mexico
- Dr. Hector Duran-Limon, University of Guadalajara, Mexico
- Dr. Hermilo Sanchez-Cruz, Autonomous University of Aguascalientes, Mexico
- Dr. Jairo Gutierrez, Auckland University of Technology, New Zealand
- Dr. Jochen Leidner, Coburg University of Applied Sciences, Germany
- Dr. Lizeth Solano-Romo, Autonomous University of Aguascalientes, Mexico
- Dr. Mahesh Raisinghani, Texas Woman´s University, USA
- Dr. Michael Reiche, Coburg University of Applied Sciences, Germany
- Dr. Olayele Adelakun, DePaul University, USA
- Dr. Sergio Galvan-Cruz, CIMAT, Zacatecas
- MSc. David Montoya-Murillo, Autonomous University of Aguascalientes, Mexico
- MSc. Dirk Bendlin, University of Oldenburg, Germany
- MSc. Gerardo Salazar-Salazar, Autonomous University of Aguascalientes, Mexico

We thank following colleagues for your important contribution to this research-oriented book:

- A. Kucewicz, Ramboll Deutschland GmbH, Hamburg, Germany
- Ahmad Alnafoosi, DePaul University, USA

- Angel Muñoz-Zavala, Autonomous University of Aguascalientes, Mexico
- Arturo Chavoya, University of Guadalajara, Mexico
- Cesar Muñoz-Chavez, Autonomous University of Aguascalientes, Mexico
- Christoph Wunck, Emden/Leer University of Applied Sciences, Germany
- David Montoya-Murillo, Autonomous University of Aguascalientes, Mexico
- Dirk Bendlin, University of Oldenburg, Germany
- Efosa Idemudia, Arkansas Tech University, USA
- Fen Wang, Central Washington University, USA
- Francisco Alvarez-Rodriguez, Autonomous University of Aguascalientes, Mexico
- Gerardo Salazar-Salazar, Autonomous University of Aguascalientes, Mexico
- Gloria Phillips-Wren, Loyola University Maryland, USA
- H. Kaddoura, Ramboll Deutschland GmbH, Hamburg, Germany
- Hector Duran-Limon, University of Guadalajara, Mexico
- Hermilo Sanchez-Cruz, Autonomous University of Aguascalientes, Mexico
- Hermilo Sanchez-Cruz, Autonomous University of Aguascalientes, Mexico
- Humberto Sossa-Azuela, Instituto Politécnico Nacional, Mexico
- Jeffrey Yi-Lin Forrest, Slippery Rock University, USA
- Jochen Leidner, Coburg University of Applied Sciences, Germany
- Jonas Kallisch, Emden/Leer University of Applied Sciences, Germany
- Jorge Marx-Gómez, University of Oldenburg, Germany
- Julio Ponce-Gallegos, Autonomous University of Aguascalientes, Mexico
- Lizeth Solano-Romo, Autonomous University of Aguascalientes, Mexico
- M. Werther Häckell, Ramboll Deutschland GmbH, Hamburg, Germany
- Mahesh Raisinghani, Texas Woman´s University, USA
- Manuel Mora, Autonomous University of Aguascalientes, Mexico
- Martha Hernandez-Ochoa, University of Guadalajara, Mexico
- Michael Reiche, Coburg University of Applied Sciences, Germany
- Olayele Adelakun, DePaul University, USA
- Paola Reyes-Delgado, Autonomous University of Aguascalientes, Mexico
- Sergio Galvan-Cruz, CIMAT, Zacatecas
- Tiko Iyamu, Cape Peninsula University of Technology, South Africa

We also thank Prof. Dr. Hamid Arabnia, Editor of the Springer Nature - Book Series: Transactions on Computational Science & Computational Intelligence, for all academic support provided for the realization of this book.

Finally, we express our gratitude to Mary James, Senior Editor at Springer Book Series, and her Editorial Staff for all guidance provided for the publication of this book.

Contents

About the Editors

Manuel Mora is a full-time Professor in the Information Systems Department at the Autonomous University of Aguascalientes (UAA), Mexico. Dr. Mora holds an M.Sc. in Computer Sciences (Artificial Intelligence area, 1989) from Monterrey Tech (ITESM), and an Eng.D. in Engineering (Systems Engineering area, 2003) from the National Autonomous University of Mexico (UNAM). He has published over 90 research papers in international top conferences, research books, and JCR indexed journals such as *IEEE-TSMC, European Journal of Operational Research, International Journal of Information Management, Engineering Management, International Journal of Information Technology and Decision Making, Information Technology for Development, International Journal in Software Engineering and Knowledge Engineering, Computer Standards & Interfaces, Software Quality Journal, Expert Systems*, and *Software and Systems Modeling*. Dr. Mora is a senior member of ACM (since 2008), an SNI at Level II, and serves in the ERB of several international journals indexed by Emergent Source Citation Index focused on decision-making support systems (DMSS) and IT services systems.

Fen Wang is a full Professor in the Information Technology and Administrative Management Department at Central Washington University (CWU). Before joining CWU, Prof. Wang was an Assistant Professor and Director of the Management Information Systems (MIS) program at the Eastern Nazarene College in Massachusetts. Prof. Wang holds a B.S. in MIS and an M.S. and a Ph.D. in Information Systems from the University of Maryland Baltimore County. Prof. Wang has brought over 10 years of professional and research experience in information technology management to her students. Her research focuses on intelligent decision support technologies and E-business strategies. These efforts have resulted in contributions to the applied literature on information technologies that have been well received in the research community. Prof. Wang has published over 30 papers in internationally circulated journals and book series, including the *International Journal of E-Business Research* (IJEBR), *International Journal of Decision Support System Technology* (IJDSST), *Intelligent Decision Technologies* (IDT), *Information*

Technology for Development (ITFD), and the *Encyclopedia of E-Commerce, E-Government and Mobile Commerce*. Prof. Wang has also consulted for a variety of public and private organizations on IT management and applications.

Jorge Marx Gómez studied Computer Engineering and Industrial Engineering at the University of Applied Sciences Berlin (Technische Fachhochschule Berlin). He was a Lecturer and Researcher at the Otto-von-Guericke-Universität Magdeburg (Germany) where he also obtained a Ph.D. degree in Business Information Systems with the work Computer-based Approaches to Forecast Returns of Scrapped Products to Recycling. From 2002 to 2003, he was a visiting Professor of Business Informatics at the Technical University of Clausthal (TU Clausthal, Germany). In 2004, he received his habilitation for the work Automated Environmental Reporting through Material Flow Networks at the Otto-von-Guericke-Universität Magdeburg. In 2005 he became a full Professor and Chair of Business Information Systems at the Carl von Ossietzky University Oldenburg (Germany). His research interests include Very Large Business Applications, Business Information Systems, Federated ERP-Systems, Business Intelligence, Data Warehousing, Interoperability, and Environmental Management Information Systems.

Hector Duran-Limon, PhD is currently a full Professor at the Information Systems Department, University of Guadalajara, Mexico. He completed a Ph.D. at Lancaster University, England, in 2002. Following this, he was a Post-doctoral Researcher until December 2003. He obtained an IBM Faculty award in 2008. His research interests include Cloud Computing and High-Performance Computing (HPC). He is also interested in Software Architecture, Software Product Lines, and Component-based Development. In 2006, he was invited to create a Ph.D. program in Information Technologies for the University of Guadalajara, becoming a member of the Academic Council. Contact him at the Information Systems Department, University of Guadalajara, Mexico; hduran@cucea.udg.mx.

Contributors

Olayele Adelakun School of Computing, DePaul University, Chicago, IL, USA

Ahmad B. Alnafoosi School of Computing, DePaul University, Chicago, IL, USA

Francisco Javier Alvarez-Rodriguez Autonomous University of Aguascalientes, Aguascalientes, Mexico

Dirk Bendlin Carl von Ossietzky University of Oldenburg, Oldenburg, Germany

Arturo Chavoya Information Systems, CUCEA, University of Guadalajara, Zapopan, Mexico

Hector A. Duran-Limon Information Systems, CUCEA, University of Guadalajara, Zapopan, Mexico

Jeffrey Yi-Lin Forrest Slippery Rock University, Slippery Rock, PA, USA

Sergio Galvan-Cruz Software Engineering Unit, CIMAT, Zacatecas, Mexico

M. Werther Häckell Ramboll Deutschland GmbH, Hamburg, Germany

Martha Hernandez-Ochoa Knowledge Fundamentals, CUNORTE, University of Guadalajara, Zapopan, Mexico

Efosa C. Idemudia Department of Management & Marketing, Arkansas Tech University, Russellville, AR, USA

Tiko Iyamu Cape Peninsula University of Technology, Cape Town, South Africa

H. Kaddoura Ramboll Deutschland GmbH, Hamburg, Germany

Jonas Kallisch Emden/Leer University of Applied Sciences, Emden, Germany

A. Kucewicz Ramboll Deutschland GmbH, Hamburg, Germany

Jochen L. Leidner Coburg University of Applied Sciences, Coburg, Germany

Jorge Marx-Gómez Department of Informatics, University of Oldenburg, Oldenburg, Germany

David Montoya-Murillo Autonomous University of Aguascalientes, Aguascalientes, Mexico

Manuel Mora Information Systems, Autonomous University of Aguascalientes, Aguascalientes, Mexico

Cesar Muñoz-Chavez Computer Science, Autonomous University of Aguascalientes, Aguascalientes, Mexico

Angel Muñoz-Zavala Autonomous University of Aguascalientes, Aguascalientes, Mexico

Gloria Phillips-Wren Loyola University Maryland, Baltimore, MD, USA

Julio Ponce-Gallegos Autonomous University of Aguascalientes, Aguascalientes, Mexico

Mahesh S. Raisinghani Department of Business & Economics, Texas Woman´s University, Denton, TX, USA

Michael Reiche Coburg University of Applied Sciences, Coburg, Germany

Paola Yuritzy Reyes-Delgado Information Systems, Autonomous University of Aguascalientes, Aguascalientes, Mexico

Gerardo Salazar-Salazar Autonomous University of Aguascalientes, Aguascalientes, Mexico

Hermilo Sanchez-Cruz Autonomous University of Aguascalientes, Aguascalientes, Mexico

Lizeth I. Solano-Romo Information Systems, Autonomous University of Aguascalientes, Aguascalientes, Mexico

Humberto Sossa-Azuela Centro de Investigación en Computación, Instituto Politécnico Nacional, Mexico City, Mexico

Fen Wang Department of Information Technology & Administrative Management, Central Washington University, Ellensburg, WA, USA

Christoph Wunck Emden/Leer University of Applied Sciences, Emden, Germany

Open Source IT for Delivering Big Data Analytics Systems as Services: A Selective Review

Manuel Mora, Paola Yuritzy Reyes-Delgado, Sergio Galvan-Cruz, and Lizeth I. Solano-Romo

1 Introduction

Big data analytics systems (BDAS) are software systems characterized fundamentally by the utilization of huge and heterogenous dataset inputs, supported by advanced information technology (IT) for ingestion, storage, processing, and end-user presentation activities, and using advanced mathematical, statistical, machine learning, and intelligent heuristics techniques with an analytics purpose – i.e., descriptive, predictive, or prescriptive – to support business decision-making [1–3].

The notion of big data can be tracked to the late 1990s and early 2000s [4–6], but until the 2010–2020 decade, big data was identified as a business strategic resource [7, 8]. In [4], the term big data was used to caution about the need of increasing the storage capacities as well as the development of new algorithms to process huge data generated for scientific visualizations. In [5], it was reported the problem of "IT infrastructure stress" which is characterized by the insufficient processing, memory, and storage levels of capacity for coping with the availability of huge multimedia data – i.e., graphs, images, audios, and videos. In [6], the concept of big data is not explicitly mentioned, but in the 3-V – volume, velocity, and variety – big data model is presented due to the expansion of e-commerce web-based platforms. In [7], the 3-V characteristics of big data were supported, and three key recommendations to cope with big data were proposed: first, to shift from processing data stocks to processing data flows; second, to hire data scientists with strong IT skills to complement business analytics users; and third, to extend the internal IT platforms with

M. Mora (✉) · P. Y. Reyes-Delgado · L. I. Solano-Romo
Information Systems, Autonomous University of Aguascalientes, Aguascalientes, Mexico
e-mail: jose.mora@edu.uaa.mx

S. Galvan-Cruz
Software Engineering Unit, CIMAT, Zacatecas, Mexico

M. Mora et al. (eds.), *Development Methodologies for Big Data Analytics Systems*, Transactions on Computational Science and Computational Intelligence, https://doi.org/10.1007/978-3-031-40956-1_1

external IT ecosystems. In [8], the concept of big data is claimed as a critical input to improving the business decision-making process due to its 3-V characteristics. Nowadays, the concept of big data has been extended to "a holistic approach to manage, process and analyze 5 Vs (i.e., volume, variety, velocity, veracity and value) in order to create actionable insights for sustained value delivery, measuring performance and establishing competitive advantages." [9; p. 235].

The concept of analytics [10] is older than the concept of big data. From 1960 to 2000, the concept of analytics was used in the economy science [10] to describe a research method where a problem was formulated with a set of equations, and their analytical and/or approximated solutions were reported including analysis of scenarios. In the next period 2001–2010, analytics concept emerged finally as a relevant approach fostered by the convergence of artificial intelligence and statistics methods used for knowledge data discovery tasks [11]. Analytics was defined in [12; p. 98] as the organizational ability to "collect, analyze, and act on data." From an IT technical perspective, this study [12] identifies the need for a data strategy, analytics software, and adequate computing IT resources. Nowadays, analytics can be defined as "the scientific process of transforming data into insight for making better decisions" [13].

Big data and analytics have converged in the last decade [1–3, 14–16], and currently big data analytics approach is fundamental for any modern business organization for supporting data-driven decision-making and creating business value [17–19]. Big data analytics refers to the approach to creating data-based business value by applying analytics techniques to complex high-volume, high-velocity, and high-variety datasets that require advanced IT for their ingestion, storage, processing, analysis, and visualization [20–23].

Notwithstanding the high business impact and value that BDAS can produce for business organizations, to develop and deliver BDAS, its realization in the business organizations require the implementation – on-premise, on the cloud, or on hybrid mode – of BDAS IT [20–23]. However, the implementation of BDAS IT demands high investments in all required IT resources – computing, storage, network, software, data, and the environment – and the selection of the cost-effective BDAS IT is a hard business managerial decision [20–23]. The utilization of open source BDAS IT is a cost-reduction alternative that business organizations can pursue to cope with the overall high investment required to implement BDAS IT [24–29].

Parallelly, IT service management (ITSM) frameworks have provided to business organizations with best processes and practices to deliver value to end users through the concept of IT services [30, 31], and the notion of BDAS as a service (BDASaaS) [32, 33] has emerged from the convergence of these three components – big data analytics, big data analytics IT platforms, and ITSM frameworks. From the perspective of ITSM managers, delivering BDASaaS implies a hard technological design effort given the variety of architectural models [34] and BDAS IT that can be used. Consequently, the selection of the rightsized technological BDAS IT is a hard ITSM decision [35, 36]. To cope with this design complexity, BDAS architectures of reference have been proposed [37]. NIST Big Data Reference Architecture (NBDRA) [37] is one of the most cited ones, and in this chapter, we used it as the theoretical framework to classify and review the main open source IT for implementing and delivering BDASaaS.

In this chapter, thus, we aim to contribute to the literature with an updated selective review of the technological landscape of the main open source IT to implement and deliver BDASaaS and to the practice with a hybrid-integrative architecture view based on the NBDRA for implementing and delivering BDASaaS using open source IT.

This chapter continues as follows. In Sect. 2, we describe the theoretical background on BDAS characteristics, on IT service implementing and delivering models, and on the NIST Big Data Reference Architecture. In Sect. 3, we report the review of the main open source IT for implementing and delivering BDASaaS. In Sect. 4, we present a discussion of the contributions to the literature and the practice from this research. Finally, in Sect. 5 we report the conclusions, limitations, and recommendations for further research.

2 Background

This section reviews the background on foundations on big data analytics systems (BDAS), on models for implementing and delivering BDAS as a services (BDASaaS), and on the new NIST Big Data Reference Architecture (NBDRA).

2.1 *Foundations of Big Data Analytics Systems*

Big data analytics systems can be defined as software systems implemented on distributed IT configured specially to manage, process, and use high-volume, high-velocity, high-variety, high-veracity, and high-value datasets with a descriptive-exploratory, predictive, or prescriptive business purpose useful for business decision-making [1, 2].

Volume, velocity, variety, veracity, and value are the 5-V core characteristics that differentiate big data analytics systems from small data analytics systems. Volume refers to huge number of business events to be registered that demands storage capabilities usually datasets in the range of terabytes (10^{12} bytes), petabytes (10^{15} bytes), exabytes (10^{18} bytes), or more. Velocity refers to a faster generation of data – i.e., higher frequency of data registering in the range of million or more events by business day. Variety implies to count with a rich diversity of data sources (internal vs. external, manual user vs. automated machine generated, real-time vs. batch ingestion engines), data structuredness (SQL vs. non-SQL), data formats (text vs. binary, analogic vs. digital, encrypted vs. non-encrypted), and data types (char, string, integer, real, image, sound, video). Veracity refers to the overall quality – characterized by objectivity, truthfulness, and credibility – of the data [38]. Value refers to the tangible (cost reductions, profitability increments, business efficiency metrics, among others) and intangible (strategy, business reputation, market value, among others) benefits produced by using big data. Value can be classified in value discovery, through exploratory actions for discovering potential valuable business insights; value creation, through the internal utilization of BDAS for incrementing business value of the firm; and value realization, through the delivery of end-user products and services enhanced with BDAS. Table 1 illustrates the usual range of characteristics between small data analytics systems (SDAS) and BDAS derived from literature [1, 2].

This structure of stages that define a BDAS process is known as the big data analytics system pipeline [1, 2]. A generic BDAS pipeline consists of the following five stages: (1) raw data sources identification, (2) raw data acquisition and preparation, (3) data storing and processing, (4) data modeling and analysis, and (5) data access and usage. In this BDAS pipeline the Big Data side corresponds to stages 1, 2 and 3, and the Analytics side to stages 4 and 5. Table 2 reports the stages, purpose, main activities, and key issues for a generic BDAS pipeline derived and adapted from the main literature [1, 2].

To summarize, the big data pipeline side is responsible for making available processed big datasets with the potential of creating business value and the analytics side for providing business value through the application of analytics procedures to the processed big datasets. Regarding the data modeling and analysis stage, there are three types of analytics procedures. Exploratory and descriptive analytics refer to procedures to report summary metrics and graphs of the big datasets that represent historical and current status of the business processes and systems related to the big datasets. Predictive analytics refer to procedures to create data-driven models that permit estimate future status of the business processes and systems related to the big datasets. Prescriptive analytics refer to procedures to create data-driven models that determine the optimal solutions or best viable alternative solutions.

2.2 *Models for Implementing and Delivering IT Services*

According to the IT service management (ITSM) literature [30, 31], IT service can be defined as a functionality enabled to IT users that generates business value, and it is delivered by an IT service system composed of IT resources, IT processes and practices, and IT people. Figure 1 – adapted from [35, 36] – illustrates the concept of IT service and IT service system.

Figure 1 shows five levels (business processes system, service level agreement (SLA) interface, IT service, IT service system, and IT suppliers). The first level accounts for the set of business processes that can be supported by one or more IT services agreed with an SLA. Second level refers to the set of SLAs that establish the expected IT service functionalities, metrics, schedule, and other relevant information on the agreed IT services for the IT customers. Third level represents the IT services provisioned to the first level – under agreed SLAs. The fourth level accounts for the system that generates and provides the IT service functionalities and pursues

Table 1 Small vs. big data analytics systems comparative profile

Attribute	Small data analytics systems	Big data analytics systems
Data volume	Number of records from thousands to millions. Size datasets from MBs to TBs. Datasets can be stored in a single data server	Number of records from millions to billions or more. Size datasets from TBs to PBs or more. Datasets must be stored in a cluster of data servers
Data variety	Datasets contain structured data (business records). Dataset sources are mainly internal business OLTP and data mart systems. Datasets are recorded using SQL and relational technologies	Datasets contain more semi-structured and unstructured (xml/json texts, text documents, binary images, binary audios, binary videos, binary streams, sensor signals) than structured data (business records). Datasets are recorded using No SQL, Graph, Key-Value, and Columnar technologies
Data velocity	Rates of generated-collected business records are in the range of hundreds to thousands per hour. Most final processed data uses batch mode	Rates of generated-collected unstructured items are in the range of thousands to millions per hour. Most final processed data uses real-time mode
Data veracity	Very high data quality due to the main utilization of structured internal business data sources and the utilization of an ETL process	From moderate to high data quality due to the main utilization of unstructured external business data sources. An ELT is used
Data value	There is an explicit and current utility value due to the need to count with the datasets for supporting business processes	There is an implicit and potential utility value since there is not mandatory to count with the datasets for supporting business processes
IT resources	Usually a centralized single or a small processing-storage server cluster	Moderate to large, distributed processing-storage server cluster
IT people	Highly skilled on analytics	Highly skilled on analytics, big data science, and big data IT services
Development process	Mainly CRISP-DM or a business intelligence dashboard designs	There is not a specific and standard development methodology
Development cost	Affordable costs (development team, development IT resources, structured datasets)	High costs (development team, development IT resources, unstructured datasets)
System lifespan	SDAS usually are used for long time spans – years – due to the stability of datasets, functional requirements, and used data processing techniques	BDAS usually are used for short time spans – months – due to the dynamism of datasets, functional requirements, and used data processing techniques
Analytics purpose	Mainly used for descriptive/exploratory and predictive purposes. Prescriptive purposes are also pursued	Firstly used for descriptive/exploratory purpose. Secondly used for predictive purpose
End users	Business managers/analysts	Big data business managers/analysts

Table 2 A generic BDAS pipeline

Stage	1. Raw data sources identification and acquisition	2. Raw data preprocessing	3. Data storage and processing	4. Data modeling and analysis	5. Data access and usage
Purpose	To identify the set of raw data sources for the big data analytics pipeline, agree legally on its accessibility, collect the agreed raw data, transmit them, and register them	To apply preprocessing procedures to raw data	To pull data of interest, apply them processing procedures, and load them in the persistent storage platforms	To elaborate data-driven models and apply those analytics procedures for specific business goals	To use data-driven models in stand-alone and/or embedded into end-user or automatic control systems for specific business goals
Main tasks	1.1 Identification of the available raw data sources. 1.2 Analysis of the available raw data sources. 1.3 Selection and legal agreement of raw data sources. 1.4 Raw data collection and transmission. 1.5 Raw data registering	2.1 Raw data preprocessing (compression/ decompression, cleaning, redundancy elimination, transformation)	3.1 Data integration, aggregation and representation. 3.2 Data replication. 3.3 Processed data ingestion/ETL	4.1 Exploratory and descriptive analytics (OLAP, descriptive statistics, descriptive charts/ graphs). 4.2 Predictive analytics (classification, regression, clustering, association). 4.3 Prescriptive analytics (optimization, simulation, heuristic methods, expert systems)	5.1 Visual interactive analytics. 5.2 Development of end-user big data analytics systems. 5.3 Development of automatic control big data analytics systems
Main issues	Variety of raw data formats (structured, text, image, audio, video, device signal). Velocity (generation rates of raw data). Volume (raw data size). Veracity (trust level of raw data). Value (business need for raw data). QoS metrics for LAN/WAN/ Internet data transmission systems. Variety, velocity, and volume of raw data	Performance metrics for preprocessing platforms. Data security issues	Performance metrics for storage server cluster, cloud storage services, and processing server clusters. Performance metrics for processing platforms. Data security and privacy issues	Performance metrics for processing server clusters and analytics platforms. Taxonomy of exploratory-descriptive, predictive, and/or prescriptive analytics procedures	QoS metrics for LAN/ WAN/Internet data transmission systems. Usability metrics. Performance metrics. Business goals metrics

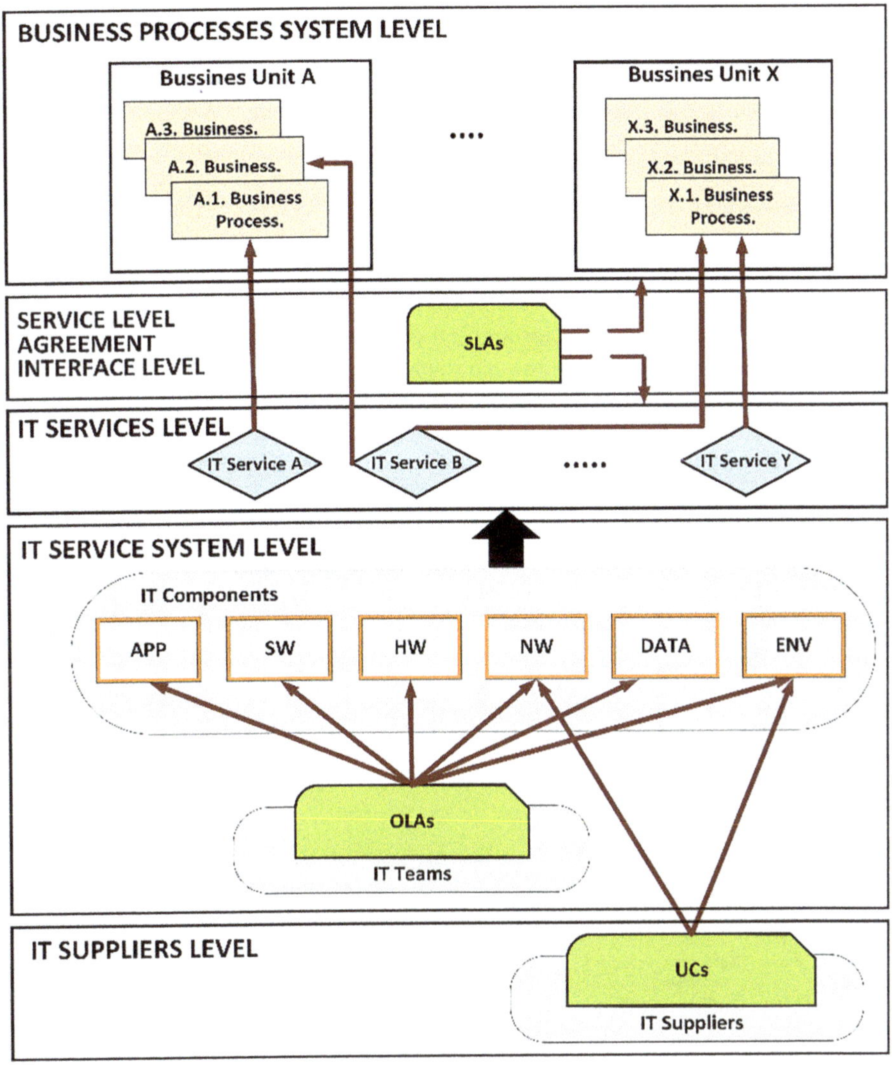

Fig. 1 IT service and IT service system concepts

to achieve the agreed IT service metrics. This level includes the IT components of software applications (APP), software base (SW), hardware equipment (HW), network equipment (NW), data (DAT), all environmental equipment and facilities (ENV), and IT service teams. This fourth level also includes the operational level agreements (OLAs) agreed internally in the diverse IT service teams and the underpinning contracts (UCs) that are agreed between the IT service organization and the external IT providers required to provide IT services. Last, fifth level represents these external IT suppliers.

Value is realized when the expected IT service utility (fit for purpose) and IT service warranty (fit for use) are achieved. The utility of an IT service refers to what the service does that is valued by the customer. The warranty for an IT service refers to how well it is delivered – i.e., how well are reached the levels of availability, capacity, continuity, and security agreed.

In the last decade, big data analytics system as service (BDASaaS) [32, 33] has emerged as an IT service. From the perspective of ITSM managers, delivering BDASaaS implies a hard design effort given the variety of architectural models that can be used. Consequently, the selection of the rightsized big data analytics system IT architecture is a hard business managerial decision.

BDASaaS can be delivered through on-premise, on cloud, or on hybrid cloud. Independently of the type of BDASaaS deployment, BDASaaS can be delivered in three different models [39]: BDASSaaS (BDAS software as a service), BDASPaaS (BDAS platform as a service), or BDASIaaS (BDAS infrastructure as a service). Figure 2 illustrates the three IT service models for BDASaaS using a hybrid functional-deployment architectural view [34] from a cloud-based IT service provider viewpoint. Figure 2 maps also the generic BDAS pipeline reported in Table 1.

BDASIaaS refers to the customer agreement for paying the utilization of physical and virtual IT resources. The cloud provider owns and hosts the physical IT resource layer, but the BDASIaaS customer remotely manages them. In this BDASIaaS provision model, the customer is free and responsible to install and manage the upper cloud layers. BDASPaaS provision model refers to the customer agreement for paying the utilization of the required cloud layers for developing BDA systems. These cloud layers are big data cluster management, big data analytics cluster management, and big data analytics development tools. The two lower

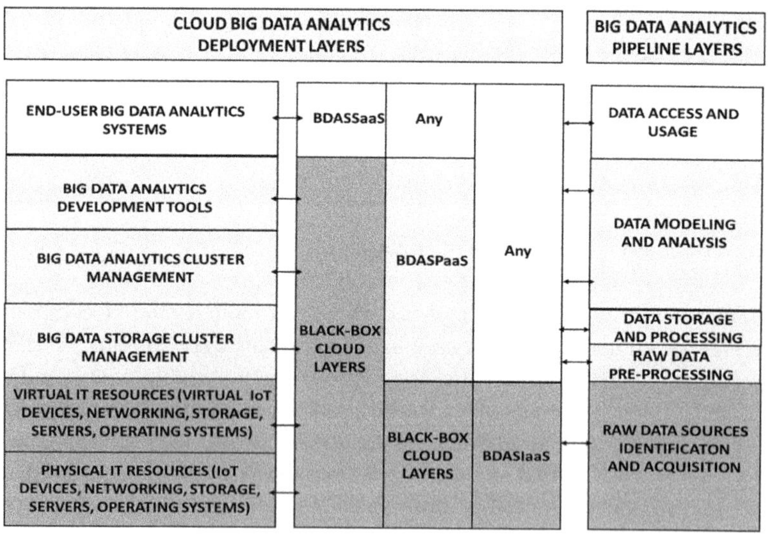

Fig. 2 Types of BDASaaS using a hybrid functional-deployment architectural view

cloud layers are considered black boxes, and the next upper layer is the responsibility of the customer. Finally, BDASSaaS refers to the customer agreement paying for the utilization of an end-user big data analytics system. All lower cloud layers are black boxes for the customer.

2.3 *The NIST Big Data Reference Architecture (NBDRA)*

The architecture of a system conveys the "fundamental concepts or properties of a system in its environment embodied in its elements, relationships, and in the principles of its design and evolution" [34; p. 2]. The architecture of a system, which is abstract, is manifested through the functional and nonfunctional properties of the system. According to [34], the architecture of a system can be designed and represented through an architecture description (AD) document. An AD document reports stakeholders and their concerns, architecture decisions and rationale, and architecture views and viewpoints. Stakeholders are any entity that will be affected by the system of interest. Concerns are the expected system properties of interest for the stakeholders. Architecture decisions and rationale are the architectural design selections done and their justifications. Architecture views are diagrams – called architecture models – governed by architecture viewpoints that depict a set of specific concerns. To guide system architects in the design of a system architecture, architecture frameworks and reference architectures (RA) have been proposed. An architecture framework "establishes a common practice for creating, interpreting, analyzing and using architecture descriptions within a particular domain of application or stakeholder community" [34; p. 9]. Reference architecture refers to "a generic architecture for a class of systems that is used as a foundation for the design of concrete architectures from this class" [40; p. 417].

Architecture frameworks and reference architectures for IT services are scarce in the literature. IT4IT [41] and TOGAF [42] can be reported as the main relevant, but they are focused on IT function [41] or enterprise architecture [42]. IT4IT [41] is a standard reference architecture for managing IT function from an IT service value stream approach. IT4IT [41] includes four IT service value streams: (1) strategy to portfolio, (2) requirement to deploy, (3) request to fulfill, and (4) detect to correct. In (2) request to fulfill value stream, it is reported that a logical service blueprint artifact must be generated, and it includes only an architecture design element, defined as "an architectural representation (diagram) of the service system components" [41; p. 74]. TOGAF [42; p. 11] is an architecture framework that "provides the methods and tools for assisting in the acceptance, production, use, and maintenance of an Enterprise Architecture." TOGAF [42] includes four domains for generating an enterprise architecture: (1) business architecture, (2) data architecture, (3) application architecture, and (4) technology architecture. Thus, TOGAF [42] can be used for generating a reference architecture for BDASaaS, but none already defined solution was found in the main literature. The ITSM literature does not provide reference architectures for IT services [30, 31, 35, 36].

For BDASaaS, several proprietary reference architectures from IT business consulting companies have been proposed. From the nonproprietary side, four main reference architectures are available [37, 43–45]. These BDASaaS reference architectures are NIST Big Data Reference Architecture V3.0 (NBDRA) [37], Reference Architecture for Big Data Systems [43], Cloud Customer Architecture for Big Data and Analytics V2.0 [44], and Open-Source Architecture for Big-Data Analytics [45]. In this chapter, NBDRA [37] is used as the reference architecture base for the review of the open source IT for implementing and delivering BDASaaS.

NIST Big Data Reference Architecture V3.0 (NBDRA) [37] consists of a vendor-neutral, technology- and infrastructure-agnostic conceptual model and two architectural views (activity view and functional view). It was designed by NITS (National Institute of Standards and Technology, USA) after several rounds of sessions in the NIST Big Data Public Working Group (NBD-PWG) with participants from industry, academia, and government agencies. According to [37; p. 3], a reference architecture provides "an authoritative source of information about a specific subject area that guides and constrains the instantiations of multiple architectures and solutions." NBDRA "is a high-level conceptual model crafted to serve as a tool for describing, discussing, and developing system specific architectures using a common framework of reference" [37; p. 3].

NBDRA addresses the requirements of interoperability, portability, reusability, extensibility, data usage, analytics, and technology infrastructure. NBDRA is structured with five main functional components ((1) system orchestrator; (2) data provider; (3) big data application provider; (4) big data framework provider (including computing and analytics processing layer, data organization and distribution layer, and infrastructures layer (computing, storage, networking)); and (5) data consumer and two fabrics (management fabric and security-privacy fabric) that provide critical internal support services for the five functional components. Figure 3 – derived from [37] – illustrates a functional architectural view of NBDRA, mapped to the BDASaaS hybrid functional-deployment architectural view from an IT service provider viewpoint (Fig. 3).

NBDRA contains two value chains: information technology (IT) and information. NBDRA IT value chain corresponds to the IT used in the five functional components and the two fabrics. NBDRA information value chain corresponds to the activities included in the big data framework provider. These activities are collection, preparation, analytics, visualization, and access. These five activities correspond to a basic NBDRA pipeline. NBDRA IT value chain cocreates value "by providing networking, infrastructure, platforms, application tools, and other IT services for hosting of and operating the Big Data in support of required data applications" [37; p. 12]. NBDRA information value chain cocreates value "by data collection, integration, analysis, and applying the results following the value chain" [37; p. 12].

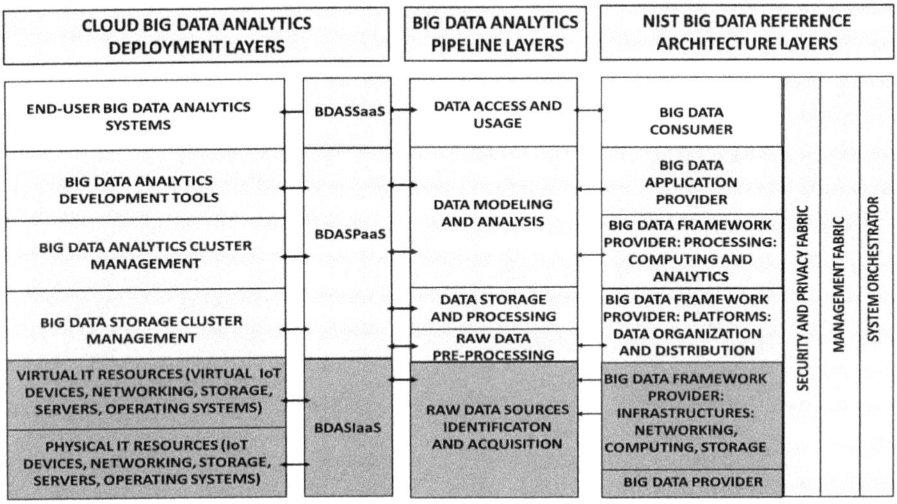

Fig. 3 NBDRA [29] mapped to the BDASaaS hybrid functional-deployment architectural view

3 Selective Review of Open Source IT for Implementing and Delivering BDASaaS

In this section, we report a selective review [46, 47] of the main open source IT that is currently available for implementing and delivering BDASaaS in their three modes: BDASIaaS, BDASPaaS, or BDASSaaS.

In the line of [46; p. 107], a conceptual review "seeks to describe, summarize, evaluate, clarify, and/or integrate the content of the primary reports" and can have an exhaustive or selective (representative) coverage and a neutral or position perspective. This conceptual review aims to describe, clarify, and integrate the information reported on open source IT tools for implementing and delivering BDASaaS collected from several studies [24–29] complemented with the review of the official documentation provided by the open source IT tool websites. Our perspective is neutral – no comparative and evaluative review is conducted on such open source IT tools.

Previous studies [24–29] provide insightful and valuable descriptive technical information on open source IT for BDASaaS. Conjointly, it can be identified from such studies the next contributions: (1) a list of the main open source IT for BDASaaS; (2) summarized descriptions of the main functionality of the open source IT for BDASaaS; and (3) an implicit ad hoc architectural layout view on how the open source IT for BDASaaS components can be integrated. This chapter takes advantage of these previous contributions and advances providing a new integrative BDASaaS hybrid framework – including generic BDASaaS pipeline, NBDRA functional architecture, and BDASaaS layers – that helps to organize the extensive list of the available open source IT for BDASaaS and to present a coherent and integrated technical architectural view. Table 3 reports the updated integrative BDASaaS hybrid framework.

Table 3 A BDASaaS hybrid framework – including generic BDAS pipeline, NBDRA functional architecture, and BDASaaS layers

BDASaaS pipeline stage	1. Raw data sources identification and acquisition	2. Raw data preprocessing	3. Data storage and processing	4. Data modeling and analysis	5. Data access and usage
BDASaaS Deployment type	BDASIaaS	BDASPaaS	BDASPaaS	BDASPaaS	BDASSaaS
NBDRA layer	Big data provider Big data framework provider: infrastructures – networking, computing, storage	Big data framework provider: platforms – data organization and distribution	Big data framework provider: platforms – data organization and distribution	Big data framework provider: processing – computing and analytics	Big data application provider Big data consumer component
Open source IT categories	Cloud platforms Processing server cluster management Streaming/CEP engines	Storage server cluster management Data lake management platforms Big data SQL databases Big data warehouses Big data non-SQL databases Big data preprocessing tools	Big data processing engines Big data SQL engines Big data non-SQL engines Big data OLAP engines	Big data analytics engines/ libraries Big data graph engines	Visual interactive analytics packages
Main functionality	To provide the physical/virtual IT infrastructure to BDAS To enable big data ingestion	To provide big data storage management To provide big data preprocessing functions	To provide big data query engines (SQL, non-SQL, and OLAP)	To provide big data analytics engines. To provide big graph analytics engines. To control batch, interactive, and streaming processing modes	To provide big data tools for interactive analytics

We conducted this selective review applying the following steps adapted from [47]: (1) to formulate the research goal; (2) to define data sources and selective criteria; (3) to collect studies; (4) to review and synthetize the findings from the collected studies; and (5) to elaborate report of findings. Table A1 (in Appendix) summarizes the five selective review steps that were applied.

Table 3 reports the updated integrative BDASaaS hybrid framework. It considers the following structural components: BDASaaS pipeline stage, BDASaaS deployment type (BDASIaaS, BDASPaaS, or BDASSaaS), NBDRA layer, open source IT categories (cloud platforms, processing server cluster management, streaming/CEP engines, storage server cluster management, data lake management platforms, big data SQL databases, big data warehouses, big data non-SQL databases, big data preprocessing tools, big data processing engines, big data SQL engines, big data non-SQL engines, big data OLAP engines, big data analytics engines/libraries, big data graph engines, and visual interactive analytics packages), and main functionality (of the BDASaaS pipeline stage).

Table 4 reports the list of the 58 open source IT tools located grouped by category of open source IT.

Table A2 (in Appendix) reports the descriptive record for each of the 58 identified open source IT tools. This descriptive record considers the next characteristics: open source IT name, self-description, BDASaaS pipeline stage, NBDRA layer, IT type, hardware requirements for small-scale production installation, and level (high, moderate, or low) of four core technical open source attributes [48]. According to [48], the most considered technical attributes for selecting an open source IT are maturity-longevity, security-reliability, documentation, and community support. Maturity-longevity refers to the period from the first release of a tool. Security-reliability refers to the extent of error-free status and hidden flaws of the tool. Documentation refers to the free-cost availability of technical and user manuals. Community support refers to the availability of technical support for tool utilization from communities of practices.

4 Discussion of Contributions

In this chapter, we have conducted a descriptive review of the landscape of BDAS as services (BDASaaS), from a high-level architectural perspective that can be useful for an IT service management manager at small business. These kinds of organizations usually lack sufficient economic, technical, and organizational resources to implement and use BDASaaS. The implementation of BDASaaS demands high costs of IT infrastructure or cloud-based fees, of licenses of proprietary big data software tools, of specialized training, and of high-tech consulting. The availability of open source IT tools for implementing BDASaaS that use moderate-cost IT infrastructure (on-premise or a cloud-based) is a path for small business.

In this landscape of BDASaaS, we report updated perspectives of the next core background topics: (1) a comparative profile between small data analytics systems and big data analytics systems (Table 1); (2) a generic BDAS pipeline of stages including main tasks and technical issues per stage (Table 2); (3) a description of the modern concept of IT service and its architectural functional view (Fig. 1); (4) the three types (software, platform, and infrastructure) of BDASaaS using a hybrid functional-deployment architectural view (Fig. 2); and (5) the NBDRA layers

Table 4 List of the 57 open source IT tools for BDASaaS grouped by big data functionality categories

BDASaaS pipeline stage	1. Raw data sources identification and acquisition	2. Raw data preprocessing	3. Data storage and processing	4. Data modeling and analysis	5. Data access and usage
Open Source IT categories and tools	Cloud platforms Apache CloudStack OpenStack Processing server cluster management Apache Mesos Apache Hadoop Yarn Apache Zookeeper Streaming/CEP engines Apache Kafka Apache Flink Apache Storm Elastic Logstash Apache IoTDB Apache Flume	Storage server cluster management Apache Hadoop/HDFS Apache Ambari Data lake management platforms Apache Hudi Delta Lakehouse Big data SQL databases PostgreSQL MySQL MariaDB Big data warehouses Apache Hive Apache Druid Big data non-SQL databases Apache HBase Apache Cassandra Big Data pre-processing tools Apache Griffin OpenRefine DataCleaner	Big data processing engines Apache Hadoop MapReduce Big data SQL engines Apache Impala Presto Spark SQL Trino Big data non-SQL engines Apache Drill Apache Pig Big data OLAP engines Apache Kylin	Big data analytics engines/libraries Apache Mahout Apache Spark core Apache Spark MLlib SparkR Sparklyr RHadoop RHive TensorFlow Keras Pytorch TorchServe ElasticSearch OpenSearch MLflow Scikit-learn Big data graph engines Apache Spark GraphX Apache Giraph Neo4j Graph Database CE Neo4j Graph Data Science CE	Visual interactive analytics packages Kibana OpenSearch Dashboards Looker Studio MS Power BI service RStudio Server Shiny Server Apache Zeppelin Apache Superset

mapped to the BDASaaS hybrid functional-deployment architectural view (Fig. 3). With these theoretical antecedents from the main literature, we report a BDASaaS hybrid framework – including the generic BDAS pipeline of stages, the NBDRA functional architecture, and the BDASaaS layers (Table 3). This BDASaaS hybrid framework is useful to guide the searching, description, and classification of available open source IT usable to implement BDASaaS (Table 4).

We consider this descriptive review provides the following contributions to the literature: (1) it organizes, integrates, and summarizes a vast literature on BDAS through the comparative profile of small vs. big data analytics systems; (2) it presents a BDAS pipeline of stages including a structure of tasks per stage; (3) it proposes a BDASaaS hybrid framework – including the generic BDAS pipeline of stages, the NBDRA functional architecture, and the BDASaaS layers – useful to understand the full stack of IT layers required to implement BDAS as services and relates these IT layers with the main available open source IT; and (4) it reports a classified list – per NBDRA layer and BDAS pipeline stage – with 58 open source IT tools.

This descriptive review also provides the following contributions to the practice: (1) it helps to ITSM managers from small organizations to a better understanding of the technical landscape of BDAS as services; (2) it makes explicit the BDAS pipeline of stages through the structure of specific tasks by stage, and it leads to create awareness on the technical challenges to implement BDASaaS in ITSM managers of small business; (3) it exhibits an essential BDASaaS hybrid framework useful for guiding ITSM managers of small business to identify the most adequate and technically viable type of service (software, platform, or infrastructure) of BDAS that the business organization can implement; and (4) it makes available an updated list of 58 open source IT tools that can be combined to implement BDASaaS.

Figure 4 illustrates a possible full-stack implementation of BDASaaS selecting open source IT tools from the reported list. NBDRA [37] indicates the functional components big data provider and big data consumer that can be internal, external,

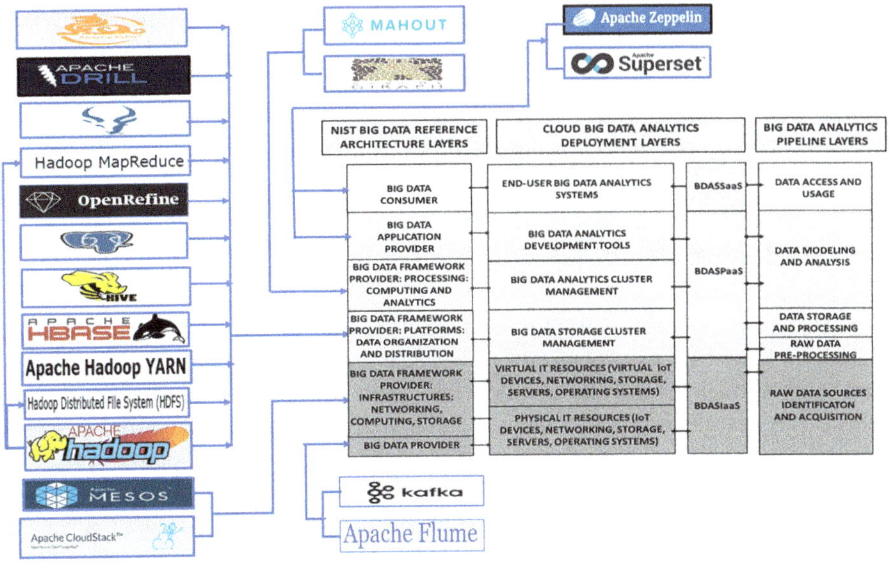

Fig. 4 A full-stack implementation of BDASaaS with open source IT tools

or combined to the business organization. Figure 4 assumes an internal perspective.

Consequently, in the NBDRA layer of big data provider are illustrated as example the Apache Kafka and Apache Flume tools to provide real-time event processing collection of data in pull and push modes, respectively. These two tools are assumed installed in a different big data framework provider – infrastructure – networking, computing, and storage layer in the business organization. In next layer, where the physical and virtualized networking, computing, and storage resources must be managed, the Apache CloudStack and Apache Mesos tools are suggested. Apache CloudStack enables the virtualization management of the physical IT resources, and Apache Mesos will manage these virtualized resources as computing and storage clusters. In the next layer of big data framework provider – Platforms – Data Organization and Distribution, there are suggested several tools: Apache Hadoop (including Core Kernel, HDFS, YARN, and MapReduce modules), Apache HBase, Apache Hive, PostgreSQL, OpenRefine, Apache Impala, Apache Drill, and Apache Kylin. Apache Hadoop provides the basic file storage management and basic MapReduce processing capabilities. Apache HBase enables a non-SQL database. Apache Hive provides a data warehouse functionality. PostgreSQL manages a SQL database. OpenRefine provides data cleaning and other preprocessing capabilities. Apache Impala provides a SQL engine. Apache Drill provides a non-SQL engine and Apache Kylin an engine to query OLAP cubes.

In the next layer of big data framework provider – Processing – Computing and Analytics, the tools of Apache Mahout and Apache Giraph are suggested. Apache Mahout provides a library of machine learning and related analytics algorithms. Apache Giraph provides processing capabilities of big data graphs. In the next layer of big data application provider, two tools are suggested: Apache Superset and Apache Zeppelin. Apache Superset is a tool for creating end-user BDAS. Apache Zeppelin provides an interactive run and test notebook for designing, testing, and exploring data-driven models. Finally, in the top layer of big data consumer, from this internal perspective, we report the same two tools that can be used for the end users to access a previously developed BDAS.

5 Conclusions

Big data analytics systems (BDAS) are high-tech and high-cost IT systems pursued mainly by large business organizations because these kinds of systems have generated value through several business benefits – better decision-making process, more reliable business decisions, automatic identification of hidden and complex business patterns of events, real-time monitoring of large quantities of critical events, and better understanding of large quantities of events through big data graphs, among others.

Small business, however, cannot afford them due the lack of economic, technical, and organizational resources. Open Source IT tools for BDAS that run over moderate-cost IT infrastructure can help small business.

In this chapter we review – from a high-level architectural perspective useful for an ITSM manager of small business – the landscape of open source IT tools for implementing BDAS as services. We located 58 software tools, and they were organized using as theoretically lenses a generic BDAS pipeline of stages derived from the main BDAS literature and the new NIST Big Data Reference Architecture (NBDRA). We identified that whereas the span of open source IT tools for BDAS is wide, this extensive variety of tools make technically complex an adequate architectural selection of functional components, due to the diverse interoperability and demanded IT resources requirements for their correct integration and performance.

Our descriptive review aims to provide essential theoretical and practical insights for implementing BDAS services, reporting a comparative profile between small data analytics systems and big data analytics systems, a generic BDAS pipeline of stages, main tasks and technical issues, a description of the modern concept of IT service and its architectural functional view, the three types (software, platform, and infrastructure) of BDASaaS using a hybrid functional-deployment architectural view, the NBDRA layers mapped to the BDASaaS hybrid functional-deployment architectural view, a high-level review of 56 open source IT tools for BDAS, and an illustrative case of a full-stack implementation of BDASaaS.

These results are limited to a selection of the main open source IT tools reported in previous studies [22, 23, 34] augmented with an updated technical search conducted by this research team. These results are also limited to a high-level descriptive review of each tool – essential description, BDASaaS pipeline stage, NBDRA layers, IT type, hardware requirements for small-scale production installation, and level (high, moderate, or low) of technical open source attributes.

Hence, we can recommend the following avenues for further research. First, we suggest refining the generic BDAS pipeline of stages to align the purpose and main tasks with the NBDRA architectural activity view of the big data application provider layer. This NBDRA layer accounts for the activities of big data collection, big data preparation, big data analytics, big data visualization, and big data access. With this refined generic BDAS pipeline of stages, a classification of open source IT tools for BDAS will fit better the NBDRA. Second, we recommend elaborating several illustrative cases of full-stack implementation of BDASaaS with open source IT tools against a set of BDAS functional architectural requirements. These illustrative cases can be useful for guiding to ITSM managers of small business to select the most convenient implementation of BDASaaS for your business organization. Third, we consider useful for practitioners to elaborate a more detailed review of the 58 open source IT tools reporting a comparative analysis of the functional capabilities between tools of the same category.

Hence, this chapter calls for further research on BDASaaS using open source tools in the context of small business and IT service management approach.

Appendix

Table A1 Selective review research steps

Step	Purpose	Outcomes	Outcomes in this research
1. To formulate the research goal.	To state the expected research goal indicating the theoretical or practical or both ones expected contributions	Research goal statement	To contribute to the literature with an updated selective review of the technological landscape of the main open source IT to implement and deliver BDASaaS and 2) to the practice with a hybrid-integrative architecture view based on the NBDRA for implementing and delivering BDASaaS using open source IT
2. To define data sources and selective criteria	To identify and agree the set of data sources to collect the studies, as well as to define the selection criteria	List of data sources Selection criteria statements	The single data source was agreed the research search engine of Google Scholar. The following joint selection criteria statements were agreed: to search in titles and abstracts the keywords "big data" *or* "data science" or "analytics" *and* "tools" *or* "platforms" *or* "technologies" in the period 2010–2022. Select only studies published in journals indexed in Scopus OR JCR index, and six studies were selected [24–29]. To include open source IT listed in such six studies *and* its website is still active. To add additional open source IT not listed in such studies, but that is relevant and has an active website
3. To collect studies	To get the studies	Set of selected studie.	Six studies [24–29] were obtained
4. To review and synthetize the findings from the collected studies	To conduct the analysis and integration of finding	Structured schema of findings	We elaborated the conceptual scheme of BDASaaS hybrid framework – including generic BDAS pipeline, NBDRA functional architecture, and BDASaaS layers – and populated a list of 58 records of open source IT tools
5. To elaborate report of findings	To produce visible results	Research results	This chapter was elaborated

Table A2 Records of the 58 open source IT tools for implementing and delivering BDASaaS

Open source IT name	Self-description	BDASaaS pipeline stage	NBDRA layer	IT type	Minimal practical HW requirements	Level of open source technical attributes
1. Apache CloudStack	Software platform "to deploy and manage large networks of virtual machines, as a highly available, highly scalable Infrastructure as a Service (IaaS) cloud computing platform" [49]	1. Raw data sources identification and acquisition	Big data framework provider: infrastructures – networking, computing, storage	Cloud platform	1 server for cluster management 4 servers for host VMs 1 NFS storage server	MatLon: high SecRel: high TechDoc: high ComSup: high
2. OpenStack	"Cloud operating system that controls large pools of compute, storage, and networking resources throughout a datacenter ... (and) provides an Infrastructure-as-a-Service (IaaS) solution." [50]	1. Raw data sources identification and acquisition	Big data framework provider: infrastructures – networking, computing, storage	Cloud platform	1 server as cluster management controller 4 servers as computing nodes to host VMs 1 block storage server 1 object storage server	MatLon: high SecRel: high TechDoc: high ComSup: high
3. Apache Mesos	"A distributed system kernel ... (that) abstracts CPU, memory, storage, and other compute resources away from machines (physical or virtual), enabling fault-tolerant and elastic distributed systems to easily be built and run effectively" [51]	1. Raw data sources identification and acquisition	Big data framework provider: infrastructures – networking, computing, storage	IT cluster management	1 server as cluster management master 4 servers as computing nodes as agents 1 storage server	MatLon: high SecRel: mod TechDoc: high ComSup: high

(continued)

Table A2 (continued)

Open source IT name	Self-description	BDASaaS pipeline stage	NBDRA layer	IT type	Minimal practical HW requirements	Level of open source technical attributes
4. Apache Hadoop Yarn	Software platform to provides a global resource manager (including a scheduler and per-application application master) and per-node NodeManager (for CPU, memory, disk, network resources) [52].	1. Raw data sources identification and acquisition	Big data framework Provider: infrastructures – networking, computing, storage	IT cluster management specific for Hadoop jobs	1 server as resource manager 4 servers as node managers/dataNodes 1 storage server	MatLon: mod SecRel: mod TechDoc: high ComSup: high
5. Apache Zookeeper	A distributed "coordination service for distributed applications. It exposes a simple set of primitives that distributed applications can build upon to implement higher level services for synchronization, configuration maintenance, and groups and naming" [53]	1. Raw data sources identification and acquisition	Big data framework provider: infrastructures – networking, computing, storage	IT cluster management (coordination service for distributed applications)	1 server as leader 4 servers as followers 1 storage server	MatLon: high SecRel: high TechDoc: high ComSup: high

6. Apache Kafka	A distributed event streaming platform for "capturing data in real-time from event sources like databases, sensors, mobile devices, cloud services, and software applications in the form of streams of events, storing these event streams durably for later retrieval; manipulating, processing, and reacting to the event streams in real-time as well as retrospectively; and routing the event streams to different destination technologies as needed" [54]	1. Raw data sources identification and acquisition	Big data provider component	Streaming/CEP engines	1 server as controller 4 servers as brokers 1 storage server	MatLon: high SecRel: high TechDoc: high ComSup: high
7. Apache Flink	"A framework and distributed processing engine for stateful computations over unbounded and bounded data stream" and includes "support for stream and batch processing, sophisticated state management, event-time processing semantics, and exactly-once consistency guarantees for state" [55]	1. Raw data sources identification and acquisition	Big data provider component	Streaming/CEP engines	1 server as master 4 servers as workers 1 storage server	MatLon: mod SecRel: high TechDoc: high ComSup: high
8. Apache Storm	"A distributed real-time computation system … to reliably process unbounded streams of data" [56]	1. Raw data sources identification and acquisition	Big data provider component	Streaming/CEP engines	1 server as master 4 servers as supervisor workers 1 storage server	MatLon: mod SecRel: high TechDoc: high ComSup: high

(continued)

Table A2 (continued)

Open source IT name	Self-description	BDASaaS pipeline stage	NBDRA layer	IT type	Minimal practical HW requirements	Level of open source technical attributes
9. Elastic Logstash	"A server-side data processing pipeline that ingests data from a multitude of sources, transforms it, and then sends it to your favorite 'stash'" [57]	1. Raw data sources identification and acquisition	Big data provider component	Streaming/CEP engines	2 Logstash nodes 1 storage server	MatLon: high SecRel: high TechDoc: high ComSup: high
10. Apache IoTDB	"IoT native database with high performance for data management and analysis, deployable on the edge and the cloud … Apache IoTDB can meet the requirements of massive data storage, high-speed data ingestion and complex data analysis in the IoT industrial fields" [58]	1. Raw data sources identification and acquisition	Big data provider component	Streaming/CEP engines	1 IoTDB datacenter server 1 IoTDB Edge server per node 1 storage server	MatLon: low SecRel: mod TechDoc: high ComSup: high
11. Apache Flume	"Flume is a distributed, reliable, and available service for efficiently collecting, aggregating, and moving large amounts of log data. It has a simple and flexible architecture based on streaming data flows. It is robust and fault tolerant with tunable reliability mechanisms and many failover and recovery mechanisms. It uses a simple extensible data model that allows for online analytic application" [59]	1. Raw data sources identification and acquisition	Big data provider component	Streaming/CEP engines	1 server as master 4 servers as workers 1 storage server	MatLon: mod SecRel: high TechDoc: high ComSup: high

11. Apache Hadoop HDFS	"The Hadoop Distributed File System (HDFS) is a distributed file system designed to run on commodity hardware" [60]	2. Raw data preprocessing	Big data framework provider: platforms – data organization and distribution	Storage server clusters	1 master nameNode server 4 dataNodes servers	MatLon: high SecRel: high TechDoc: high ComSup: high
12. Apache Ambari	"Apache Ambari is a tool for provisioning, managing, and monitoring Apache Hadoop clusters. Ambari consists of a set of RESTful APIs and a browser-based management interface" [61]	2. Raw data preprocessing	Big data framework provider: platforms – data organization and distribution	Storage server clusters	1 master namenode server 4 dataNodes servers	MatLon: high SecRel: high TechDoc: high ComSup: high
13. Apache Hudi	"A data lake platform. Apache Hudi brings core warehouse and database functionality directly to a data lake. Hudi provides tables, transactions, efficient upserts/deletes, advanced indexes, streaming ingestion services, data clustering/compaction optimizations, and concurrency all while keeping your data in open source file formats" [62]	2. Raw data preprocessing	Big data framework provider: platforms – data organization and distribution	Storage server clusters: data lake management platforms	It uses the existent HDFS infrastructure	MatLon: mod SecRel: mod TechDoc: high ComSup: high

(continued)

Table A2 (continued)

Open source IT name	Self-description	BDASaaS pipeline stage	NBDRA layer	IT type	Minimal practical HW requirements	Level of open source technical attributes
14. Delta Lakehouse	"Delta Lake is an open source project that enables building a Lakehouse architecture on top of data lakes. Delta Lake provides ACID transactions, scalable metadata handling, and unifies streaming and batch data processing on top of existing data lakes, such as S3, ADLS, GCS, and HDFS" [63]	2. Raw data preprocessing	Big data framework provider: platforms – data organization and distribution	Storage server clusters: data lake management platforms	It uses the existent Spark infrastructure	MatLon: mod SecRel: mod TechDoc: high ComSup: high
15. Apache Griffin	"Apache Griffin is an open source Data Quality solution for Big Data, which supports both batch and streaming mode. It offers a unified process to measure your data quality from different perspectives, helping you build trusted data assets, therefore boost your confidence for your business" [64]	2. Raw data preprocessing	Big data rramework provider: platforms – data organization and distribution	Big data preprocessing tools	It uses the Spark IT infrastructure	MatLon: mod SecRel: mod TechDoc: mod ComSup: mod
16. OpenRefine	"OpenRefine (previously Google Refine) is a powerful tool for working with messy data: cleaning it; transforming it from one format into another; and extending it with web services and external data" [65]	2. Raw data preprocessing	Big data framework provider: platforms – data organization and distribution	Big data preprocessing tools	1 server	MatLon: high SecRel: high TechDoc: high ComSup: high

17. DataCleaner	"DataCleaner is a strong data profiling engine for discovering and analyzing the quality of your data. Find the patterns, missing values, character sets and other characteristics of your data values" [66]	2. Raw data preprocessing	Big data framework provider: platforms – data organization and distribution	Big data preprocessing tools	1 server	MatLon: high SecRel: high TechDoc: high ComSup: high
18. PostgreSQL	"PostgreSQL is a powerful, open source object-relational database system that uses and extends the SQL language combined with many features that safely store and scale the most complicated data workloads" [67]	3. Data storage and processing	Big data framework provider: platforms – data organization and distribution	Storage server clusters: big data SQL databases	1 cluster of processing servers 1 cluster of storage servers	MatLon: high SecRel: high TechDoc: high ComSup: high
19. MySQL	MySQL is an "Open Source SQL database management system, is developed, distributed," relational and "works in client/server or embedded systems" [68]	3. Data storage and processing	Big data framework provider: platforms – data organization and distribution	Storage server clusters: big data SQL databases	1 cluster of processing servers 1 cluster of storage servers	MatLon: high SecRel: high TechDoc: high ComSup: high

(continued)

Table A2 (continued)

Open source IT name	Self-description	BDASaaS pipeline stage	NBDRA layer	IT type	Minimal practical HW requirements	Level of open source technical attributes
20. Apache Hive	"Hive is a data warehousing infrastructure based on Apache Hadoop. It provides SQL which enables users to do ad-hoc querying, summarization and data analysis easily" [69]	3. Data storage and processing	Big data framework provider: platforms – data organization and distribution	Storage server clusters: big data warehouses	It uses the existent Hadoop infrastructure	MatLon: high SecRel: high TechDoc: high ComSup: high
21. Apache Druid	"Apache Druid is a real-time analytics database designed for fast slice-and-dice analytics ('OLAP' queries) on large data sets. Most often, Druid powers use cases where real-time ingestion, fast query performance, and high uptime are important" [70]	3. Data storage and processing	Big data framework provider: platforms – data organization and distribution	Storage server clusters: big data warehouses	1 Master server 1 Query server 2 Data servers 2 NFS servers or 2 HDFS servers	MatLon: high SecRel: high TechDoc: high ComSup: high
22. Apache HBase	"Apache HBase is an open-source, distributed, versioned, non-relational database" used "when you need random, real-time read/write access to your Big Data" [71]	3. Data storage and processing	Big data framework provider: platforms – data organization and distribution	Storage server clusters: big data non-SQL database	It uses the existent Hadoop infrastructure 1 master server (HDFS nameNode) 1 master backup server 4 region servers (including Zookeeper) (HDFS dataNodes)	MatLon: high SecRel: high TechDoc: high ComSup: high

23. Apache Cassandra	Apache Cassandra is a masterless, lightweight, non-relational, and NoSQL distributed database. "NoSQL databases enable rapid, ad-hoc organization and analysis of extremely high-volume, disparate data types" [72]	3. Data storage and processing	Big data framework provider: platforms – data organization and distribution	Storage server clusters: big data non-SQL database	4 node servers with 1 TB of storage each server	MatLon: high SecRel: high TechDoc: high ComSup: high
24. Apache Hadoop MapReduce	"Hadoop MapReduce is a software framework for easily writing applications which process vast amounts of data (multi-terabyte data-sets) in-parallel on large clusters (thousands of nodes) of commodity hardware in a reliable, fault-tolerant manner" [73]	3. Data storage and processing.	Big data framework provider: platforms – data organization and distribution	Big data processing engines	1 master ResourceManager server 4 worker nodeManagers servers	MatLon: high SecRel: high TechDoc: high ComSup: high
25. Apache Impala	Impala is a high-performance SQL engine for Hadoop platform. "With Impala, you can query data, whether stored in HDFS or Apache HBase – including SELECT, JOIN, and aggregate functions – in real time. Furthermore, Impala uses the same metadata, SQL syntax (Hive SQL), ODBC driver, and user interface (Hue Beeswax) as Apache Hive, providing a familiar and unified platform for batch-oriented or real-time queries" [74]	3. Data storage and processing.	Big data framework provider: platforms – data organization and distribution	Big data SQL engines	1 cluster of processing servers 1 cluster of storage servers	MatLon: mod SecRel: mod TechDoc: high ComSup: high

(continued)

Table A2 (continued)

Open source IT name	Self-description	BDASaaS pipeline stage	NBDRA layer	IT type	Minimal practical HW requirements	Level of open source technical attributes
26. Presto	"Presto is an open source SQL query engine that's fast, reliable, and efficient at scale. Use Presto to run interactive/ad hoc queries at sub-second performance for your high volume apps" [75]	3. Data storage and processing.	Big data framework provider: platforms – data organization and distribution	Big data SQL engines	1 cluster of processing servers 1 cluster of storage servers	MatLon: high SecRel: high TechDoc: high ComSup: high
27. Apache Spark SQL	"Spark SQL is a Spark module for structured data processing … Spark SQL can also be used to read data from an existing Hive installation" [76]	3. Data storage and processing.	Big data framework provider: platforms – data organization and distribution	Big data SQL engines	1 Spark context server 1 Spark cluster manager server Processing servers 4 worker node servers	MatLon: high SecRel: high TechDoc: high ComSup: high
28. Trino	"Trino is a distributed SQL query engine designed to query large data sets distributed over one or more heterogeneous data sources" [77]	3. Data storage and processing.	Big data framework provider: platforms – data organization and distribution	Big data SQL engines	1 cluster of processing servers 1 cluster of storage servers	MatLon: high SecRel: high TechDoc: high ComSup: high

29. Apache Drill	Apache Drill is a "schema-free SQL Query Engine for Hadoop, NoSQL and Cloud Storage … Drill supports a variety of NoSQL databases and file systems, including HBase, MongoDB, MapR-DB, HDFS, MapR-FS, Amazon S3, Azure Blob Storage, Google Cloud Storage, Swift, NAS and local files" [78]	3. Data storage and processing.	Big data framework provider: platforms – data organization and distribution	Big data non-SQL engines	1 cluster of processing servers 1 cluster of storage servers	MatLon: high SecRel: high TechDoc: high ComSup: high
30. Apache Pig	"Apache Pig is a platform for analyzing large data sets. Pig's language, Pig Latin, lets you specify a sequence of data transformations such as merging data sets, filtering them, and applying functions to records or groups of records. Pig comes with many built-in functions but you can also create your own user-defined functions to do special-purpose processing" [79]	3. Data storage and processing.	Big data framework provider: platforms – data organization and distribution	Big data non-SQL engines	1 cluster of processing servers 1 cluster of storage servers	MatLon: high SecRel: high TechDoc: high ComSup: high
31. Apache Kylin	Apache Kylin is a "distributed Analytical Data Warehouse for Big Data … Kylin is able to achieve near constant query speed regardless of the ever-growing data volume" [80]	3. Data storage and processing.	Big data framework provider: platforms – data organization and distribution	Big data OLAP engines	1 cluster of processing servers 1 cluster of storage servers	MatLon: high SecRel: high TechDoc: high ComSup: high

(continued)

Table A2 (continued)

Open source IT name	Self-description	BDASaaS pipeline stage	NBDRA layer	IT type	Minimal practical HW requirements	Level of open source technical attributes
32. Apache Mahout	"Apache Mahout(TM) is a distributed linear algebra framework and mathematically expressive Scala DSL designed to let mathematicians, statisticians, and data scientists quickly implement their own algorithms" [81]	4. Data modeling and analysis	Big data framework provider: processing – computing and analytics	Big data analytics engines/libraries	1 cluster of processing servers 1 cluster of storage servers	MatLon: high SecRel: high TechDoc: high ComSup: high
33. Apache Spark core	"Apache Spark is a unified analytics engine for large-scale data processing. It provides high-level APIs in Java, Scala, Python and R … It also supports a rich set of higher-level tools including Spark SQL for SQL and structured data processing, pandas API on Spark for pandas workloads, MLlib for machine learning, GraphX for graph processing, and Structured Streaming for incremental computation and stream processing" [82]	4. Data modeling and analysis	Big data framework provider: processing – computing and analytics	Big data analytics engines/libraries	1 Spark context server 1 Spark cluster manager server Processing servers 4 worker node servers	MatLon: high SecRel: high TechDoc: high ComSup: high
34. Apache Spark MLlib	"MLlib is Spark's machine learning (ML) library. Its goal is to make practical machine learning scalable and easy" [83]	4. Data modeling and analysis	Big data framework provider: processing – computing and analytics	Big data analytics engines/libraries	1 Spark context server 1 Spark cluster manager server Processing servers 4 worker node servers	MatLon: high SecRel: high TechDoc: high ComSup: high

35. Apache SparkR	"SparkR is an R package that provides a light-weight frontend to use Spark from R" [84]	4. Data modeling and analysis	Big data framework provider: processing – computing and analytics	Big data analytics engines/libraries	1 Spark context server 1 Spark cluster manager server Processing servers 4 worker node servers	MatLon: high SecRel: high TechDoc: high ComSup: high
36. Sparklyr	"Sparklyr provides bindings to Spark's distributed machine learning library. In particular, sparklyr allows you to access the machine learning routines provided by the spark.ml package. Together with sparklyr's dplyr interface, you can easily create and tune machine learning workflows on Spark, orchestrated entirely within R" [85]	4. Data modeling and analysis	Big data framework provider: processing – computing and analytics	Big data analytics engines/libraries	1 Spark context server 1 Spark cluster manager server Processing servers 4 worker node servers	MatLon: high SecRel: high TechDoc: high ComSup: high
37. RHadoop	"RHadoop is a collection of five R packages that allow users to manage and analyze data with Hadoop." It includes the packages RHDFS and RHbase (for working with HDFS and HBASE data stores, respectively, in R) and the RMR package to provide a way for data analysts to access massive [86]	4. Data modeling and analysis	Big data framework provider: processing – computing and analytics	Big data analytics engines/libraries	1 Spark context server 1 Spark cluster manager server Processing servers 4 worker node servers	MatLon: mod SecRel: mod TechDoc: mod ComSup: mod

(continued)

Table A2 (continued)

Open source IT name	Self-description	BDASaaS pipeline stage	NBDRA layer	IT type	Minimal practical HW requirements	Level of open source technical attributes
38. RHive	"RHive is an R extension facilitating distributed computing via HIVE query. RHive allows easy usage of HQL (Hive SQL) in R, and allows easy usage of R objects and R functions in Hive" [87]	4. Data modeling and analysis	Big data framework provider: processing – computing and analytics	Big data analytics engines/libraries	1 Spark context server 1 Spark cluster manager server Processing servers 4 worker node servers	MatLon: mod SecRel: mod TechDoc: mod ComSup: mod
39. TensorFlow	TensorFlow is a deep learning ML library to "build and train models by using the high-level Keras API … For large ML training tasks, use the Distribution Strategy API for distributed training on different hardware configurations without changing the model definition" [88]	4. Data modeling and analysis	Big data framework provider: processing – computing and analytics	Big data analytics engines/libraries	1 computing cluster of 4 servers 1 data server	MatLon: high SecRel: high TechDoc: high ComSup: high
40. Keras	"Keras is a deep learning API written in Python, running on top of the machine learning platform TensorFlow … While TensorFlow is an infrastructure layer for differentiable programming, dealing with tensors, variables, and gradients, Keras is a user interface for deep learning, dealing with layers, models, optimizers, loss functions, metrics, and more" [89]	4. Data modeling and analysis	Big data framework provider: processing – computing and analytics	Big data analytics engines/libraries	1 computing server 1 data server	MatLon: high SecRel: high TechDoc: high ComSup: high

41. PyTorch	"PyTorch is an optimized tensor library for deep learning using GPUs and CPUs" [90]	4. Data modeling and analysis	Big data framework provider: processing – computing and analytics	Big data analytics engines/libraries	1 computing cluster of 4 servers 1 data server	MatLon: high SecRel: high TechDoc: high ComSup: high
42. TorchServe	"TorchServe takes a Pytorch deep learning model and it wraps it in a set of REST APIs. Currently it comes with a built-in web server that you run from command line. This command line call takes in the single or multiple models you want to serve, along with additional optional parameters controlling the port, host, and logging" [91]	4. Data modeling and analysis	Big data framework provider: processing – computing and analytics	Big data analytics engines/libraries	1 computing server 1 data server	MatLon: low SecRel: low TechDoc: mod ComSup: mod
43. ElasticSearch	"Elasticsearch is a distributed, RESTful search and analytics engine capable of addressing a growing number of use cases. As the heart of the Elastic Stack, it centrally stores your data for lightning fast search, fine-tuned relevancy, and powerful analytics that scale with ease" [92]	4. Data modeling and analysis	Big data framework provider: processing – computing and analytics	Big data analytics engines/libraries	1 computing server 1 data server	MatLon: high SecRel: high TechDoc: high ComSup: high

(continued)

Table A2 (continued)

Open source IT name	Self-description	BDASaaS pipeline stage	NBDRA layer	IT type	Minimal practical HW requirements	Level of open source technical attributes
44. OpenSearch	"OpenSearch is a distributed search and analytics engine based on Apache Lucene. After adding your data to OpenSearch, you can perform full-text searches on it with all of the features you might expect: search by field, search multiple indices, boost fields, rank results by score, sort results by field, and aggregate results" [93]	4. Data modeling and analysis	Big data framework provider: processing – computing and analytics	Big data analytics engines/libraries	1 computing server 1 data server	MatLon: high SecRel: high TechDoc: high ComSup: high
45. MLflow	"MLflow is a platform to streamline machine learning development, including tracking experiments, packaging code into reproducible runs, and sharing and deploying models. MLflow offers a set of lightweight APIs that can be used with any existing machine learning application or library (TensorFlow, PyTorch, XGBoost, etc), wherever you currently run ML code (e.g. in notebooks, stand-alone applications or the cloud)" [94]	4. Data modeling and analysis	Big data framework provider: processing – computing and analytics	Big data analytics engines/libraries	1 computing server 1 data server	MatLon: high SecRel: high TechDoc: high ComSup: high

46. Scikit-learn	"Scikit-learn is an open source machine learning library that supports supervised and unsupervised learning. It also provides various tools for model fitting, data preprocessing, model selection, model evaluation, and many other utilities" [95]	4. Data modeling and analysis	Big data framework provider: processing – computing and analytics	Big data analytics engines/libraries	1 computing server 1 data server	MatLon: high SecRel: high TechDoc: high ComSup: high
47. Apache Spark GraphX	GraphX is a Spark component for graphs and graph-parallel computation. "GraphX extends the Spark RDD by introducing a new Graph abstraction: a directed multigraph with properties attached to each vertex and edge. To support graph computation, GraphX exposes a set of fundamental operators (e.g., subgraph, joinVertices, and aggregateMessages) as well as an optimized variant of the Pregel API. In addition, GraphX includes a growing collection of graph algorithms and builders to simplify graph analytics tasks" [96]	4. Data modeling and analysis	Big data framework provider: processing – computing and analytics	Big graph engines/libraries	1 Spark context server 1 Spark cluster manager server Processing servers 4 worker node servers	MatLon: high SecRel: high TechDoc: high ComSup: high
48. Apache Giraph	"Apache Giraph is an iterative graph processing framework, built on top of Apache Hadoop. The input to a Giraph computation is a graph composed of vertices and directed edges … Each vertex stores a value, so does each edge. The input, thus, not only determines the graph topology, but also the initial values of vertices and edges" [97]	4. Data modeling and analysis	Big data framework provider: processing – computing and analytics	Big graph engines/libraries	1 master namenode server 4 dataNodes servers	MatLon: high SecRel: high TechDoc: high ComSup: high

(continued)

Table A2 (continued)

Open source IT name	Self-description	BDASaaS pipeline stage	NBDRA layer	IT type	Minimal practical HW requirements	Level of open source technical attributes
49. Neo4j Graph Database CE	Neo4j Graph Database is "a native graph data store built from the ground up to leverage not only data but also data relationships" [98]	4. Data modeling and analysis	Big data framework provider: processing – computing and analytics	Big graph engines/libraries	1 computing cluster of 4 servers 1 data server	MatLon: high SecRel: high TechDoc: high ComSup: high
50. Neo4j Graph Data Science CE	"Neo4j Graph Data Science is a connected data analytics and machine learning platform that helps you understand the connections in big data to answer critical questions and improve predictions." It includes "65+ pretuned graph algorithms and machine learning (ML) modeling to analyze your connected data" [99]	4. Data modeling and analysis	Big data framework provider: processing – computing and analytics	Big data analytics engines/libraries Big graph engines/libraries	1 computing cluster of 4 servers 1 data server	MatLon: high SecRel: high TechDoc: high ComSup: high
51. Kibana	"Kibana is a free and open user interface that lets you visualize your Elasticsearch data and navigate the Elastic Stack. Do anything from tracking query load to understanding the way requests flow through your apps" [100]	5. Data access and usage	Big data consumer component	Visual interactive analytics packages	1 computing server	MatLon: high SecRel: high TechDoc: high ComSup: high

52. OpenSearch Dashboards	"OpenSearch Dashboards is an open-source, integrated visualization tool that makes it easy for users to explore their data in OpenSearch. From real-time application monitoring, threat detection, and incident management to personalized search, OpenSearch Dashboards gives you the data visualizations needed to graphically represent trends, outliers, and patterns in your data. The image below shows a sample of data visualizations in OpenSearch Dashboards" [101]	5. Data access and usage	Big data consumer component	Visual interactive analytics packages	1 computing server	MatLon: high SecRel: high TechDoc: high ComSup: high
53. Looker Studio	"Looker Studio is a free tool that turns your data into informative, easy to read, easy to share, and fully customizable dashboards and reports" [102]	5. Data access and usage	Big data consumer component	Visual interactive analytics packages	Free cloud-based access	MatLon: high SecRel: high TechDoc: high ComSup: high
54. MS Power BI Service	"Power BI is a collection of software services, apps, and connectors that work together to help you create, share, and consume business insights in the way that serves you and your business most effectively. The Microsoft Power BI service (app.powerbi.com), sometimes referred to as Power BI online, is the SaaS (Software as a Service) part of Power BI" [103]	5. Data access and usage	Big data consumer component	Visual interactive analytics packages	Free cloud-based access	MatLon: high SecRel: high TechDoc: high ComSup: high

(continued)

Table A2 (continued)

Open source IT name	Self-description	BDASaaS pipeline stage	NBDRA layer	IT type	Minimal practical HW requirements	Level of open source technical attributes
55. RStudio	"RStudio is an integrated development environment (IDE) for R. It includes a console, syntax-highlighting editor that supports direct code execution, as well as tools for plotting, history, debugging and workspace management" [104]	5. Data access and usage	Big data consumer component	Visual interactive analytics packages	1 computing server	MatLon: high SecRel: high TechDoc: high ComSup: high
56. Shiny Server	It hosts Shiny web applications and interactive documents online. "Shiny is an R package that makes it easy to build interactive web apps straight from R. You can host standalone apps on a webpage or embed them in R Markdown documents or build dashboards. You can also extend your Shiny apps with CSS themes, htmlwidgets, and JavaScript actions" [105].	5. Data access and usage	Big data consumer component	Visual interactive analytics packages	1 computing server	MatLon: high SecRel: high TechDoc: high ComSup: high
57. Apache Zepellin	"Web-based notebook that enables data-driven, interactive data analytics and collaborative documents with SQL, Scala, Python, R and more" [106]	5. Data access and usage	Big data consumer component	Visual interactive analytics packages	1 computing server	MatLon: low SecRel: low TechDoc: mod ComSup: mod

58. Apache SuperSet	"Apache Superset is a modern, enterprise-ready business intelligence web application. It is fast, lightweight, intuitive, and loaded with options that make it easy for users of all skill sets to explore and visualize their data, from simple pie charts to highly detailed deck. gl geospatial charts" [107]	5. Data access and usage	Big data consumer component	Visual interactive analytics packages	1 computing server	MatLon: low SecRel: low TechDoc: mod ComSup: mod

References

1. Watson, H.J.: Tutorial: Big data analytics: concepts, technologies, and applications. Commun. Assoc. Inf. Syst. **34**, 1247–1268 (2014)
2. Phillips-Wren, G., Lakshmi, S.I., Uday, K., Ariyachandra, T.: Business analytics in the context of big data: a roadmap for research. Commun. Assoc. Inf. Syst. **37**(1), 448–472 (2015)
3. Kune, R., Konugurthi, P.K., Agarwal, A., Chillarige, R.R., Buyya, R.: The anatomy of big data computing. Softw. Pract. Exp. **46**(1), 79–105 (2016)
4. Cox, M., Ellsworth, D.: Managing big data for scientific visualization. In: ACM SIGGRAPH Proceedings, pp. 3–8, Los Angeles, CA (1997, Aug)
5. Mashey, J.R.: Big data and the next wave of {InfraStress} problems, solutions, opportunities. Paper presented at the 1999 USENIX Annual Technical Conference, Monterrey, CA, June 6-11, 1999
6. Laney, D.: 3D Data Management: Controlling Data Volume, Velocity and Variety. META Group Research Note 6 (2001)
7. Davenport, T.H., Barth, P., Bean, R.: How Big Data Is Different. Sloan Manag. Rev. **54**(1), 22–24 (2012)
8. McAfee, A., Brynjolfsson, E.: Big data: the management revolution. Harvard Bus. Rev. **90**(10), 1–9 (2012)
9. Wamba, S.F., Akter, S., Edwards, A., Chopin, G., Gnanzou, D.: How 'big data' can make big impact: findings from a systematic review and a longitudinal case study. Int. J. Prod. Econ. **165**, 234–246 (2015)
10. Tedford, J.R.: Analytics of decision making. J. Farm Econ. **46**(5), 1353–1362 (1964)
11. Kohavi, R., Neal, J.R., Simoudis, E.: Emerging trends in business analytics. Commun. ACM. **45**(8), 45–48 (2002)
12. Davenport, T.H.: Competing on analytics. Harvard Bus. Rev. **84**(1), 98–107 (2006)
13. INFORMS.: Best definition of analytics. https://www.informs.org/About-INFORMS/News-Room/O.R.-and-Analytics-in-the-News/Best-definition-of-analytics (2019). Accessed 1 Mar 2019
14. Russom, P.: Big data analytics. TDWI Best Pract. Rep. **19**(4), 1–34 (2011)
15. Tsai, C.W., Lai, C.F., Chao, H.C., Vasilakos, A.V.: Big data analytics: a survey. J. Big Data. **2**(1), 1–32 (2015)
16. Sun, Z., Huo, Y.: The spectrum of big data analytics. J. Comput. Inform. Syst. **61**(2), 154–162 (2021)
17. Monino, J.L.: Data value, big data analytics, and decision-making. J. Knowl. Econ. **12**(1), 256–267 (2016)
18. Saggi, M.K., Jain, S.: A survey towards an integration of big data analytics to big insights for value-creation. Inf. Proc. Manag. **54**(5), 758–790 (2018)
19. Dong, J.Q., Yang, C.: Business value of big data analytics: a systems-theoretic approach and empirical test. Inform. Manag. **57**, 103124 (2020)
20. Eckerson, W.: Big Data Analytics: Profiling the Use of Analytical Platforms in User Organizations. TDWI (2011, Sept)
21. Alharthi, A., Krotov, V., Bowman, M.: Addressing barriers to big data. Bus. Horizons. **60**(3), 285–292 (2017)
22. Baig, M.I., Shuib, L., Yadegaridehkordi, E.: Big data adoption: state of the art and research challenges. Inf. Proc. Manag. **56**(6), 102095 (2019)
23. Hu, H., Wen, Y., Chua, T.S., Li, X.: Toward scalable systems for big data analytics: a technology tutorial. IEEE Access. **2**, 652–687 (2014)
24. Barlas, P., Lanning, I., Heavey, C.: A survey of open source data science tools. Int. J. Intell. Comput. Cybern. **8**(3), 232–226 (2015)
25. Singh, D., Reddy, C.K.: A survey on platforms for big data analytics. J. Big Data. **2**(1), 1–20 (2015)

26. Grover, P., Kar, A.K.: Big data analytics: a review on theoretical contributions and tools used in literature. Global J. Flexible Syst. Manag. **18**(3), 203–229 (2017)
27. Oussous, A., Benjelloun, F.Z., Lahcen, A.A., Belfkih, S.: Big Data technologies: a survey. J. King Saud. Univ. Comput. Inf. Sci. **30**(4), 431–448 (2018)
28. Ajah, I.A., Nweke, H.F.: Big data and business analytics: trends, platforms, success factors and applications. Big Data Cognit. Comput. **3**(2), 1–32 (2019)
29. Ikegwu, A.C., Nweke, H.F., Anikwe, C.V., Alo, U.R., Okonkwo, O.R.: Big data analytics for data-driven industry: a review of data sources, tools, challenges, solutions, and research directions. Cluster Comput. **25**, 3343–3387 (2022)
30. TSO.: ITIL 4, Create, Deliver, Support. The Stationary Office, London (2018)
31. ISO/IEC: ISO/IEC 20000-2:2019, Information Technology — Service Management — Part 2: Guidance on the Application of Service Management Systems. International Organization for Standardization, Geneva (2019)
32. Delen, D., Demirkan, H.: Data, information and analytics as services. Decis. Support. Syst. **55**(1), 359–363 (2013)
33. Wang, X., Yang, L.T., Liu, H., Deen, M.J.: A big data-as-a-service framework: state-of-the-art and perspectives. IEEE Trans. Big Data. **4**(3), 325–340 (2017)
34. ISO/IEC/IEEE: ISO/IEC/IEEE: 42010: 2011 Systems and Software Engineering, Architecture Description. International Organization for Standardization, Geneva (2011)
35. Hunnebeck, L.: Service Design. The Stationary Office, London (2011)
36. Mora, M., Raisinghani, M., O'Connor, R.V., Marx Gomez, J., Gelman, O.: An extensive review of IT service design in seven international ITSM processes frameworks: part I. Int. J. Inf. Technol. Syst. Appr. **7**(2), 83–107 (2014)
37. NIST: NIST Big Data Interoperability Framework: Volume 6, Reference Architecture Version 3. NIST Special Publication 1500-6r2. National Institute of Standards and Technology, Gaithersburg (2019)
38. Lukoianova, T., Rubin, V.L.: Veracity roadmap: is big data objective, truthful and credible? Adv. Classif. Res. Online. **24**(1), 4–15 (2014)
39. Mell, P., Grance, T.: The NIST Definition of Cloud Computing. Special Publication 800-145. National Institute of Standards and Technology, Gaithersburg (2011)
40. Angelov, S., Grefen, P., Greefhorst, D.: A framework for analysis and design of software reference architectures. Inf. Softw. Technol. **54**(4), 417–431 (2012)
41. The Open Group.: The Open Group IT4IT™ Reference Architecture, Version 2.1. The Open Group, Berkshire (2017)
42. The Open Group.: The TOGAF® Standard, Version 9.2. Berkshire, The Open Group, Berkshire (2018)
43. Pääkkönen, P., Pakkala, D.: Reference architecture and classification of technologies, products and services for big data systems. Big Data Res. **2**(4), 166–186 (2015)
44. Cloud Standards Consumer Council: Cloud Customer Architecture for Big Data and Analytics V2.0. Cloud Standards Consumer Council, Massachusetts (2017)
45. Gökalp, M.O., Kayabay, K., Zaki, M., Koçyiğit, A., Eren, P.E., Neely, A.: Big-Data Analytics Architecture for Businesses: a Comprehensive Review on New Open-Source Big-Data Tools. Cambridge Service Alliance, Cambridge (2017)
46. Cooper, H.M.: Organizing knowledge syntheses: a taxonomy of literature reviews. Knowl. Soc. **1**(1), 104–126 (1988)
47. Templier, M., Paré, G.: A framework for guiding and evaluating literature reviews. Commun. Assoc. Inf. Syst. **37**(1), 112–137 (2015)
48. Mora, M., Marx Gómez, J., O'Connor, R.V., Gelman, O.: An MADM risk-based evaluation-selection model of free-libre open source software tools. Int. J. Technol. Policy Manag. **16**(4), 326–354 (2016)
49. Apache Organization.: Cloudstack. https://docs.cloudstack.apache.org (2022). Accessed 1–26 Aug 2022

50. OpenStack Organization.: OpenStack. https://docs.openstack.org (2022). Accessed 1–26 Aug 2022
51. Apache Organization.: Mesos. https://mesos.apache.org/documentation (2022). Accessed 1–26 Aug 2022
52. Apache Organization.: Hadoop Yarn. https://hadoop.apache.org/docs/stable/hadoop-yarn (2022). Accessed 1–26 Aug 2022
53. Apache Organization.: Zookeeper. https://zookeeper.apache.org/doc/ (2022). Accessed 1–26 Aug 2022
54. Apache Organization.: Kafka. https://kafka.apache.org/documentation (2022). Accessed 1–26 Aug 2022
55. Apache Organization.: Flink. https://flink.apache.org/flink-architecture.html (2022). Accessed 1–26 Aug 2022
56. Apache Organization.: Storm. https://storm.apache.org (2022). Accessed 1–26 Aug 2022
57. Elastic Organization.: ElasticSearch. https://www.elastic.co/guide/en/logstash/ (2022). Accessed 1–26 Aug 2022
58. Apache Organization.: IoTDB. https://iotdb.apache.org (2022). Accessed 1–26 Aug 2022
59. Apache Organization.: Flume. https://flume.apache.org/documentation.html (2022). Accessed 1–26 Aug 2022
60. Apache Organization.: Hadoop. https://hadoop.apache.org/docs/ (2022). Accessed 1–26 Aug 2022
61. Apache Organization.: Ambari. https://ambari.apache.org/ (2022). Accessed 1–26 Aug 2022
62. Apache Organization.: Hudi. https://hudi.apache.org/docs/ (2022). Accessed 1–26 Aug 2022
63. Delta Lake.: Delta Lake. https://docs.delta.io/latest/index.html (2022). Accessed 1–26 Aug 2022
64. Apache Organization.: Griffin. https://griffin.apache.org/ (2022). Accessed 1–26 Aug 2022
65. OpenRefine Organization.: OpenRefine. https://openrefine.org/ (2022). Accessed 1–26 Aug 2022
66. DataCleaner.: DataCleaner. https://datacleaner.github.io/docs/ (2022). Accessed 1–26 Aug 2022
67. PostgreSQL Organization.: PostgreSQL. https://www.postgresql.org/docs/ (2022). Accessed 1–26 Aug 2022
68. MySQL Organization.: MySQL. https://dev.mysql.com/doc/ (2022). Accessed 1–26 Aug 2022
69. Apache Organization.: Hive. https://hive.apache.org/ (2022). Accessed 1–26 Aug 2022
70. Apache Organization.: Druid. https://druid.apache.org/docs/ (2022). Accessed 1–26 Aug 2022
71. Apache Organization.: Hbase. https://hbase.apache.org/book.html (2022). Accessed 1–26 Aug 2022
72. Apache Organization.: Cassandra. https://cassandra.apache.org/doc/ (2022). Accessed 1–26 Aug 2022
73. Apache Organization.: Hadoop MapReduce https://hadoop.apache.org/docs/stable/hadoop-mapreduce-client/hadoop-mapreduce-client-core/MapReduceTutorial.html (2022). Accessed 1–26 Aug 2022
74. Apache Organization.: Impala. https://impala.apache.org/docs/ (2022). Accessed 1–26 Aug 2022
75. Presto.: Presto. https://prestodb.io/docs/current/ (2022). Accessed 1–26 Aug 2022
76. Apache Organization.: Spark SQL. https://spark.apache.org/docs/latest/sql-programming-guide.html (2022). Accessed 1–26 Aug 2022
77. Trino Organization.: Trino. https://trino.io/docs/current/ (2022). Accessed 1–26 Aug 2022
78. Apache Organization.: Drill. https://drill.apache.org/ (2022). Accessed 1–26 Aug 2022
79. Apache Organization.: Pig. https://pig.apache.org/ (2022). Accessed 1–26 Aug 2022
80. Apache Organization.: Kylin. https://kylin.apache.org/docs31/ (2022). Accessed 1–26 Aug 2022
81. Apache Organization.: Mahout. https://mahout.apache.org/ (2022). Accessed 1–26 Aug 2022

82. Apache Organization.: Spark. https://spark.apache.org/docs/latest/index.html (2022). Accessed 1–26 Aug 2022
83. Apache Organization.: Spark MLlib. https://spark.apache.org/docs/latest/ml-guide.html (2022). Accessed 1–26 Aug 2022
84. Apache Organization.: Spark R. https://spark.apache.org/docs/latest/api/R/index.html (2022). Accessed 1–26 Aug 2022
85. RStudio.: RStudio CE. https://spark.rstudio.com/ (2022). Accessed 1–26 Aug 2022
86. Revolution Analytics.: RHadoop. https://github.com/RevolutionAnalytics/RHadoop/ wiki (2022). Accessed 1–26 Aug 2022
87. Nexr.: RHive. https://github.com/nexr/RHive/wiki/User-Guide (2022). Accessed 1–26 Aug 2022
88. Tensorflow Organization.: Tensorflow. https://www.tensorflow.org/learn (2022). Accessed 1–26 Aug 2022
89. Keras IO.: Keras. https://keras.io/getting_started/ (2022). Accessed 1–26 Aug 2022
90. Pytorch Organization.: Pytorch. https://pytorch.org/docs/stable/index.html (2022). Accessed 1–26 Aug 2022
91. Pytorch Organization.: PytorchServe. https://github.com/pytorch/serve/blob/master/ docs/ server.md (2022). Accessed 1–26 Aug 2022
92. Elastic Organization.: Elastic. https://www.elastic.co/elasticsearch/ (2022). Accessed 1–26 Aug 2022
93. Opensearch Organization.: Opensearch. https://opensearch.org/docs/latest/opensearch/ index/ (2022). Accessed 1–26 Aug 2022
94. Pypi Organization.: Pypi. https://pypi.org/project/mlflow/ (2022). Accessed 1–26 Aug 2022
95. Scikit Organization.: Scikit. https://scikit-learn.org/ (2022). Accessed 1–26 Aug 2022
96. Apache Organization.: Spark GraphX. https://spark.apache.org/docs/latest/graphx-programming-guide.html#overview (2022). Accessed 1–26 Aug 2022
97. Apache Organization.: Giraph. https://giraph.apache.org/intro.html (2022). Accessed 1–26 Aug 2022
98. Neo4j.: Neo4j Graph Database CE. https://neo4j.com/product/neo4j-graph-database/ (2022). Accessed 1–26 Aug 2022
99. Neo4j.: Neo4j Graph Data Science CE. https://neo4j.com/product/graph-data-science/ (2022). Accessed 1–26 Aug 2022
100. Elastic.: Kibana. https://www.elastic.co/kibana/ (2022). Accessed 1–26 Aug 2022
101. Opensearch Organization.: OpenSearch Dashboards. https://opensearch.org/docs/1.0/ dashboards/index/ (2022). Accessed 1–26 Aug 2022
102. Google Company.: Lookerstudio. https://lookerstudio.google.com/overview (2022). Accessed 1–26 Aug 2022
103. Microsoft Company.: PowerBI. https://learn.microsoft.com/en-us/power-bi/fundamentals/ power-bi-service-overview (2022). Accessed 1–26 Aug 2022
104. RStudio Company.: RStudio Server CE. https://www.rstudio.com/products/rstudio/ #rstudio-server (2022). Accessed 1–26 Aug 2022
105. RStudio Company.: Shiny Server CE. https://shiny.rstudio.com/tutorial/ (2022). Accessed 1–26 Aug 2022
106. Apache Organization.: Zeppelin. https://zeppelin.apache.org/ (2022). Accessed 1–26 Aug 2022
107. Apache Organization.: Superset. https://superset.apache.org/docs/intro (2022)

The Role of Machine Learning in Big Data Analytics: Current Practices and Challenges

Hector A. Duran-Limon, Arturo Chavoya, and Martha Hernández-Ochoa

1 Introduction

Huge amounts of data are generated on a daily basis by diverse application domains. Social media, mobile phones, sensors, and medical imaging among others are examples of data sources. The exponential growth of both the Internet and data digitalization has fueled the generation of high volumes of data. According to the International Data Corporation, such generation of data will increase from 33 zettabytes in 2018 to 175 zettabytes in 2025 [1]. For instance, regarding social media data generated in 1 minute in October 2021 [2], we have that 694 million songs were streamed in the USA, there were 4.2 million Google searches, 210 million emails were sent, and 21 million snaps were created.

Big data analytics (BDA) enables the extraction of valuable information from large datasets that are obtained from multiple sources. Such valuable information involves patterns and correlations that can help organizations to make better decisions [3–5]. Laney defined big data in terms of Volume, Velocity, and Variety [6]. Two more Vs were added later: Value and Veracity [7]. Currently, the 5 Vs are the most widely accepted conceptualization of big data. Volume refers to large volumes of data that increase exponentially with time. Velocity regards the speed at which data are generated and processed. The diverse number of data sources and heterogeneity of the data denote Variety. In addition, Value refers to the extracted patterns

H. A. Duran-Limon (✉) · A. Chavoya
Information Systems, CUCEA, University of Guadalajara, Guadalajara, Mexico
e-mail: hduran@cucea.udg.mx; achavoya@cucea.udg.mx

M. Hernández-Ochoa
Knowledge Fundamentals, CUNORTE, University of Guadalajara, Guadalajara, Mexico
e-mail: martha.ochoa@cunorte.udg.mx

M. Mora et al. (eds.), *Development Methodologies for Big Data Analytics Systems*, Transactions on Computational Science and Computational Intelligence, https://doi.org/10.1007/978-3-031-40956-1_2

and correlations that can help to make better decisions. Lastly, Veracity involves the level of confidence on the data.

The main elements of BDA include descriptive analytics, predictive analytics, and prescriptive analytics. Descriptive analytics is in charge of describing what has happened, where past and current patterns can be identified and highlighted. In contrast, predictive analytics identify correlations among different variables whereby the value of a variable can be forecasted when other variables suffer changes. On the other hand, prescriptive analytics helps to find the best option or recommendation under uncertainty conditions.

The data can be in different formats, namely, unstructured, semi-structured, and structured. Unstructured data do not have a structural organization and comprise videos, audios, pictures, and online text. Semi-structured data is data that is partially structured; an example of data in this format is XML data in the web, which employ an informal tag-type format for organizing the data. Lastly, structured data can be normally extracted from relational databases and spreadsheets. Crucially, most of the data is either unstructured or semi-structured.

Traditional approaches such as data warehousing and the use of a classic relational database management system (RDBMS) have become impractical to analyze such unstructured and semi-structured data [8]. On the other hand, machine learning (ML) algorithms have proven to be successful in analyzing vast amounts of data [4, 7, 9–11].

Machine learning is part of the artificial intelligence field, which involves algorithms and statistical models that are able to learn and adapt without following explicit instructions to do so [12]. ML algorithms can be categorized in three main classes: unsupervised learning, supervised learning, and reinforcement learning. Unsupervised learning is used to find a hidden structure on unlabeled data. This kind of algorithm groups data into clusters. Unsupervised learning can be used, for example, for customer segmentation and pattern classification. In contrast, with supervised learning, the data must be already labeled or structured. Algorithms of this kind infer a function from the labeled data that enables them to make either predictions or decisions. There are two subcategories of supervised learning: classification and regression. The former is used to identify the class of a data point. Classification can be used for speech recognition, image recognition, and fraud detection, among others. On the other hand, regression algorithms are employed for prediction. The value of the dependent variable is predicted from a continuous dataset. The independent variables are used for modeling or training. Regression has been applied, for example, for weather forecasting or predicting the value of a stock in the stock market. Lastly, reinforcement learning is an approach whereby each type of action is given a different reward. By using trial and error, the actions that have the greatest reward are learned. The goal of reinforcement learning is to find the policy that maximizes the reward function. This type of method is commonly applied in gaming and robotics.

Some of the most important domain areas of BDA include health and human welfare, weather forecasting, customer transactions, customer preferences, financial analysis, and social networking and the Internet. In this chapter, we focus on describing some of the most widely used ML algorithms and platforms for BDA, as well as analyzing the role that the use of ML has played in some of these domain areas. More specifically, we present the use of ML for BDA in the areas of healthcare, weather forecasting, and social networking and the Internet.

The chapter is organized as follows. We first present some of the most widely used ML algorithms in BDA. Then, we present the most commonly used distributed platforms for processing big data. This section is followed by a review of a selection of three important domain areas where BDA is employed. Finally, some concluding remarks are drawn.

2 Machine Learning Techniques

In this section, we present some of the machine learning techniques that are frequently used in BDA [4].

2.1 *Support Vector Machines*

Support vector machines (SVMs) [13] are supervised learning models that are employed for binary classification and regression analysis of data. The training algorithm is a non-probabilistic binary linear classifier that classifies the trained data into one of two categories. The SVM maps training data into points in space whereby the width of the gap between the two categories is maximized. New data is then mapped to the same space in which it is predicted to which category it belongs and what position in space it takes. A data point involves a number of features. A hyperplane is used to classify these data points into two classes. It is desirable that the margin or distance from the hyperplane to the nearest point of each side is maximized, as shown in Fig. 1.

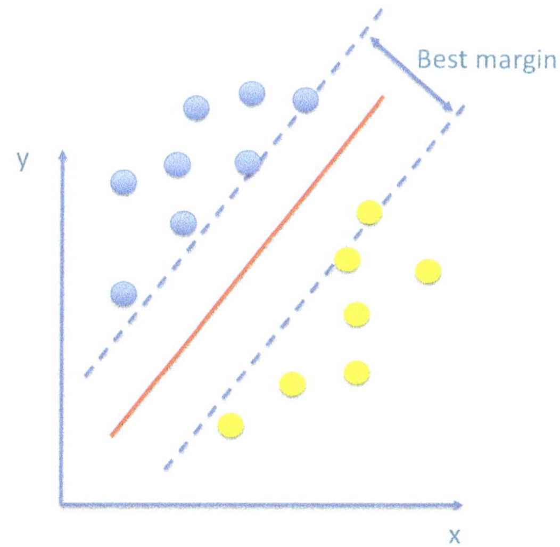

Fig. 1 Linear SVM

In many cases, it is not always possible to separate perfectly both classes. Soft margin classifiers (also called support vector classifiers) allow that certain data points be in the incorrect side, so that the distance of the hyperplane is maximized from the majority of the data points of both sides, obtaining a more robust classifier with a better predictive capacity when applied to new data points. In case the separation between the groups is nonlinear, the dimensions of the space can be expanded. In fact, the dimension of the hyperplane depends on the number of features (i.e., data inputs characterizing each data point). A kernel function can be used to efficiently map the input data into high-dimensional spaces.

Support vector regression (SVR) is a variant of SVM. This variant employs a regression scheme used for predicting values. In SVM, the margins do not include data, whereas in SVR the margin lines are chosen so that they cover all data (hard margin) or permit some violation (soft margin). These margins involve a tolerance error (epsilon). The aim here is to find the function that represents a line that is between the two margins, as shown in red in Fig. 2. SVR also allows a nonlinear regression analysis.

The parameter needed by SVMs is the so-called soft margin parameter, which is normally indicated with C, and the kernel function. In the case of SVRs, an additional parameter called ϵ is needed. In cases where there is much noise, ϵ must be selected accordingly to reflect the variance of noise. In cases where no noise is present, we have an interpolation problem and ϵ corresponds to the preset interpolation accuracy. The larger the value of ϵ, the smaller the number of support vectors required, and vice versa. In addition, a procedure of cross validation is commonly employed for the selection of both the kernel function and the optimal value of C.

A parallel implementation of SVM that employs MapReduce to reduce the training time is presented in [14].

Fig. 2 Linear SVR

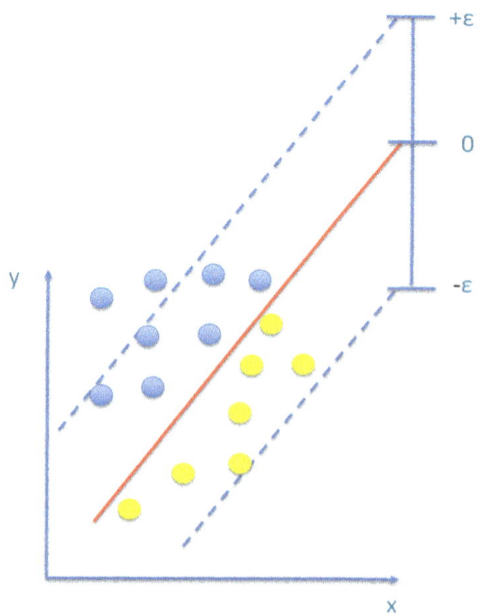

2.2 Decision Trees

Decision trees are nonparametric supervised learning models that are used for classification and regression analysis of data. Decision trees are able to carry out a multi-class classification on a dataset. There are various methodologies for generating decision trees. The classification and regression trees (CART) algorithm [15] is the most widely used, which is described below.

A decision tree is a binary tree that can be constructed by splitting the data input into subsets of data based on an attribute evaluation. There are two kinds of nodes: decision nodes and leaf nodes. Decision nodes contain a condition to split the data, whereas leaf nodes help to decide the class of a new data point. Decision trees that classify data into categories are called classification trees, whereas decision trees that predict values are called regression trees. In the case of classification trees, the best split is found using the Gini impurity index, which is equivalent to using the entropy or information-gain criterion. On the other hand, in the case of regression trees, the best split is the split that minimizes the residual sum of squares (RSS) of the observed and predicted values.

A recursive partitioning is carried out in which this splitting process is performed on each derived branch. A decision tree is split down from the root to leaf nodes. The data points are located in axis-parallel (hyper-) rectangles, as shown in Fig. 3. In case of overfitting, there are mechanisms that help address this issue. Pruning is one of such mechanisms, which involves the process of removing a branch from a decision node.

Once the decision tree is constructed, the predictions are carried out on the leaves where the mode is taken for classification, whereas the mean is used for regression.

One of the main advantages of regression trees over other ML approaches is that the graphical model of a regression tree helps to understand the phenomenon represented in the data. That is, the features located in the upper nodes of the tree play a more important role in the prediction process. For instance, regarding weather

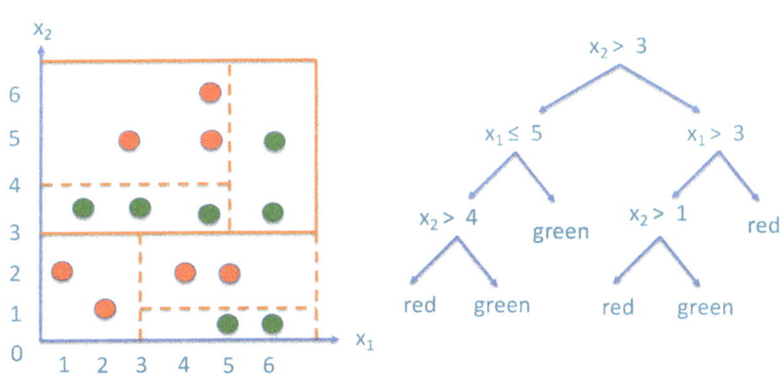

Fig. 3 Data points space of a classification tree

forecast, in case of having as an upper node, let us say wind speed, and as a lower node moisture, this would indicate that wind speed has a higher impact on the temperature than moisture.

There are a number of parallel versions of decision trees implemented with MapReduce, such as [16, 17].

2.3 Clustering Algorithms

Clustering algorithms create clusters of datasets, whose members are more closely related to each other than to members of other clusters. The main idea of these algorithms is to distribute input data into clusters without requiring labels for the training set [18]. The behavior of these algorithms is shown in Fig. 4. In this kind of algorithm, two requirements are met: (1) each cluster must have a set and (2) at least one element must exist in the cluster [19].

One popular clustering algorithm is *k*-means, which is described as follows. *K*-means groups datasets based on closeness to each other using the Euclidean distance [20], where the aim is to minimize the distance between the elements within the cluster, and *k* is the number of clusters. The *k*-means algorithm consists of assigning each element from the dataset to the defined *k*-th cluster closer to this element. For each iteration, the *k*-th cluster is calculated once the associated elements are observed to the related cluster. This process is iterative until all elements from the dataset are assigned to the clusters [21]. For *n* elements and a dimension *d*, the *k*-means algorithm complexity is $O(k*n*d)$, so it is computationally efficient [22]. The steps of the algorithm are defined as follows:

1. Define the number of clusters *k*.
2. Select *k* random elements from the dataset as centroids. In other words, select one element (called centroid) for each cluster.
3. All the elements are assigned to the closest cluster centroid.
4. Recalculate the *k*-th cluster once all the elements are associated with related clusters.
5. Repeat until one of the next criteria is met.

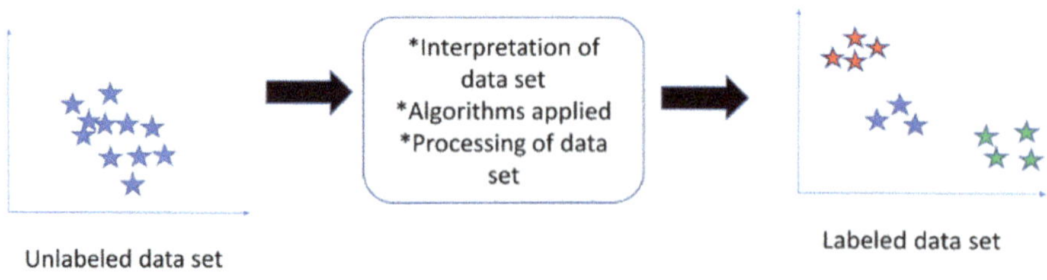

Fig. 4 Behavior of clustering algorithms

(a) The k-th value is reached.
(b) The elements remain in the same cluster.
(c) Once the new cluster is defined, the centroid is the same.

Some advantages of k-means are as follows: it is based on mathematical ideas, it is easily implemented, and it has fast convergence [23]. However, there are some drawbacks such as the following: when a global cluster is used, it is not effective; also, the size and density of the cluster is not handled by the algorithm [20]; with the traditional k-means algorithm, it is difficult to analyze a massive dataset; and prediction of a k value is hard. K-means is utilized for document classification, insurance fraud detection, customer segmentation, rideshare data analysis, automatic clustering of IT alerts, and call record details analysis, among other applications [22]. There are several improvements of this algorithm that have been made in different research works [21, 24–27], for example, "spectral clustering," which uses standard linear algebra methods. This algorithm is built on graph Laplacian matrices [21].

The main steps are described as follows:

1. Create a similarity graph to cluster between N objects.
2. Compute the first k eigenvectors of its Laplacian matrix.
3. Run k-means to separate objects into k classes.

Distributed clustering algorithms are classified into homogeneous and heterogeneous, based on the type of dataset they process. Most distributed clustering algorithms are focused on homogenous datasets. Some distributed clustering algorithms are described next. In [28], the authors proposed a distributed dynamic clustering algorithm (DDCA), which is based on k-means and a tree topology. Another article presented a Noise-based k-means that has better results for urban hotspots over k-means [29]. In [30] a new approach for very large spatial heterogeneous datasets is proposed, which is based on the k-means algorithm but generates clusters dynamically.

2.4 Artificial Neural Networks

Some of the most used ML techniques in BDA are the different variants of artificial neural networks (ANNs) [31]. ANNs are a family of models inspired in biological neural networks and consist of at least one input layer and one output layer of nodes, where each node corresponds to an artificial neuron and the nodes in one layer are connected to the nodes in the adjacent layer. The nodes in the input layer receive the values introduced in the model, and the nodes in the output layer produce the response of the model. There can also be one or more intermediate layers, which are known as "hidden" layers. The role of the hidden layers is to discover features that are informative for the desired goal. The connection among two nodes is represented by a function, whose parameters need to be adjusted by training the network with input data. An ANN that only contains an input layer and an output layer and all the nodes in one layer are connected to all nodes in the other layer is known as a perceptron [32].

Deep learning models are a special kind of ANNs that use a "deep" architecture, that is, one that contains more than one hidden layer. Deep learning methods are very effective when dealing with a large number of training samples. The current success of deep learning is due to a great extent to three factors: (1) recent advances in the development of high-performance central processing units (CPUs) and graphics processing units (GPUs), (2) the availability of big data, and (3) recent developments in ML algorithms. Unlike shallow architectures that depend on the availability of expert human knowledge to train the supervised models, deep models can discover useful features from data in a hierarchical way from fine to abstract in an unsupervised manner, where each layer in the network discovers new characteristics of the data in an incremental way. Deep learning models can be classified as either multilayer neural networks that take nonstructured vector values as input or convolutional neural networks (CNNs) that take multidimensional structured values as input. Within the first category, three widely used deep models are stacked autoencoders, deep belief networks, and deep Boltzmann machines. These models differ in the way the connections among layers are made, whether they are directed or undirected, and the direction of the connection (toward the output layer or toward the input layer). On the other hand, CNNs use the spatial and configurational information of adjacent data points that cannot exist in the vectorized data used by the multilayered neural networks. This characteristic makes CNNs especially suitable to analyze 2D or 3D data (such as images) to discover patterns of interest [32].

3 Open-Source Platforms for Big Data Analytics

We present some of the most used open-source platforms for BDA after a brief introduction to the MapReduce model.

3.1 MapReduce

MapReduce [33] is a programming model developed by Google for processing big data on a distributed platform. The data is processed in batches in parallel by using either clusters or grid systems.

This programming model involves two main operations: Map and Reduce. The former involves splitting and mapping the data, whereas the latter performs a summary operation. The Map function takes input key-value pairs (K1, V1), which are transformed to different key-value pairs (K2, V2). Afterward, a shuffling process is carried out, whereby all pairs with the same key (K2) are collected and grouped according to their key value. MapReduce then uses the Reduce function to process the data of each group, which is transformed into different key-value pairs (K3, V3).

Both the Map and the Reduce functions are run in parallel. Data inputs and data outputs are stored in a distributed file system.

The performance and scalability of MapReduce may be negatively impacted when there are large amounts of data that need to be written by the Map operation. Also, the communication costs commonly overcome computation costs given that many MapReduce implementations employ a distributed storage in order to address crash recovery.

MapReduce is useful for different kinds of applications, such as distributed pattern-based searching, distributed sorting, web access log stats, ML, and document clustering, among others. There are a number of frameworks implementing MapReduce such as Hadoop and Spark, which are presented below.

3.2 Apache Hadoop

Apache Hadoop is a parallel computing framework whose main function is to store and process large datasets across clusters of computers [34]. Hadoop is designed to scale up from single servers to thousands of nodes, each one having its own storage and processing. It consists of four modules: (1) Hadoop Common, which includes utilities to support the other Hadoop modules; (2) Hadoop Distributed File System (HDFS), which is a distributed file system that gives high-throughput access to application data; (3) Hadoop YARN (Yet Another Resource Negotiator), which is a framework that provides cluster resource management and job scheduling for managing the extensive storage resources and keeping track of the processing workload across clusters; and (4) Hadoop MapReduce, which is a YARN-based system for the parallel processing of large datasets contained on HDFS clusters; during a Map step, the master node divides the job into smaller tasks and distributes the resources depending on the task, and after the computations, the Reduce step aggregates all the partial results to produce an integrated solution to the problem [34, 35].

Even though Apache Hadoop is extensively used, it has some drawbacks. One problem with Hadoop is that it is strictly a batch computing platform, and as such, it is not suitable for real-time streaming applications where immediate results are expected. Another problem with Hadoop is the skew problem, which happens when during a Map and Reduce operation there is an imbalance in the time between a Map step and the corresponding Reduce step, which can cause a delay in the execution of one of the steps [34]. Some of the problems with Hadoop are solved with Spark, which is better suited for real-time data processing, but Hadoop is still considered more suitable for BDA in terms of cost, security, and fault tolerance when batch processing is involved [36].

3.3 Apache Spark

Spark is the topmost used tool (34.88%) for BDA among experts in this field, according to [4]. It is a parallel and open-source cluster computing framework developed as an Apache project. Spark was created in 2009 in UC Berkeley's AMPLab [37]. Spark runs on top of an HDFS (Hadoop Distributed File System) infrastructure. Spark also supports SparkQL, Spark Streaming, MLib, and GraphX libraries for ML and data mining. Multi-language and analytics are also supported. Spark is deployed in a big data hybrid (batch and real time) processing model [38]. Spark can access Hadoop Distributed File System (HDFS), Hbase, and Cassandra [39].

Some benefits of using this platform are that it is easier to use, programs run faster (up to 100 times quicker than Hadoop MapReduce [39]), and it has high processing speed. Also, Spark is highly efficient with massive amounts of data and has fault tolerance without replication, reducing read disk, write disk, and the network I/O cost and employing in-memory computation operations. Furthermore, it covers batch, streaming, interactive, and iterative workloads [40]. In Spark, resilient distributed datasets (RDDs) are the main abstraction and provide a way to treat all distributed RAM as a single memory, which provides robustness against data loss. In [38] the authors figured out that Spark performs better than MapReduce for all datasets due to its in-memory computation, less overhead in setting up jobs for every iteration, and lower network I/O cost. For these reasons, Spark is in general the preferred choice by experts in big data.

On the other hand, some drawbacks are that Spark consumes more memory in operation than Hadoop, and as such the cost is very high and the latency is higher, so results have lower throughput and iteration processing. Other problems involve that there is no file management system and that there are small file issues, among others.

3.4 Other Open-Source Platforms and Tools

Even though Hadoop and Spark are the main big data platforms currently in use, there are other platforms that can be useful under specific circumstances, and some of them can even interact with Hadoop or Spark. Some of these platforms are listed next.

Apache Storm can be an alternative to Hadoop MapReduce when there is a heavy need for real-time big data processing. The main difference between Hadoop and Storm is that the former runs jobs, whereas the latter runs topologies, and while a MapReduce job can finish, a topology continues processing incoming data until the user terminates the process [34].

Apache Flink is a platform that provides real-time processing of data streams, and at the same time it can process historical batch data. Flink offers many libraries, including support for ML, a graph API, and a table API to process SQL operations, among others [34].

On the other hand, Apache Flume is an agent-based platform that provides reliable, distributed, and accessible web services from various sources to collect, aggregate, and transfer large amounts of streaming data to a centralized data store [41].

Regarding storage systems for BDA, traditional SQL database systems are not suitable for storing large quantities of unstructured data, such as text documents. Consequently, in these cases, there has been a need to transition to NoSQL databases for storing this kind of data to be processed by BDA systems. In a recent systematic literature review, the NoSQL storage tools most cited in the publications were MongoDB, Hbase, CouchDB, Cassandra, and Neo4J, although other storage systems such as BigTable, HyperTable, and SimpleDB were also cited [42].

As for other tools used for big data, WEKA (Waikato Environment for Knowledge Analysis) is an open-source software that contains, among other things, a collection of implementations of well-known ML algorithms [43]. Another popular tool for the analysis of big data is the R language, which provides a wide variety of statistical and graphics techniques [44]. Finally, some ML algorithms cited in the literature are implemented using general-purpose programming languages, such as Python and C++.

4 Domain Areas of Big Data Analytics

We present some of the main domain areas to which BDA is currently applied. The articles presented in this section were considered based on some inclusion-exclusion criteria, which are described next. The inclusion criteria were as follows: (i) the article is written in the English language; (ii) the article must relate to ML algorithms, BDA, and related platforms and/or tools; (iii) the article was published between the years 2012 and 2022; (iv) the article was published in a journal or conference; (v) the article addresses one of the considered domains; and (vi) the article was selected from a subset of high quality journals and conferences, such as those supported by IEEE or ACM. On the other hand, the exclusion criteria were (i) articles not published within the period 2012–2022, (ii) papers not published in a journal or conference, and (iii) papers with not enough relevance to the main topic of this chapter.

4.1 Healthcare

Healthcare systems produce the largest and fastest growing datasets corresponding mainly to electronic medical records (EMRs) and imaging data, which are considered clinical data [45]. Other types of healthcare-related data are patient behavior and sentiment data such as those coming from wearable sensors and social sites;

administration and cost activity data, such as financial and operational data, and patient profiles including dietary habits, exercise patterns, and environmental factors; and pharmaceutical and research and development data, including mechanism of action of drugs, and their side effects and toxicity [46].

Collected patient information is growing both in volume and complexity. For instance, neuroimaging currently produces more than 10 petabytes (10^{15}) of data each year, and genomic sequencing data is expected to reach exabyte (10^{18}) proportions per year within the next decade, exceeding other big data fields such as astronomy [47]. Given that healthcare is a data-intensive field and that health data comes from numerous sources and in different formats, traditional software systems are not able to handle this kind of data [34]. It is therefore justified to use the tools provided by BDA to collect, organize, analyze, and evaluate massive datasets from healthcare systems in order to identify patterns and other information of interest that can lead as an ultimate goal to improve human welfare [48].

Health BDA has mainly four challenges:

1. Data aggregation, as health big data come from different sources, it has to be put together from warehouses located in different places and in real time.
2. Data maintenance and storage, which require both SQL and NoSQL databases systems, as the data are growing at an exponential rate and come in different formats.
3. Data integration and interoperability, as data come in structured, semi-structured, and unstructured formats, and a way has to be found to standardize all these data so that systems can operate together.
4. Data analysis, as the time and resource requirements increase exponentially as the number of records increases, the hardware and software needed to analyze health data have to grow in size and complexity to provide robust analytical tools to perform analyses that extract knowledge from the data [34].

The three types of analytics are of interest in healthcare: descriptive, predictive, and prescriptive [46]. In a 2020 review of 804 articles that applied BDA to healthcare data, almost half of the articles used predictive analytics, approximately a third used prescriptive analytics, and nearly a quarter used descriptive analytics [46]. These results emphasize the fact that in healthcare, predicting outcomes is more valuable than building an explanatory model, as delaying action waiting for a complete model can cost lives [49]. In this same review, 70% of studies used clinical data, many articles (40%) included experiments with the hope that the proposed predictive and prescriptive models be incorporated in systems used by decision-makers in healthcare organizations, and nearly 65% of the articles focused on ML and data mining techniques applied to the field of health, such as the classification of medical data and symptoms and diagnosis and prediction of diseases [46]. In general, ML and statistical methods such as data mining are among the main approaches used in predictive analysis in order to make informed decisions on patient care by examining current and historical facts to predict future outcomes [50].

Machine learning techniques can be valuable for the prediction of disease occurrences or their complications. Although many ML algorithms can be applied to solve health related problems, each type of problem might best be solved using a particular technique or a certain combination of techniques. For instance, deep learning has been successfully applied to the classification of medical images and videos, frequently in combinations with the processing of EMRs [47]. In healthcare, the following ML algorithms have been used on big data [34, 48]: K-nearest neighbor, support vector machines, neural networks, k-means clustering techniques, ensemble learning, Markov decision process, decision trees, and naïve Bayes.

Regarding the platforms, the following big data platforms are popular in health informatics: Hadoop, Spark, High Performance Computing (HPC) cluster, Flink, and Storm [34]. The Hadoop ecosystem has been used in the following applications: treatment of cancer and genomics; monitoring of patient vitals; collection of real-time data related to patient care; processing of large datasets related to drugs, diseases, symptoms, and other factors to extract meaningful information for insurance companies; and prevention and detection of frauds [50].

The selection of the big data platform as solution for a specific healthcare problem depends on a number of factors, such as real-time requirements, speed, data size, scalability, and throughput, among others. Some applications, such as EMR collection, might not require real-time processing, and a platform that does not require live streaming such as Hadoop MapReduce will suffice, but for other applications, a real-time response will be a must, such as the analysis of an electrocardiogram in order to determine a possible intervention. For other applications such as diagnosis suggestion support, scalability and storage of huge amounts of data is a necessity, in which case a scaling system like Spark would be the right choice [34].

Issues and future directions concerning big data in healthcare involve the increased volumes of health data at an intense rate, which demand an increment in IT infrastructure to allow healthcare organizations and researchers to safely manage and exploit the ever-increasing quantities of datasets and enable clinical decision-making in real time based on personalized data from patients [46]. Another concern in healthcare is the high heterogeneity of data sources, the noise introduced in high-throughput experiments, and the variety of experimental techniques and environmental conditions; these heterogeneous data must frequently be collected and preprocessed before applying the data mining methods to extract valuable knowledge. Big data privacy and security of healthcare data are also two important issues that must be addressed in BDA software, by, for example, using advanced encryption algorithms and pseudo-anonymization of the personal data; these software solutions must offer security on the network level and authentication of all users handling these data, as well as appropriate governance standards and practices [51]. Given the sensitive nature of healthcare data, attempts to protect medical and clinical data have been provided by legal provisions such as the Health Insurance Portability and Accountability Act (HIPAA) in the USA, which safeguards the collection, storage, and disclosure of identifiable healthcare data. However, this protection is provided only for so-called covered entities, such as insurance companies and healthcare facilities, but does not cover firms that own social networks such as Facebook, Google, and Twitter, which in some cases have been known to make illegal use of personal information from users. Data protection laws should be extended beyond healthcare settings and encompass systems—such as social network services—that allow the amassment, storage, and analysis of personal information [52].

Regarding the application of ML techniques on big data in various fields within healthcare, some examples are shown in Table 1.

The sample applications from Table 1 were chosen to cover different healthcare fields from a number of datasets from patients from various parts of the world. A more detailed description of the examples given in Table 1 follows.

Gulshan et al. [53] used a deep convolutional neural network for the detection of diabetic retinopathy and macular edema in US patients. The CNN was trained using a dataset of 128,175 retinal images that were classified in a scale of 3 to 7 for diabetic retinopathy and macular edema by a set of 54 US ophthalmologists. The trained neural network was validated using two separate datasets of 9963 and 1748 images. At an operating point selected for high sensitivity for the detection of diabetic retinopathy and diabetic macular edema, the algorithm had a sensitivity of 97.5% and 96.1% and a specificity of 93.4% and 93.9% for the two respective

Table 1 Some applications of machine learning algorithms on big data in healthcare

Ref.	Desired goal	Platform and/or tools	Machine learning algorithm	Datasets	Issues
[53]	Detection of diabetic retinopathy and diabetic macular edema in US patients	An implementation of the Inception-v3 architecture	Deep convolutional neural network	128,175 retinal images	Research needed on applicability in clinical settings
[54]	Prediction of diabetes on Indian populations	Hadoop, R	Decision tree, naïve Bayes, random forest	Data from 75,664 patients with 13 attributes	A higher number of Hadoop nodes could be useful
[55]	Prediction of multimodal cerebral infarction risk in Chinese populations	Not mentioned	Convolutional neural networks	20,320,848 records from patients containing structured and unstructured text data	The accuracy of the algorithms depends on the feature description of the disease
[56]	Prediction of obesity in children in the USA	WEKA	Decision tree algorithm ID3	Data from 7519 patients	Clinical data can have missing or erroneous values
[57]	Disease detection in populations from Saudi Arabia	Spark	Naïve Bayes, logistic regression	18.9 million tweets	Privacy risks concerning public data need to be addressed

validation datasets. The authors state that the feasibility of using the algorithm in a clinical setting for the detection of these diseases requires further research.

In another study, Yuvaraj and SriPreethaa [54] compared three ML algorithms for their ability to predict diabetes using data from Indian populations. A dataset from 75,664 patients obtained from the Indian National Institute of Diabetes was used, with each record having 13 attributes related to diabetes. From this dataset, 70% of the data was used for training the algorithm, and the remaining 30% was used for validation of the model. The following ML algorithms were compared against each other in terms of precision, recall, F-measure, and accuracy on a Hadoop cluster with four nodes running R language scripts: decision tree, naïve Bayes, and random forest. Under the conditions tested, the random forest algorithm yielded a better precision for predicting diabetes by at least 3% than the other two algorithms for all evaluation measurements. The authors propose to use a Hadoop cluster with more nodes to speed up the process and to compare other ML algorithms.

Chen et al. [55] used a convolutional neural network algorithm to predict the risk of cerebral infarction using data from 31,919 hospitalized patients in Central China from the years 2013 to 2015. The data consisted of 20,320,848 records in total and was composed of structured and unstructured data. The structured data included laboratory data and the basic information from the patient, such as age, gender, and life habits, whereas the unstructured text data included the patients' narration of their illness, as well as the doctors' notes on the case. A CNN-based multimodal (using both structured and unstructured data) disease risk prediction algorithm was designed based on a unimodal (using only unstructured text data) CNN prediction algorithm. The multimodal disease risk prediction algorithm achieved 94.8% accuracy and a faster convergence speed than the unimodal disease risk prediction algorithm. The authors found out that the accuracy of the algorithms depended on the quality of the descriptions of the diseases in the data available.

Dugan et al. [56] compared six ML algorithms to predict obesity in children from the USA after the age of 2 using only data collected before this age. The ML techniques analyzed were the WEKA implementations of the random tree, random forest, J48, ID3, naïve Bayes, and Bayes algorithms. The data was collected from a US pediatric clinical support system and consisted of records from 7519 patients. Results showed that the decision tree algorithm ID3 accurately predicted obesity in children after the age of 2. These authors emphasized that clinical data might have missing or erroneous values that can affect the accuracy of the prediction.

In another study, Alotaibi et al. [57] developed a symptoms and disease detection tool using Twitter data in Arabic and proposed its use by the healthcare system in the Kingdom of Saudi Arabia. The data consisted of 18.9 million tweets collected from November 2018 to September 2019. The proposed tool implemented the naïve Bayes and the logistic regression algorithms and ran on a Spark platform. The tool detected that the top 5 diseases in Saudi Arabia according to the available Twitter data were dermal diseases, heart diseases, hypertension, cancer, and diabetes. The results were evaluated using numerical criteria (Accuracy and F1-score) and validated against available healthcare statistics. The data obtained by the proposed system could be used by healthcare officials, among other things, to create awareness in the public about the top diseases and how to prevent them. On the other hand, the availability of healthcare data in public social networks raises privacy concerns that need to be addressed.

From these examples, we can see that the main focus of the analysis of big data using ML techniques lies on the detection of present diseases or the prediction of future diseases. Another commonality is that these studies consist of proposals to be used in clinical settings, rather than descriptions of working systems currently in use in healthcare facilities. Furthermore, distributed platforms such as Hadoop and Spark in these reports are not as widely used as they should in order to process the large amounts of data required by BDA systems. The above suggests that the use of ML in BDA is still mainly in an exploratory phase before its adoption in real-world applications in the healthcare field. On the other hand, although the ML algorithm used depends largely on the kind of application desired in healthcare, it is noticeable that deep learning algorithms are steadily being used more frequently in BDA in these and other works, instead of the more traditional ML algorithms. Finally, a concern that is emphasized is the matter of privacy of healthcare data, since the records and other clinical data from patients frequently require to be processed in a different location from the one where it was produced and can also require to be accessed by different people in the BDA systems.

In general, it can be concluded that although there is still room for improvements in a number of aspects, ML techniques will be indispensable tools in the extraction of knowledge from big data derived from healthcare systems in order to improve the well-being of humans at the individual and the population level.

4.2 Weather Forecasting

Weather forecasting has gained attention in the last decades due to its potential to save lives. For instance, forecasting hurricanes, cyclones, heavy rains, and tornados can help in implementing evacuation plans more efficiently. Weather forecasting is also important in agriculture as it allows farmers to prepare their lands for any anticipated weather changes. Furthermore, social events and sport events can be organized based on weather predictions.

Currently, weather forecasting primarily relies on model-based methods, in which the atmosphere is modeled as a fluid. Partial differential equations of fluid dynamics and thermodynamics [58] are solved using numerical methods. Sample measurements of the current state of the atmosphere are taken in order to approximate the future states by solving such equations. Solving these equations can be computationally expensive depending on the size and granularity of the modeled area. There are different numerical weather prediction models. The Weather Research and Forecasting (WRF) model [59, 60] is currently the world's most used model mainly due to its open-source nature as well as its higher resolution and accuracy. WRF was developed in the 1990s and it was openly released in the year 2000 [60].

Data-driven computer modeling systems, including BDA, can be used as an alternative to numerical weather prediction methods. One of the advantages of the data-driven approaches is obtaining a higher accuracy for short-term forecasts [61]. Several ML approaches have been applied to weather forecasting. Below we present approaches that employ ANN.

An ensemble of neural networks is proposed by Ahmadi et al. [62] for weather prediction. The authors' approach outperformed other similar approaches. One of the main disadvantages of this solution is that the ensemble creates a redundancy. Patil et al. [63] used neural networks to forecast sea surface temperature, whereas Rodríguez-Fernández et al. [64] applied neural networks to predict soil moisture. On the other hand, Sharaff and Roy [65] presented a comparative analysis of regression methods and the back propagation neural network for temperature forecasting. The authors concluded that the back propagation network achieves better accuracy than linear regression and regression trees.

One of the first attempts to employ deep ANNs to the domain area of weather forecasting was carried out by Liu [66], which presented a deep neural network-based feature representation for weather forecasting. The results showed that deep ANNs achieved a higher accuracy than traditional methods such as support vector regression (SVR). Also, a deep neural network was used for ultrashort-term wind speed prediction by Dalto et al. [67]. The authors' results show that deep neural networks outperformed shallow neural networks. In addition, Shi et al. [68] presented a deep learning approach with long short-term memory (LSTM) for precipitation nowcasting. The authors' approach uses a convolutional long short-term memory (LSTM) prediction of rain intensity over local areas. The accuracy of the fully connected LSTM approach is overtaken by the convolutional LSTM. Moreover, Hossain et al. [69] showed that their deep learning approach was able to obtain a higher accuracy than traditional ANNs for predicting temperature. Besides, Yonekura et al. [70] employed a deep learning neural network to predict short-term local temperature and rain. The deep learning approach obtained a higher accuracy than other ML methods.

Apart from ANNs, some other ML models have been used. Voyant et al. [71] presented a comparison of different traditional ML algorithms for radiation forecasting. The authors concluded that ANN and ARIMA are equivalent in terms of accuracy and that SVR, random forests, and regression trees obtained promising results. In addition, Rasel et al. [72] showed that SVR outperformed ANNs in rainfall prediction. However, ANNs obtained better results than SVR for temperature forecast. Mahmood et al. [73] employed a cumulative distribution function for the prediction of extreme weather changes. Moreover, Zhan et al. [74] carried out a correlation analysis of both meteorological and hydrological data whereby a correlation matrix is obtained. The authors then used an SVR model for horizontal comparison in order to obtain a higher accuracy. A random forest model was used for the same purpose; however, the SVR model obtained better results. Lastly, Maliyeckel et al. [75] proposed a hybrid ML model for rainfall prediction. The authors employed algorithms of the LightGBM framework together with an SVR model. The former is a gradient booting framework that uses tree-based learning algorithms. The authors reported that the hybrid model obtained better results than each of the individual models.

The algorithm that is currently more widely used for weather forecasting is an artificial neural network. Recently, deep learning networks have received special attention in the area of weather forecasting. It has been shown that deep learning networks achieve better accuracy than traditional ML methods. In particular, deep networks are able to model complex data with fewer elements than shallow networks. The reason is that the extra layers enable the composition of features from lower layers. One disadvantage of deep networks is that a larger computation time is required for training. Other algorithms that have been successfully employed are SVR, decision trees, and random forest.

Regarding distributed platforms, Hadoop and Spark are the most widely used systems for processing big data related to weather forecasting [76]. In addition, some of the most used languages to develop ML algorithms in the area of weather forecasting include Python and MATLAB [76].

There are a number of issues that need to be addressed regarding the use of BDA for weather forecasting. First of all, most of the works mentioned above do not use a distributed computing model such as MapReduce to manage large amounts of data. Rather, most of these works focus on developing and evaluating different ML algorithms for weather forecasting but miss to evaluate the scalability of their approaches. As a consequence, many proposals report good accuracy for short-term weather predictions. However, further research is needed to evaluate the accuracy of the ML models for larger-term forecasts where a larger amount of data is required. Another issue that requires further attention is that most works do not use a development process methodology for implementing BDA in the area of weather forecasting, giving place to ad hoc practices that can make this task far more complicated.

In Table 2, we selected a sample of works that cover different aspects of the weather forecasting domain. More concretely, we selected works aiming to forecast different aspects of weather such as temperature, rainfall, thunderstorms, wind speed, and severe convective weather.

We present next a more detailed description of the applications shown in Table 2. Hewage et al. [61] used two variants of recurrent neural networks (RNN) called long short-term memory (LSTM) and temporal convolutional networks (TCNs) for weather forecasting. The authors developed a multi-input multi-output (MIMO) model and a multi-input single-output (MISO) model. The former is fed with ten surface parameters (i.e., surface temperature, surface pressure, X component of wind, Y component of wind, humidity, convective rain, non-convective rain, snow water equivalent, soil temperature, and soil moisture) and predicts the same parameters; thus, only one model is needed to predict all the parameters. The latter is fed with ten surface parameters and predicts a single parameter; hence, ten models are required for predicting all the parameters. The authors employed 675,924 records to develop the models. Also, the Keras tool (a Python library) was employed for developing and evaluating the models. The LSTM and TCN models outperformed classic ML approaches such as standard regression, SVR, and random forest. The proposed models also produced better prediction results than WRF in the case of short-term forecasting. However, WRF produces better forecasting results in the case of long-term forecasting.

Table 2 Some applications of machine learning algorithms on big data in the weather forecasting domain

Ref.	Desired goal	Platforms and/or tools	Machine learning algorithm	Datasets	Issues
[61]	Use deep learning for weather forecasting. Proposed model outperforms traditional methods such as regression, SVM, and random forest. Also, better results than WRF for short-term forecasting	Python	Deep learning network with long short-term memory	675,924 records	The approach has not been tested for long-term forecasting
[77]	Deep learning approach to predict severe convective weather	Python	A convolutional network model	4,582,577 thunderstorm samples; 3,609,185 heavy rain samples; 1,468,158 hail samples	Deep CNN training time is much longer than simpler algorithms. There are still inadequacies in the proposed algorithm as it issues too many false alarms of hail
[78]	Deep learning approach to predict severe convective weather such as heavy rain and thunderstorms	Not mentioned	Deep convolutional neural network	Temperature prediction data from 2009 to 2015; wind prediction data from 2000 to 2010	The approach has only been tested for short-term forecasting
[79]	A regression tree model for very short-term wind speed prediction	R	Regression tree	3061 hourly samples of wind speed	Only supports very short-term predictions
[80]	An SVR model to forecast rainfall of landslides	Spark	SVR	The data was taken from September 15, 2016, to February 28, 2017	The study considered only a small dataset involving 4008 records

The work of Zhou et al. [77] proposes a deep learning approach for severe convective weather involving heavy rain, hail, and thunderstorms. The authors employed 5 years of severe weather observation involving 4,582,577 thunderstorm samples, 3,609,185 heavy rain samples, and 1,468,158 hail samples. The results of this work show that the six-layer convolutional neural network obtained better results than SVR, random forest, and other traditional ML approaches. The proposed deep learning model is currently used in the National Meteorological Center of China to provide guidance on the operational forecast of severe convective weather events in China. Unfortunately, although the authors employ big data to develop their models, their work does not report on the computing approach taken to deal with big data.

Mehrkanoon [78] proposed a convolutional neural network to predict temperature and wind speed, both involving short-term forecasts. The author shows that the two-layer and three-layer networks outperform shallow networks. The datasets employed for developing the models include data from 2009 to 2015 for temperature prediction, whereas data from 2000 to 2010 was used for wind speed prediction. The authors used large amounts of data to develop their models; nevertheless, their work does not mention what tools and platforms were employed.

Troncoso et al. [79] evaluated the accuracy of different types of regression tree models employed in very short-term forecasts of wind speed. The authors also show that regression trees are able to outperform—for this specific problem—other ML approaches such as SVR and neural networks. The package CORElearn (a library of R) was used to generate the models. The authors used a sample of 3061 samples of hourly wind speed measures taken by eight towers.

Lee et al. [80] proposed an SVR model to forecast rainfall of landslides on the Apache Spark platform. This platform was configured in the standalone mode in which the worker node employed the SVR model. The model was developed with data taken from September 15, 2016, to February 28, 2017.

We can see that these works have aimed at forecasting different variables of the weather. For example, Troncoso et al. focus on forecasting wind speed, whereas Lee et al. target forecasting rainfall of landslides. Other works focus on severe convective weather, such as the work by Zhou et al. and the work by Mehrkanoon. Other efforts have taken a more holistic approach in which multiple variables are predicted, such as the case of the work by Hewage et al. On the other hand, the most popular tools employed are Python, R, and Apache Spark. Crucially, most of the reviewed works do not pay attention to the issue of efficiently processing big data; rather, the authors focus on showing which ML algorithm is more accurate. In fact, apart from the works of Zhou et al. and Hewage et al., most of the reviewed approaches do not include large amounts of data in their experiments. Therefore, further work is still required to investigate the accuracy of the proposed ML methods in the case of large datasets and long-term forecasting.

4.3 Social Networks and the Internet

Social networking and the Internet handle a large amount of passive data. These data involve user information, historical data, comments, interaction, blogs, etc. from websites and social media networks. Some examples of websites and social media networks are: Twitter Inc.'s microblogging site twitter.com, Google Inc.'s video platform youtube.com, Meta Corp.'s instagram.com, facebook.com, the associated Meta WhatsApp messaging service, and devices apps. In other words, these websites and social media networks catch information flows from web-based life or applications that predict end-user behavior patterns. The main goal of ML is to enable data-driven decision-making. This decision must be accurately based on analyzed data. However, these data should have privacy, security, accuracy, and confidentiality for this domain, as the increase of everyday data generated by humans is 2.5 quintillion bytes [37].

Some properties and data types, from social networking and the Internet, are time, GPS coordinates, user ID, texts, videos, velocity, address, posts, SMS, and IP social media, among others [89] that need to be analyzed by ML algorithms in this domain. A wide variety of unstructured data is produced mainly from email conversations and social networking sites as graphics and text [19]. Data evolve rapidly in a highly connected society, which is generated by data sources such as social media, mobile devices, and the Internet of Things (IoT) [81].

There are some successive phases to manage organization data processes such as data generation, data acquisition, data preprocessing, data storage, data analysis, data visualization, and data exposition, which are defined in [81]. In data generation, the data is generated from different sources (e.g., IoT, social media, operational and commercial data); therefore, data acquisition has three subphases: data identification, data collection, and data transfer. In data analysis, ML models are applied to predict future events and drive proactive decisions. The most common ML algorithms are clustering, graph analysis, decision trees, classification, and regression and association analysis in ML analysis [81].

Table 3 shows some applications of ML algorithms on big data obtained from the social networks and the Internet domain.

Some of the current ML algorithms for social networking and the Internet are found in the state-of-the-art literature. For instance, Nti et al. [4] studied the applications of the decision tree, neural network, and support vector machine algorithms, and the platforms that they used were Hadoop, MapReduce, and Spark, with SQL (structured query language) as language. The authors' aim was to make data-driven decisions to accomplish the desired goals. On the other hand, Kaur and Lal [19] used k-means and hierarchical clustering algorithms, along with the SparkR platform and the R language; their main aim was to improve clustering and reduce CPU utilization through ML. In another work proposed by Latif and Afzal [82], logistic regression, simple logistic multilayer, perceptron J48, and naive Bayes PART implemented with WEKA and Java were used; they concluded that efficient models could predict a movie's popularity for social networking. Lakshmanaprabu et al. [83] used

Table 3 Some applications of machine learning algorithms on big data in the social networks and the Internet domain

Ref.	Desired goal	Platform and/or tools	Machine learning algorithm	Datasets	Issues
[4]	Proposal of a taxonomy with a keyword search and using the appropriate tool or platform for the right task	Hadoop, MapReduce, Spark	Decision trees, neural networks, and support vector machines	Data from 1512 published articles	The high number of free BDA tools, platforms, and data mining tools makes it challenging to select the appropriate one for the right task
[19]	Improvement of clustering and reduction of CPU utilization through ML	SparkR	*K*-means, hierarchical clustering	622 datasets	There is a need to use these algorithms for heterogeneous data such as image, video, streams, etc.
[82]	Construction of efficient models that can predict the popularity of a movie	WEKA	Logistic regression, simple logistic multilayer perceptron J48, and naive Bayes PART	2000 data points	It is necessary to select specific attributes related to a movie
[83]	Reduction of noise and unwanted data from the database to improve the efficiency of their algorithm	Hadoop, MapReduce	Linear kernel, SVM	34,042 data instances	Large amounts of data are required to analyze social networking systems involving Internet of Things
[84]	Reduction of redundant and irrelevant datasets	Not mentioned	Random forest, K-means, and support vector machines	Approximately 33,000 attack accounts, NSL-KDD	It is necessary to monitor the network and analyze the incoming traffic dynamically

a linear kernel support vector, Hadoop MapReduce, and Java to reduce noise and unwanted data from a database to improve the efficiency of their algorithm. Finally, Patgiri et al. [84] used random forest, support vector machine (SVM), and NSL-KDD to reduce redundant and irrelevant datasets.

Considering other works, in [85] the authors used classification, regression, dimensionality reduction, clustering, and density estimation to classify the good, the bad, and the ugly use for cybersecurity and cyber physical systems. On the other hand, the authors in [86] analyzed big data for social transportation. The authors concluded that social data contain abundant information and evolve with time. The authors in [87] developed a model for fake news detection using SVM and NB. Other approaches such as [88] focused on traffic management. The authors used online learning, which was handled by an online adaptive clustering algorithm and incremental learning. Incremental learning is based on the incremental knowledge acquisition and self-learning (IKASL) algorithm, decremental learning, and concept drift detection. Finally, the authors in [19] used k-means and hierarchical clustering algorithms for analyzing social networking using SparkR. The authors' models were fed with social media involving YouTube datasets.

The algorithm more commonly used for social networking is SVM, which is internally deployed for tracking and classifying key metrics—e.g., likes, loyalty, and value information. Other algorithms that have been employed for processing social media data are naive Bayes, decision trees, and the clustering algorithm k-means. Therefore, the selection of the most appropriate algorithm for the social network and Internet domain depends on the goal of the application; for instance, SVM is a supervised algorithm, whereas k-means is an unsupervised algorithm, and both have been used for this domain. Furthermore, from the previous examples, it can be seen that authors frequently used a combination of two or more ML algorithms to achieve a higher performance.

Some of the challenges in this application domain are the large number of free BDA tools, platforms, and data mining tools that are available, making it difficult to select the appropriate one for the right task. Another challenge is the large diversity of heterogeneous data formats that exist for social media, such as image, video, and text, among others.

5 Conclusions

Given the huge amounts of data that are currently produced in practically all domains of human knowledge and social interaction and that this quantity of data will most likely continue to increase exponentially in the foreseeable future, the use of automated computational tools to extract meaningful information from these data is no longer an option but a necessity. Big data analytics in conjunction with machine learning algorithms are poised to fill this need, as machine learning techniques have been developed precisely to computationally automate the extraction of knowledge from data.

Concerning the domain areas mentioned in the previous section, and the three kinds of big data analytics (descriptive, predictive, and prescriptive), healthcare and weather forecasting benefit especially from predictive analytics. That is, in general it is more important, for instance, to predict the appearance of new disease outbreaks, or to predict bad weather conditions, than it is to explain previous occurrences of events.

As for the machine learning techniques currently in use in big data analytics, practically all kinds of algorithms can be applied depending on the specific goals desired. However, deep learning algorithms have proven to still have room for improvements and for being applied in other application domains in big data analytics systems.

Regarding the open-software platforms currently used, both Apache Hadoop and Apache Spark continue to be the preferred choice for big data analytics systems, with a preference for Spark when speed and real-time processing are needed, and a predilection for Hadoop when processing data in batches and a speedy response is not an issue.

In the literature review that we made, we found that many researchers used big data analytics at a small scale only to demonstrate the feasibility of a machine learning approach to solve a problem, but without the application to actual big data for solving real-world problems. Furthermore, many reports concentrated on achieving high accuracy on their proposals for integrating machine learning with big data analytics systems, but without concern for building computationally efficient systems. We also found that in the reviewed literature, the use of distributed systems using Hadoop or Spark is not as widespread as it should be in big data analytics systems. Thus, we consider that research and applications of big data analytics in conjunction with machine learning will continue to grow in the years to come, both in academia and industry.

References

1. Reinsel, D., Gantz J., Rydning, J.: The Digitalization of The World: From Edge to Core (2018), https://www.seagate.com/files/www-content/our-story/trends/files/idc-seagate-dataage-whitepaper.pdf
2. Rahman, M.S., Reza, H.: A systematic review towards Big Data analytics in social media. Big Data Min. Anal. **5**, 228–244 (2022). https://doi.org/10.26599/BDMA.2022.9020009
3. Fisher, D., DeLine, R., Czerwinski, M., Drucker, S.: Interactions with Big Data analytics. Interactions. **19**, 50–59 (2012). https://doi.org/10.1145/2168931.2168943
4. Nti, I.K., Quarcoo, J.A., Aning, J., Fosu, G.K.: A mini-review of machine learning in big data analytics: applications, challenges, and prospects. Big Data Min. Anal. **5**, 81–97 (2022). https://doi.org/10.26599/BDMA.2021.9020028
5. Wixom, B., Ariyachandra, T., Douglas, D., Goul, K., Gupta, B., Iyer, L., Kulkarni, U., Mooney, B.J.G., Phillips-Wren, G., Turetken, O.: The current state of business intelligence in academia: the arrival of big data. Commun. Assoc. Inf. Syst. **34**, 1–13 (2014). https://doi.org/10.17705/1cais.03401
6. Laney, D.: 3D data management: Controlling data volume velocity and variety, https://studylib.net/doc/8647594/3d-data-management%2D%2Dcontrolling-data-volume%2D%2Dvelocity%2D%2Dan... (2001)
7. Qiu, J., Wu, Q., Ding, G., Xu, Y., Feng, S.: A survey of machine learning for big data processing. EURASIP J. Adv. Signal Proc. **2016**, 1–16 (2016)
8. EMC (ed.): Data Science and Big Data Analytics: Discovering, Analyzing, Visualizing and Presenting Data. Wiley Publishing (2015)
9. Grover, P., Kar, A.K.: Big Data analytics: a review on theoretical contributions and tools used in literature. Global J. Flex. Syst. Manag. **18**, 203–229 (2017). https://doi.org/10.1007/s40171-017-0159-3
10. Mikalef, P., Pappas, I.O., Krogstie, J., Giannakos, M.: Big data analytics capabilities: a systematic literature review and research agenda. Inf. Syst. E-Bus. Manag. **16**, 547–578 (2018). https://doi.org/10.1007/s10257-017-0362-y
11. Zhou, L., Pan, S., Wang, J., Vasilakos, A.V.: Machine learning on big data: opportunities and challenges. Neurocomputing. **237**, 350–361 (2017). https://doi.org/10.1016/j.neucom.2017.01.026

12. Russell, S., Norvig, P.: Artificial Intelligence: a Modern Approach. Prentice Hall (2010)
13. Cristianini, N., Shawe-Taylor, J.: An Introduction to Support Vector Machines and Other Kernel-Based Learning Methods. Cambridge University Press (2000)
14. Sun, Z.Q., Fox, G.C.: Study on parallel SVM based on MapReduce. In: International Conference on Parallel and Distributed Processing Techniques and Applications, pp. 495–561, Las Vegas, NV, USA (2012)
15. Breiman, L., Friedman, J., Stone, C.J., Olshen, R.A.: Classification and Regression Trees. Taylor & Francis (1984)
16. Dai, W., Ji, W.-Z.: A MapReduce implementation of C4.5 Decision Tree algorithm. Int. J. Database Theory Appl. **7**, 49–60 (2014)
17. Purdilă, V., Pentiuc, Ş.-G.: MR-Tree-A scalable MapReduce algorithm for building decision trees. J. Appl. Comput. Sci. Math. **8**, 16–19 (2014)
18. Mahdavinejad, M.S., Rezvan, M., Barekatain, M., Adibi, P., Barnaghi, P., Sheth, A.P.: Machine learning for internet of things data analysis: a survey. Digit. Commun. Netw. **4**, 161–175 (2018). https://doi.org/10.1016/j.dcan.2017.10.002
19. Kaur, N., Lal, N.: Clustering of social networking data using SparkR in Big Data. In: Mayank, S., Gupta, P.K., T.V, F.J, Ö.T (eds.) Advances in Computing and Data Sciences, pp. 217–226. Springer Singapore, Singapore (2018)
20. Arora, P., Deepali, Varshney, S.: Analysis of K-means and K-Medoids algorithm for Big Data. In: International Conference on Information Security & Privacy (ICISP2015), pp. 507–512 (2016)
21. Prabhu, C.S.R., Chivukula, A.S., Mogadala, A., Ghosh, R., Livingston, L.M.J.: Big Data Analytics: Systems, Algorithms, Applications. Springer, Singapore (2019)
22. Ray, S.: A quick review of Machine Learning algorithms. In: 2019 International Conference on Machine Learning, Big Data, Cloud and Parallel Computing (COMITCon), pp. 35–39 (2019)
23. Yuan, C., Yang, H.: Research on K-value selection method of K-means clustering algorithm. J (Basel). **2**, 226–235 (2019). https://doi.org/10.3390/j2020016
24. Narayanan, B.N., Djaneye-Boundjou, O., Kebede, T.M.: Performance analysis of machine learning and pattern recognition algorithms for Malware classification. In: 2016 IEEE National Aerospace and Electronics Conference (NAECON) and Ohio Innovation Summit (OIS), pp. 338–342 (2016)
25. Narayanan, B.N., Hardie, R.C., Kebede, T.M.: Performance analysis of a computer-aided detection system for lung nodules in CT at different slice thicknesses. J. Med. Imag. **5**, 14504 (2018). https://doi.org/10.1117/1.JMI.5.1.014504
26. Narayanan, B.N., Hardie, R.C., Kebede, T.M., Sprague, M.J.: Optimized feature selection-based clustering approach for computer-aided detection of lung nodules in different modalities. Pattern Anal. Appl. **22**, 559–571 (2019). https://doi.org/10.1007/s10044-017-0653-4
27. Al-Yaseen, W.L., Othman, Z.A., Nazri, M.Z.A.: Multi-level hybrid support vector machine and extreme learning machine based on modified K-means for intrusion detection system. Expert Syst. Appl. **67**, 296–303 (2017). https://doi.org/10.1016/j.eswa.2016.09.041
28. Ge, Y., Tang, K.: Distributed dynamic cluster algorithm for wireless sensor networks. In: 6th International Conference on Wireless, Mobile and Multi-Media (ICWMMN 2015), pp. 23–27 (2015)
29. Ran, X., Zhou, X., Lei, M., Tepsan, W., Deng, W.: A novel K-means clustering algorithm with a noise algorithm for capturing urban hotspots. Appl. Sci. (Switzerland). **11** (2021). https://doi.org/10.3390/app112311202
30. Bendechache, M., Kechadi, M.-T.: Distributed clustering algorithm for spatial data mining. In: 2015 2nd IEEE International Conference on Spatial Data Mining and Geographical Knowledge Services (ICSDM), pp. 60–65 (2015)
31. Chiroma, H., Abdullahi, U.A., Abdulhamid, S.M., Abdulsalam Alarood, A., Gabralla, L.A., Rana, N., Shuib, L., Targio Hashem, I.A., Gbenga, D.E., Abubakar, A.I., Zeki, A.M., Herawan, T.: Progress on artificial neural networks for Big Data analytics: a survey. IEEE Access. **7**, 70535–70551 (2019). https://doi.org/10.1109/ACCESS.2018.2880694

32. Shen, D., Wu, G., Suk, H.-I.: Deep learning in medical image analysis. Annu. Rev. Biomed. Eng. **19**, 221–248 (2017). https://doi.org/10.1146/annurev-bioeng-071516-044442
33. Dean, J., Ghemawat, S.: MapReduce: simplified data processing on large clusters. Commun. ACM. **51**, 107–113 (2008). https://doi.org/10.1145/1327452.1327492
34. Harerimana, G., Jang, B., Kim, J.W., Park, H.K.: Health Big Data analytics: a technology survey. IEEE Access. **6**, 65661–65678 (2018). https://doi.org/10.1109/ACCESS.2018.2878254
35. Apache Software Foundation: Apache Hadoop, https://hadoop.apache.org/
36. Ketu, S., Mishra, P.K., Agarwal, S.: Performance analysis of distributed computing frameworks for Big Data analytics: Hadoop vs Spark. Computación y Sistemas. **24**, 669–686 (2020). https://doi.org/10.13053/CyS-24-2-3401
37. Mohd, A.B., Banu, A., Yakub, M.: Evolution of big data and tools for big data analytics. J. Interdiscipl. Cycle Res. **12**, 309–316 (2020)
38. Gupta, P., Sharma, A., Jindal, R.: Scalable machine-learning algorithms for big data analytics: a comprehensive review. WIREs Data Min. Knowl. Discov. **6**, 194–214 (2016). https://doi.org/10.1002/widm.1194
39. Raza, M.U., XuJian, Z.: A comprehensive overview of BIG DATA technologies: a survey. In: Proceedings of the 5th International Conference on Big Data and Computing, pp. 23–31. Association for Computing Machinery, New York, NY, USA (2020)
40. Venkatram, K., Geetha, M.A.: Review on Big Data & analytics – concepts, philosophy, process and applications. Cybern. Inf. Technol. **17**, 3–27 (2017). https://doi.org/10.1515/cait-2017-0013
41. Ikegwu, A.C., Nweke, H.F., Anikwe, C.V., Alo, U.R., Okonkwo, O.R.: Big data analytics for data-driven industry: a review of data sources, tools, challenges, solutions, and research directions. Cluster Comput. (2022). https://doi.org/10.1007/s10586-022-03568-5
42. Faridoon, A., Imran, M.: Big data storage tools using NoSQL databases and their applications in various domains: a systematic review. Comput. Inf. **40**, 489–521 (2021). https://doi.org/10.31577/cai_2021_3_489
43. Witten, I.H., Frank, E., Hall, M.A., Pal, C.J., DATA, M.: Practical machine learning tools and techniques. In: Data Mining (2005)
44. R Core Team: R.: A Language and Environment for Statistical Computing, https://www.R--project.org/ (2022)
45. Kambatla, K., Kollias, G., Kumar, V., Grama, A.: Trends in big data analytics. J. Parallel Distrib. Comput. **74**, 2561–2573 (2014). https://doi.org/10.1016/j.jpdc.2014.01.003
46. Galetsi, P., Katsaliaki, K.: A review of the literature on big data analytics in healthcare. J. Oper. Res. Soc. **71**, 1511–1529 (2020). https://doi.org/10.1080/01605682.2019.1630328
47. Cirillo, D., Valencia, A.: Big data analytics for personalized medicine. Curr. Opin. Biotechnol. **58**, 161–167 (2019). https://doi.org/10.1016/j.copbio.2019.03.004
48. Akundi, S.H., Soujanya, R., Madhuri, P.M.: Big Data analytics in healthcare using Machine Learning algorithms: a comparative study. Int. J. Online Biomed. Eng. (iJOE). **16**, 19–32 (2020). https://doi.org/10.3991/ijoe.v16i13.18609
49. Agarwal, R., Dhar, V.: Editorial—Big Data, data science, and analytics: the opportunity and challenge for IS research. Inf. Syst. Res. **25**, 443–448 (2014). https://doi.org/10.1287/isre.2014.0546
50. Sunil Kumar, M.S.: Big Data analytics for healthcare industry: impact, applications, and tools. Big Data Min. Anal. **2**, 48 (2019). https://doi.org/10.26599/BDMA.2018.9020031
51. Ristevski, B., Chen, M.: Big Data analytics in medicine and healthcare. J. Integr. Bioinform. **15** (2018). https://doi.org/10.1515/jib-2017-0030
52. Gostin, L.O., Halabi, S.F., Wilson, K.: Health data and privacy in the digital era. JAMA. **320**, 233–234 (2018). https://doi.org/10.1001/jama.2018.8374
53. Gulshan, V., Peng, L., Coram, M., Stumpe, M.C., Wu, D., Narayanaswamy, A., Venugopalan, S., Widner, K., Madams, T., Cuadros, J., Kim, R., Raman, R., Nelson, P.C., Mega, J.L., Webster, D.R.: Development and validation of a Deep Learning algorithm for detection of

diabetic retinopathy in retinal fundus photographs. JAMA. **316**, 2402–2410 (2016). https://doi.org/10.1001/jama.2016.17216
54. Yuvaraj, N., SriPreethaa, K.R.: Diabetes prediction in healthcare systems using machine learning algorithms on Hadoop cluster. Cluster Comput. **22**, 1–9 (2019). https://doi.org/10.1007/s10586-017-1532-x
55. Chen, M., Hao, Y., Hwang, K., Wang, L., Wang, L.: Disease prediction by machine learning over big data from healthcare communities. IEEE Access. **5**, 8869–8879 (2017). https://doi.org/10.1109/ACCESS.2017.2694446
56. Dugan, T.M., Mukhopadhyay, S., Carroll, A., Downs, S.: Machine learning techniques for prediction of early childhood obesity. Appl. Clin. Inform. **06**, 506–520 (2015)
57. Alotaibi, S., Mehmood, R., Katib, I., Rana, O., Albeshri, A.: Sehaa: a Big Data analytics tool for healthcare symptoms and diseases detection using Twitter, Apache Spark, and machine learning. Appl. Sci. **10** (2020). https://doi.org/10.3390/app10041398
58. Richardson, L.F., Lynch, P.: Weather Prediction by Numerical Process. Cambridge University Press (2007)
59. NCAR/UCAR.: WRF model users site, http://www2.mmm.ucar.edu/wrf/users/
60. Powers, J.G., Klemp, J.B., Skamarock, W.C., Davis, C.A., Dudhia, J., Gill, D.O., Coen, J.L., Gochis, D.J., Ahmadov, R., Peckham, S.E., Grell, G.A., Michalakes, J., Trahan, S., Benjamin, S.G., Alexander, C.R., Dimego, G.J., Wang, W., Schwartz, C.S., Romine, G.S., Liu, Z., Snyder, C., Chen, F., Barlage, M.J., Yu, W., Duda, M.G.: The weather research and forecasting model: overview, system efforts, and future directions. Bull. Am. Meteorol. Soc. **98**, 1717–1737 (2017). https://doi.org/10.1175/BAMS-D-15-00308.1
61. Hewage, P., Trovati, M., Pereira, E., Behera, A.: Deep learning-based effective fine-grained weather forecasting model. Pattern Anal. Appl. **24**, 343–366 (2021). https://doi.org/10.1007/s10044-020-00898-1
62. Ahmadi, A., Zargaran, Z., Mohebi, A., Taghavi, F.: Hybrid model for weather forecasting using ensemble of neural networks and mutual information. In: 2014 IEEE Geoscience and Remote Sensing Symposium, pp. 3774–3777 (2014)
63. Patil, K., Deo, M.C.: Basin-scale prediction of sea surface temperature with artificial neural networks. In: 2018 OCEANS - MTS/IEEE Kobe Techno-Oceans (OTO), p. 1–5 (2018)
64. Rodriguez-Fernandez, N.-J., de Rosnay, P., Albergel, C., Aires, F.: SMOS Neural Network Soil Moisture Data Assimilation. (2017)
65. Sharaff, A., Roy, S.R.: Comparative analysis of temperature prediction using regression methods and back propagation neural network. In: 2018 2nd International Conference on Trends in Electronics and Informatics (ICOEI), pp. 739–742 (2018)
66. Liu, J.N.K., Hu, Y.-X., You, J.J., Chan, P.W.: Deep neural network based feature representation for weather forecasting. In: The 2014 World Congress in Computer Science, Computer Engineering, and Applied Computing (2014)
67. Dalto, M., Matuško, J., Vašak, M.: Deep neural networks for ultra-short-term wind forecasting. In: 2015 IEEE International Conference on Industrial Technology (ICIT), pp. 1657–1663 (2015)
68. Shi, X., Chen, Z., Wang, H., Yeung, D.-Y., Wong, W., Woo, W.: Convolutional LSTM network: a machine learning approach for precipitation Nowcasting. In: Proceedings of the 28th International Conference on Neural Information Processing Systems, pp. 802–810. MIT Press, Cambridge, MA (2015)
69. Hossain, M., Rekabdar, B., Louis, S.J., Dascalu, S.: Forecasting the weather of Nevada: a deep learning approach. In: 2015 International Joint Conference on Neural Networks (IJCNN), p. 1–6 (2015)
70. Yonekura, K., Hattori, H., Suzuki, T.: Short-term local weather forecast using dense weather station by deep neural network. In: 2018 IEEE International Conference on Big Data (Big Data), pp. 1683–1690 (2018)
71. Voyant, C., Notton, G., Kalogirou, S., Nivet, M.-L., Paoli, C., Motte, F., Fouilloy, A.: Machine learning methods for solar radiation forecasting: a review. Renew. Energy. **105**, 569–582 (2017). https://doi.org/10.1016/j.renene.2016.12

72. Rasel, R.I., Sultana, N., Meesad, P.: An application of data mining and machine learning for weather forecasting. In: Meesad, P., Sodsee, S., Unger, H. (eds.) Recent Advances in Information and Communication Technology 2017, pp. 169–178. Springer International Publishing, Cham (2018)
73. Mahmood, M.R., Patra, R.K., Raja, R., Sinha, G.R.: A novel approach for weather prediction using forecasting analysis and data mining techniques. In: Saini, H.S., Singh, R.K., Kumar, G., Rather, G.M., Santhi, K. (eds.) Innovations in Electronics and Communication Engineering, pp. 479–489. Springer Singapore, Singapore (2019)
74. Zhan, Y., Zhang, H., Liu, Y.: Forecast of meteorological and hydrological features based on SVR model. In: 2021 4th International Conference on Advanced Electronic Materials, Computers and Software Engineering (AEMCSE), pp. 579–583 (2021)
75. Maliyeckel, M.B., Sai, B.C., Naveen, J.: A comparative study of LGBM-SVR hybrid machine learning model for rainfall prediction. In: 2021 12th International Conference on Computing Communication and Networking Technologies (ICCCNT), p. 1–7 (2021)
76. Fathi, M., Haghi Kashani, M., Jameii, S.M., Mahdipour, E.: Big Data analytics in weather forecasting: a systematic review. Arch. Comput. Methods Eng. **29**, 1247–1275 (2022). https://doi.org/10.1007/s11831-021-09616-4
77. Zhou, K., Zheng, Y., Li, B., Dong, W., Zhang, X.: Forecasting different types of convective weather: a deep learning approach. J. Meteorolog. Res. **33**, 797–809 (2019). https://doi.org/10.1007/s13351-019-8162-6
78. Mehrkanoon, S.: Deep shared representation learning for weather elements forecasting. Knowledge-Based Syst. **179**, 120–128 (2019). https://doi.org/10.1016/j.knosys.2019.05.009
79. Troncoso, A., Salcedo-Sanz, S., Casanova-Mateo, C., Riquelme, J.C., Prieto, L.: Local models-based regression trees for very short-term wind speed prediction. Renew. Energy. **81**, 589–598 (2015). https://doi.org/10.1016/j.renene.2015.03.071
80. Lee, Z.-J., Lee, C.-Y., Yuan, X.-J., Chu, K.-C.: Rainfall forecasting of landslides using support vector regression. In: 2020 3rd IEEE International Conference on Knowledge Innovation and Invention (ICKII), pp. 1–3 (2020)
81. Faroukhi, A.Z., Alaoui, I., Gahi, Y., Amine, A.: An adaptable big data value chain framework for end-to-end big data monetization. Big Data Cogn. Comput. **4**, 1–27 (2020). https://doi.org/10.3390/bdcc4040034
82. Latif, M.H., Afzal, H.: Prediction of movies popularity using machine learning techniques. Int. J. Comput. Sci. Netw Secur. **16**, 127–131 (2016)
83. Lakshmanaprabu, S.K., Shankar, K., Khanna, A., Gupta, D., Rodrigues, J.J.P.C., Pinheiro, P.R., de Albuquerque, V.H.C.: Effective features to classify big data using social internet of things. IEEE Access. **6**, 24196–24204 (2018)
84. Patgiri, R., Varshney, U., Akutota, T., Kunde, R.: An investigation on intrusion detection system using machine learning. In: Proceedings of the 2018 IEEE Symposium Series on Computational Intelligence, SSCI 2018, p. 1684–1691. Institute of Electrical and Electronics Engineers Inc. (2019)
85. Liang, F., Hatcher, W.G., Liao, W., Gao, W., Yu, W.: Machine learning for security and the Internet of Things: the good, the bad, and the ugly. IEEE Access. **7**, 158126–158147 (2019). https://doi.org/10.1109/ACCESS.2019.2948912
86. Zheng, X., Chen, W., Wang, P., Shen, D., Chen, S., Wang, X., Zhang, Q., Yang, L.: Big Data for social transportation. IEEE Trans. Intell. Transp. Syst. **17**, 620–630 (2016). https://doi.org/10.1109/TITS.2015.2480157
87. Jain, A., Shakya, A., Khatter, H., Gupta, A.K.: A smart system for fake news detection using machine learning. In: 2019 International Conference on Issues and Challenges in Intelligent Computing Techniques (ICICT), p. 1–4 (2019)
88. Nallaperuma, D., Nawaratne, R., Bandaragoda, T., Adikari, A., Nguyen, S., Kempitiya, T., de Silva, D., Alahakoon, D., Pothuhera, D.: Online incremental machine learning platform for Big Data-driven smart traffic management. IEEE Trans. Intell. Transp. Syst. **20**, 4679–4690 (2019). https://doi.org/10.1109/TITS.2019.2924883

The Data Value Chain Ontology

Dirk Bendlin, Jorge Marx Gómez, H. Kaddoura, A. Kucewicz, and M. Werther Häckell

1 Introduction

In 2012, the authors of [10] called data scientist the "sexiest job of the twenty-first century." However, 10 years later, what is left of that appeal? On the one hand, decision-makers are still not data scientists and still find it difficult to evaluate data-analytic approaches; on the other hand, data scientists lack the expertise to assess the value of their analyses for a specific domain. This chapter presents an ontology derived from a literature review to describe a data value chain (DVC) and its visualization. The resulting model aims to bridge the gap between decision-makers and data scientists on both sides of the DVC. The ontology is based on the existing business model canvas (BMC) ontology from [33] and describes how a DVC can be modeled, visualized, and analyzed. The DVC ontology can serve as guidance to identify potential areas of improvement for individuals and organizations in deriving value from data.

D. Bendlin (✉)
Carl von Ossietzky Universität Oldenburg, Oldenburg, Germany
e-mail: dirk.bendlin@uni-oldenburg.de

J. M. Gómez
University of Oldenburg, Oldenburg, Germany

H. Kaddoura · A. Kucewicz · M. W. Häckell
Ramboll Deutschland GmbH, Hamburg, Germany

M. Mora et al. (eds.), *Development Methodologies for Big Data Analytics Systems*, Transactions on Computational Science and Computational Intelligence, https://doi.org/10.1007/978-3-031-40956-1_3

2 Problem Identification and Motivation

Data science builds the foundation for data-driven decision-making [35]. Data-driven decision-making is bridging the gap between data scientists and decision-makers. But what prevents organizations from getting the most value out of their data?

According to a survey by [2], more than half of interviewed organizations stated that 50% or more of their decisions had been based primarily on "intuition or experience." At the same time, "big data" presents challenges due to its sheer volume, velocity, and variability [7]. This complexity requires methodological and technical expertise. Much of raw data is not suitable for direct analysis; thus, data scientists and analysts must invest considerable time into data preparation [21].

Due to these challenges, data represents a significant and mostly underutilized asset for organizations. To recover the value of data, a procedure is needed to make the flow of data within a company visible as a valuable, measurable resource. Based on literature, this chapter suggests a data value chain (DVC) ontology and visualization. Inductively from relevant literature, a basic DVC ontology REV0 was created following the design science research methodology process by [34]. Industry partners assessed this approach at workshops of the WiSA Big Data research project and suggested simplifications. These ideas and a deductive literature study in relevant online databases resulted in the DVC ontology REV1.

3 Definition of Objective and Solution

This chapter presents an ontology for a DVC based on a literature review. Ontologies originate from philosophy, in which they describe the "study of the kind of things that exist" [8]. Ontologies, however, have become well-known and significant in the field of information systems since the 1980s [17]. The language of many areas can be reflected in computational ontologies [17]. Ontologies capture not only vocabulary but also conceptualizations [8]. Reference ontologies provide a thorough explanation of a domain [38]. Because ontologies may need to be updated as domain knowledge evolves, they must be descriptive to preserve available domain knowledge [37].

The definition of value creation from data can be very broad; therefore, we have limited this review to canvas models based on the BMC ontology by [33]. According to [24], the BMC is an established model to map and examine products in a business model. At the center of the business models are (data) products (DPs), which are defined for this work as relevant questions which can be answered utilizing data (analysis).

The BMC ontology consists of pillars and elements. A pillar describes a main concept of the DVC ontology; it consists of multiple elements. As an example, the customer interface (pillar) consists of the elements "customer relationship, distribution channels, and the target group." Each of the elements has been further described by [33] (see Table 1).

Table 1 Description of the business model elements [33]

Name of element	Name
Definition	Provides a precise description of the business model element
Part of	Defines to which pillar of the ontology the element belongs or of which element it is a sub-element
Related to	Describes to which other elements of the ontology an element is related
Set of	Indicates into which sub-elements an element can be decomposed
Cardinality	Defines the number of allowed occurrences of an element or sub-element inside the ontology
Attributes	Lists the attributes of the element or sub-element. The allowed values of an attribute are indicated between accolades {VALUE1, VALUE2}. Their occurrences are indicated in brackets (e.g., 1–*n*) Each element or sub-element has two standard attributes, NAME and DESCRIPTION, that contain a chain of characters {abc}
References	Indicates the main references related to the business model element

For visualization, the DVC can be supplemented by an evaluation model consisting of a series of key performance indicators (KPIs) derived from the literature. These indicators enable the assessment of the potential profit and related costs of the DPs. Offshore wind turbines generate electricity that can be sold. Wind turbines provide a range of operational data. Deeper understanding of this data can enable wind turbine owners to increase their energy production and increase their revenues. A DP could, for example, be the question when a certain component would fail. This information creates a lot of value, which could be an argument for investing in optimizing the data value chain, for example, by installing additional sensors or using better methods for data analysis.

4 Methodology

For creating the ontologies from a literature review, the approaches of [32] and [39] were combined and organized according to the design science research methodology process by [34] (see Fig. 1).

The BMC ontology by [33] was chosen as a baseline. First, the BMC was extended using a backward search of relevant prior work [14, 24]. The resulting DVC ontology REV0 was combined with an evaluation model which is a novelty of this approach. This ontology was evaluated in workshops with the industry partners Deutsche Windtechnik, Vattenfall, and Ocean Breeze of the WiSA Big Data research project (funded by the German Federal Ministry of Economics and Climate Protection FKZ: 03EE3016). Workshop participants suggested a potential for simplification of the REV0 ontology by reducing the number of elements. For this simplification, a four-step literature review was performed as described by [15]. First, relevant publications from previous work were identified. Second, these

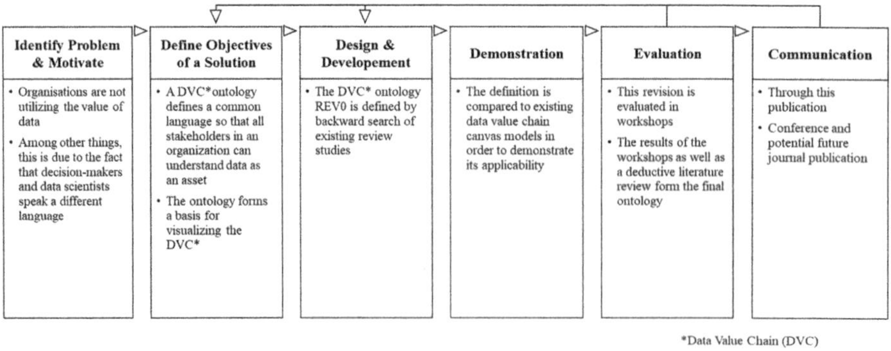

Fig. 1 Process for developing the DVC ontology

publications were reviewed, and third, inclusion criteria and search algorithms were defined. Fourth, the search algorithm was performed in the relevant online libraries [26]. Finally, the results were graded as A (important), B (less important, possibly additional material), and C (not important) according to [34]. With these results, the derived DVC ontology REV1 can serve as a starting point in future evaluations of the value chain for creating the highest value out of data for decision-making purposes.

4.1 Results of Initial Literature Review and Workshop

In the first step, sources and keywords were derived based on related work addressing canvas models for DPs [14, 24]. Through this preliminary work, the pillars and elements presented in Table 2 were derived and coded in accordance with [29].

Related to DPs, the definitions of the nine elements of the original BMC were refined and supplemented by additional elements as well as sub-elements. All elements were coded (as defined in Table 2) for use in the evaluation process.

4.2 Results of the Evaluation Model

To represent the value of data, [36] suggests to follow the goodwill approach. Goodwill is an American accounting concept with which organizations can assign a value to intangible assets.

The business value of information (BVI) was chosen for this chapter, as it assesses the usability of data for its own processes by evaluating data quality aspects. The evaluated data (p) can be assessed with various extendable variables (Eq. 1). The result can additionally be refined individually [28] based on further dimensions of data quality.

Table 2 Initial coding of elements for DVC ontology REV0 based on an initial literature review

Pillar	Element	Sub-elements
Customer interface	CR: customer relationship – what is the relationship between the customer (user), in this case the organization or the user, and the DP? In which process (step) is it used and by whom?	CR1: relationship DP < > stakeholders
		CR2: relevant process step for the DP
		CR3: acceptance criteria for the result
	CH: (distribution) channels – in which format and frequency is the DP finally communicated?	CH1: frequency of use
		CH2: form of use of the DP
		CH3: visualization
	TG: target groups – who should use the DP in the company?	TG1: user story (Who is informed? Who is helped by the DP?)
		TG2: Who decides? Who is the end user?
		TG3: user story – who does what?
(Data) product	VP: value proposition – what is the (potential) value of the DP?	VP1: What is the monetary value of this decision?
		VP2: What decision improves or enables the DP?
		VP3: requirement (nonfunctional)
Infrastructure management	KA: key activities – which activities must be performed "internally" within data analysis and DP development? --> Data science canvas and data science workflow	KA1: problem statement (requirement)
		KA2: data integration
		KA3: data analysis methods and model selection
		KA4: data acquisition (How?)
		KA5: model construction (train test)
		KA6: data preparation
		KA7: data analysis and result (KPIs)
	KR: key resources – what is needed to establish and continuously deliver and use the DP?	KR1: required resources
		KR2: data sources
		KR3: influencing constraints
	KP: key partners – which partners are involved in the development and continuous delivery?	KP1: internal and external partners?
		KP2: How is support provided?
		KP3: What is the role of the partner in the data science team (e.g., DP owner)? KP4: Which software is needed to create the DP?

(continued)

Table 2 (continued)

Pillar	Element	Sub-elements
Financial aspects	C$: cost structure – what are the costs associated with providing the DP?	C$1: What is the cost or effort to generate the analysis result of the DP (incl. CAPEX for measurement devices)?
		C$2: risks
	R$: income sources (benefits) – what benefits can be derived from the use of the DP? How is this to be evaluated?	R$1: What income or cost savings can be achieved by the DP?
		R$2: data value

$$\text{BVI} = \sum_{p=1}^{n} \left(\text{Relevance}_p\right) * \text{Accuracy} * \text{Completeness} * \text{Timeliness} \tag{1}$$

Here, p stands for the business processes or functions (DPs) and n stands for the number of processes. Relevance (0 to 1) evaluates the usefulness of the data for one or more business processes. Accuracy reflects the percentage of records deemed correct. Completeness assesses the percentage of total data compared to the available amount of data for an analysis. Timeliness indicates how quickly new or updated data is available.

According to the DVC definition described in this chapter, the BVI is divided into its different components. For this chapter, the KPIs of the evaluation model have been limited to data acquisition (DAQ), data governance (DAGO), data management (DAMA), data analysis (DANA), and decision-making (DEC).

In economics, the return on investment (RoI) is considered to be "one of the most important return ratios" [42]. The RoI relates the profit to the costs. Different performance indicators can be used to evaluate the RoI, which can impede a uniform comparison of this indicator [43]. The RoI is calculated as follows [43]:

$$\text{RoI} = \frac{\text{Return}}{\text{Capital invested}} \times 100 \tag{2}$$

For this description, we call this indicator return on information (ROI). The ROI simultaneously assigns cost or effort factors to the components of the BVI (BVI_{DAQ}, BVI_{DAGO}, BVI_{DAMA}, BVI_{DANA}, and BVI_{DEC}) as shown in Fig. 2. The ROI for the different parts of the value chain is calculated by dividing the data quality (potential income) by the cost (effort of implementation). The various indicators can be recorded in three different ways: as a pure ratio, in the form of a utility value analysis as weighted target fulfillment levels, or in the form of real costs and benefits in euros.

The DVC ontology must be understood as a fundamental model for evaluating, visualizing, and optimizing DVCs. Due to the high individuality of DVCs and strategies for creating value from data, adaptability and extendibility of the ontology is necessary to achieve continuous validity.

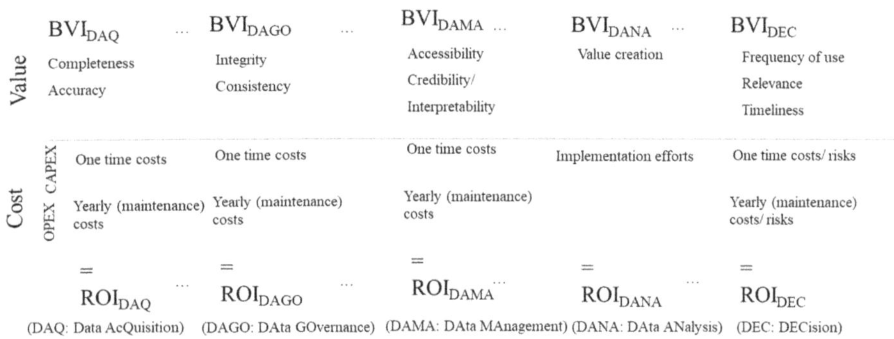

Fig. 2 Hierarchy of KPIs for evaluating the DVC

4.3 Results of the Extended Literature Review

The literature review was performed in the next step. Multiple search algorithms were created to produce a better output (see Table 3). Following [15], the results of the different search algorithms were classified into categories A, B, and C.

Table 3 lists the total number of papers, the number of relevant papers, and the search algorithm for each online library. Five online libraries (HBR, MDPI, SAGE journals, Wiley Online Library, and Google Scholar) have been summarized under "others" because their results did not exhibit high relevance for the research. Of the 36 relevant "A" and "B" sources, 13 were "A" sources. The analysis of these sources resulted in 21 models, which are shown in Table 4. Not all of these models were sufficiently detailed or fulfilled enough criteria for an ontology according to [32]. However, they served as helpful additions to design the DVC ontology REV1. The elements were scored equally according to the frequency of use in the literature and in the workshops. Only elements that scored more than 50% on average were considered for the DVC ontology REV1.

5 Ontology Derived from the Results of the Literature Review

The DVC ontology aims to create a better understanding of how value can be extracted from data. To do so, all components of a DVC must be considered, from the data source to the processing or acquisition of data to the data analysis and decision. The basis of the DVC ontology is the existence or possibility of collecting data to answer a core question. We hope that this work can support future implementation and research by assessing individual DVCs in different domains.

Table 3 Algorithms applied for the literature review

Library	Total [A, B, C]	Relevant [A, B]	Search algorithm
IEEE Xplore including IET	233	7	(("All Metadata":Business Model) or ("All Metadata":Big Data) or ("All Metadata":AI) or ("All Metadata":Data Science) or ("All Metadata":Data Product) or ("All Metadata": Research Project)) and ("All Metadata":Canvas)
ACM Digital Library	84	3	[[All: "business model canvas"]] or [[All: "ai canvas"]] or [[All: "data science canvas"]] or [[All: "big data canvas"]] or [[All: "data product canvas] or [All: „research project canvas"]]]
Springer Link	36	4	Adaptation not required
Science Direct	583	6	(("Data Science") or ("Big Data") or ("Data Product")) and ("canvas")
AISeL[a]	434	6	("Business" and "Model") or ("AI") or ("Data" and "Science") or ("Big" and "Data") or ("Data" and "Product") or ("Research" and "project") and "canvas")
Others (e.g., MDPI, Wiley)	199	10	Individual adaptation required
Total	1579	36	–

[a]*AISeL* Association for Information Systems eLibrary

5.1 Delimitation

Because DVCs for different domains can vary, it can be assumed that no universal model exists. The literature review identified 21 different canvas models for creating value from data. Different perspectives and focus areas were identified; thus, the model should be general and expandable to be more flexible in its use. A modular approach is needed to build adjustable DVC evaluation models according to the needs of different applications and research areas.

5.2 Ontology Structure and Taxonomy

Based on the BMC ontology described by [33], the DVC ontology model REV0 was defined based on previous literature (top-down) and extended with the results of the literature reviews (bottom-up) and the workshop results. Analogously to [33], we first defined the pillars of the DVC (see Table 5).

The explanatory model in Fig. 3 visualizes how the DVC ontology can be embedded into the evaluation process of a DVC. The classification defines an appropriate set of categories that correspond to the research question and the domain in which the DVC is located. They form a basis for the evaluation to identify data products.

Table 4 Classification of canvas models

	CR1	CR2	CR3	CH1	CH2	CH3	TG1	TG2	TG3	VP1	VP2	VP3	KA1	KA2	KA3	KA4	KA5	KA6	KA7	KR1	KR2	KR3	KP1	KP2	KP3	KP4	C$1	C$2	R$1	R$2
1											X				X		X				X									
2	X	X					X	X		X						X				X	X		X				X		X	
3											X			X	X	X		X		X	X									
4	X	X					X	X	X					X	X	X		X			X		X	X	X	X				
5	X	X									X												X							
6										X	X			X	X	X														X
7				X																			X							
8	X																				X		X							
9	X						X	X	X	X	X	X		X		X		X	X		X	X	X					X		
10	X				X	X				X	X	X			X	X					X							X		
11	X									X						X				X	X	X				X		X		
12			X			X					X		X	X	X	X	X	X	X	X	X				X	X	X		X	X
13																														
14	X				X	X		X						X	X	X		X	X		X						X	X	X	X
15																														
16	X	X	X				X				X	X		X	X	X	X	X	X	X	X	X								
17	X	X	X				X	X	X		X	X	X		X		X		X		X	X	X	X	X		X			
18	X					X					X			X	X	X		X	X			X	X	X						
19										X	X				X	X	X				X		X							
20										X	X				X	X											X	X		
21					X					X			X			X		X	X			X	X			X	X		X	
Lit	50%	23%	14%	5%	14%	18%	23%	23%	14%	36%	55%	18%	14%	36%	55%	64%	23%	36%	32%	23%	59%	27%	45%	14%	14%	18%	27%	23%	18%	14%
Ws	18%	55%	91%	91%	91%	64%	100%	91%	91%	100%	82%	27%	55%	0%	36%	36%	9%	9%	27%	45%	100%	55%	73%	64%	36%	0%	27%	0%	55%	18%
Av.	34%	39%	52%	48%	52%	41%	61%	57%	52%	68%	68%	23%	34%	18%	45%	50%	16%	23%	29%	34%	80%	41%	59%	39%	25%	9%	27%	11%	37%	16%

1: AI canvas [1]; 2: AI project canvas [31]; 3: analytics canvas [25]; 4: big data management canvas [31]; 5: business process canvas [22]; 6: data insight generator [31]; 7: data canvas [31]; 8: data collection map [19]; 9: data innovation board [31]; 10: data product canvas [11]; 11: data project canvas [31]; 12: data science canvas [41]; 13: data service card [6]; 14: data value map [30]; 15: data-driven business value matrix [5]; 16: deep learning AI canvas [31]; 17: enterprise AI canvas [31]; 18: key activity canvas [16]; 19: machine learning canvas [31]; 20: research project canvas [27]; 21: the open data value canvas [12]

Table 5 Pillars of the DVC ontology

BMC ontology	DVC ontology
Product	Data product
Customer interface	User story
Infrastructure management	DVC infrastructure
Financial aspects	Evaluation model

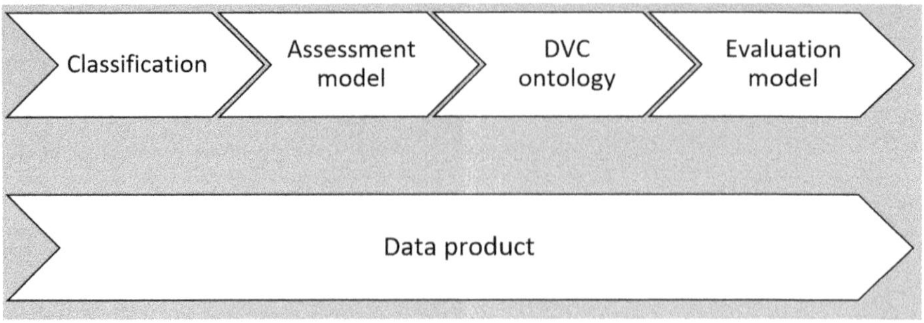

Fig. 3 Explanatory model of the evaluation process of the DVC

The assessment model is performed after the classification. DPs are first prioritized by assessing the potential data value created and the implementation efforts.

The DVC ontology describes the DP details and how they are related. These characteristics help to place the DP into the organization's value chain. In the evaluation model, each step should be built logically on the previous one, making it straightforward to recall. To meet these requirements, indicators (such as the ROI_{DAQ}) are introduced to evaluate the DVC.

The results of the literature review and the evaluation model are used to produce further detail in the pillars. This detailing enables later visualization and analysis of the DVC. At the same time, the evaluation model should remain extendable to match the pace of development in business informatics and to enable long-term general applicability. Based on the findings of the DVC ontology REV0, the original structure did not seem to be fully suitable. [31] classifies their data science canvas into 12 elements. This work aims to remain as close as possible to the BMC model, as this is regarded as an established model to capture the value chain of (data) products. The DVC ontology REV1 remains extendable, as indicated by the "…" in Table 6.

The relationship of the pillars and elements is shown in Fig. 4. Pillars are represented in the background and elements are represented with boxes. Relationships of the elements of the DVC ontology are indicated with arrows. This visualization follows the approach by [33].

Our canvas model has focused on DPs in the center as a value proposition which has been created using a company's DVC infrastructure. On the left side, the DVC infrastructure is represented by the key partners, key activities, and key resources, similar to the description in [33]. The user story on the right side helps to understand the user interface, their specific needs, how the decision is visualized, and through

Table 6 Pillars and elements for the DVC ontology

Pillar	Elements	Description
Data product	Value proposition	Describes the (potential) value of the DP for the user in form of a relevant question, which can be answered utilizing data (analysis)
User interface	Channels	Determines how the DP is used and visualized (e.g., dashboards, reports, etc.). Visualization is a sub-element to channels
	User story	"As a … [who?] I want to … [what?] so that … [why?]" The decision is a sub-element to the user story [9]
	Acceptance criteria	Describes the acceptance criteria of the users for the DP
DVC infrastructure	Key resources	Combines key partners and data source; key partners are a sub-element to key resources
	Key partners	Describes which internal or external partners are involved in creating, maintaining, or improving the DP
	Key activities	Includes the key activities for the DP along the DVC
Evaluation model	ROI_{DAQ}	ROI_{DAQ} is the KPI for data acquisition, including the respective cost and revenue aspects; it describes how the data acquisition is performed
	ROI_{DAGO}	ROI_{DAGO} is the KPI for data governance, including the respective cost and revenue aspects; it describes how the data analysis is conducted and evaluated
	ROI_{DAMA}	ROI_{DAMA} is the KPI for data management, including the respective cost and revenue aspects; it describes how the data analysis is conducted and evaluated
	ROI_{DANA}	ROI_{DANA} is the KPI for data analysis, including the respective cost and revenue aspects; it describes how the data analysis is conducted and evaluated
	$ROI_{...}$	$ROI_{...}$ (…placeholder for potential extension of the DVC ontology)
	ROI_{DEC}	ROI_{DEC} is the KPI for the decision, including the respective cost and revenue aspects

which channels it is communicated. Data sources are part of the key resources and the key partners and support the creation of value from the data. The following DVC ontology and Canvas REV1 was derived accordingly and is shown in Figs. 4 and 5.

In the next section, different elements including the relationship, cardinality, and attributes are detailed following [33].

The functionality of the ontology can be illustrated by using an example from the literature selection of data-driven methods for improving offshore wind operation and maintenance planning collected and evaluated by [13]. Not all criteria of the DVC were specified from the literature; thus, the results can only be shown partly as an example. Table 7 demonstrates how such data products can be embedded into this ontology and evaluated using the developed ROI indicators (here, ROI_{DANA}).

The ontology can be applied on a variety of DPs found in literature or in organizations, making it an extensively applicable model.

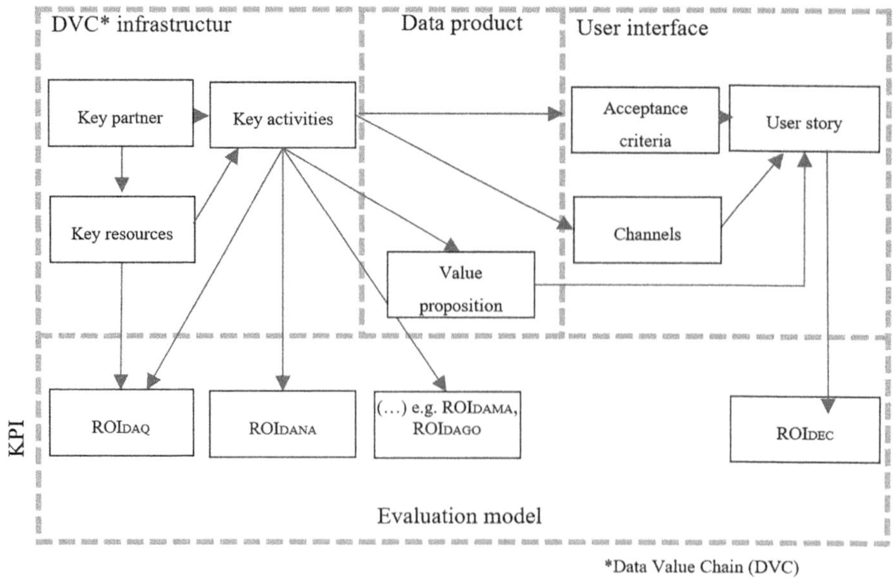

Fig. 4 DVC ontology REV1

<table>
<tr>
<td rowspan="2">KP (key partners): Which partners are involved in the development and continuous data provision?</td>
<td>KA (key activities): Which activities must be carried out “internally” within the data evaluation and DP creation?</td>
<td rowspan="2">DP (data product): A relevant question, which can be answered utilizing data (analysis).</td>
<td>AC (acceptance criteria): How does the data scientist relate to the user of the DP?</td>
<td rowspan="2">US (user story): As a … [who?] I want… [what?] to know in order to … decide … [why?]. Decision is a sub-element to the user story.</td>
</tr>
<tr>
<td>KR (key resources): What is necessary to set up the DP and continuously provide it, as well as to use it? (Data source, data acquisition, data analysis)</td>
<td>CH (channels): In what format and at what frequency will the DP finally be communicated?</td>
</tr>
<tr>
<td colspan="3">C$ (cost structure): What costs are associated with the provision of the DP?</td>
<td colspan="2">R$ (revenue streams): What benefits can be derived from the use of the DP? How is this to be evaluated?</td>
</tr>
</table>

Fig. 5 DVC Canvas REV1

Table 7 Exemplary data products derived from literature

DVC infrastructure	Data product	User interface	Evaluation model
Key partners: wind park operator, safety planner, decision-makers, maintenance team Key activities: Perform a k-nearest neighbor, support vector classification, or linear discriminant analysis	Value proposition: risk-based inspection	Acceptance criteria: model results User story: Detect and identify faults in the wind park Channels: via monitoring procedures	ROI_{DANA} Excellent accuracy and efficiency Excellent prediction accuracy
Key resource: datasets			Not robust to data noise Not suitable for high-dimensional datasets

5.3 *Data Product*

The classification according to [33] shows an overrepresentation of infrastructure elements, while the value proposition portion as well as the financial aspects (here called evaluation model) are underrepresented (see Table 8).

5.4 *Infrastructure Management*

Eight canvases were classified as mainly focused on infrastructure management (see Table 10). Although the elements included in the canvases varied greatly, most of the canvases contained more than 14 elements and were thus extensive. Canvases of the category infrastructure had their focus on infrastructure management, in which great emphasis was placed on data integration, data acquisition, and data analysis methods as well as model selection (see Table 4; ratings ≥50). One of the core elements – the key resources – is described in Table 9.

The element key partners describe which internal or external partners are needed for the DVC infrastructure (Table 10). For example, they could refer to external data providers or data analysts. Some canvas models, such as the key activity canvas [16] or the analytics canvas [25], are specialized in describing key activities and how they relate to the key partners.

Key activities refer to all the actions that are necessary within the DVC to prepare, evaluate, visualize, or provide results and for the maintenance of this process (Table 11). Some of the canvas models found in the literature review, for example, the key activity canvas, place a primary focus on the different activities the key partners need to perform [16].

Table 8 Description of the element value proposition

Name of element	Value proposition (VP)
Definition	The value proposition (VP) describes how and why users should benefit from a DP
Part of	Pillar: data product
Cardinality	1–*n*
Attributes	Description string{abc}
References	[33]

Table 9 Description of the key resources element

Name of element	Key resources (KR)
Definition	The key resources (KR) element answers the question of what is required to establish and operate a DP. The subcategories are the resources (internal and external partners) required, the data sources, and the methods. As a minimum, a potential data source should be named
Part of	DVC infrastructure
Related to	Data source, key activities (KA), data analysis (DANA) method
Cardinality	1–*n*
Attributes	Description string{abc}
References	[33]

Table 10 Description of the key partners element

Name of element	Key partners (KP)
Definition	Key partners (KP) includes internal and external partners (KP1), how they provide support (KP2), and what role partners have in the data science team. Some DPs might not require partners to be built (e.g., if the DP already exists)
Part of	DVC infrastructure
Related to	Data acquisition (DAQ), …, data analysis (DANA)
Set of	Key resources (KR)
Cardinality	0–*n*
Attributes	Description string{abc}
References	[33]

Table 11 Description of the key activities element

Name of element	Key activities (KA)
Definition	The key activities describe the essential resources needed for creating and maintaining the DP along the DVC
Part of	Infrastructure (data value creation)
Related to	ROI_{DAQ}, ROI_{DANA} method, key resources (KR)
Set of	Key resources (KR)
Cardinality	0–*n*
Attributes	Description string{abc}
References	[33]

5.5 User Interface

The majority of canvases (9 out of 21) focused on the pillar user interface. Even though the variation of elements considered by the canvases in this category was larger, they placed great emphasis on the relationship between the DP and the stakeholders, as well as on internal and external partners (e.g., AI project canvas [31] or data collection map [18]). Aspects such as data source and data acquisition were also important in many canvases. In general, most of the canvases contained approximately 11 elements. Table 12 describes the main element of the user interface, the user story.

The channels describe how a DP is communicated to the user, for example, via newsletter, emails, dashboards, reports, or other means (Table 13). [12] describes channels as "online websites, data lakes, APIs,[1] or dedicated apps". The data science canvas [41] also refers to data storytelling for the way data is presented to the user of a specific DP.

Acceptance criteria play an important role for a DP to be accepted by its users (Table 14). These criteria can be different KPIs [23] or special requirements for the visualization.

5.6 Evaluation Model

At the beginning, the evaluation model was developed using a qualitative literature research and expert interviews. The evaluation model should be straightforward to understand and generally applicable in order to enable possible transferability. Indicators are necessary in the evaluation model process to make optimization potentials apparent. The indicators begin with the ROI_{DAQ} (Table 15).

ROI_{DAGO} (Table 16) describes the processes by which an organization manages data. Some canvas models such as the big data management canvas [18] refer to this process as data-knowledge management or data-knowledge engineering.

The ROI_{DAMA}, which is an important aspect of the DVC, is evaluated in the next step (Table 17). Data management is sometimes accommodated using data integration (coupling different data sources for better results) [31].

ROI_{DANA} (Table 18) evaluates the value of the data analysis. The data analysis is most prominent in three of the studied canvases, namely, the AI canvas [1], the deep learning canvas [31], and the machine learning canvas [31]. These canvases are focused on the data source, data analysis methods, and model selection. The deep learning canvas, with 15 different elements, is more holistic and complex and thus considers further aspects which also relate to infrastructure.

[1] API: *Application Programming Interface*

Table 12 Description of the customer (user) relationship element

Name of element	User story (US)
Definition	The user interface (UI) describes how the user interacts with the DP and how it is distributed
Part of	User interface
Related to	User story (US), channels (CH), visualization, decision (DEC)
Cardinality	1–n
Attributes	DESCRIPTION for the decision that contains a string {abc} DESCRIPTION of the user story that contains a string {abc}
References	[33]

Table 13 Description of the channel element

Name of element	Channel (CH)
Definition	Channels describe how the results of the DVC are distributed to the users (e.g., via dashboards, reports, etc.)
Part of	User interface
Related to	User story (US), visualization, decision (DEC)
Set of	User story
Cardinality	0–n
Attributes	DESCRIPTION of the used channels as a string {abc} DESCRIPTION of the used visualization as a string {abc}
References	[33]

Table 14 Description of the acceptance criteria (AC) element

Name of element	Acceptance criteria (AC)
Definition	The acceptance criteria describe which criteria from the users exist to access the presented results of the DP
Part of	User interface, decision (DEC)
Related to	User story (US), visualization, decision (DEC)
Set of	User story (US)
Cardinality	1–n
Attributes	Description string{abc}
References	[33]

Table 15 Description of the ROI_{DAQ} element

Name of element	ROI_{DAQ} (return on information, data acquisition)
Definition	Describes data acquisition in plain language and assesses indicators as outlined in the assessment model description
Part of	Financial aspects
Related to	ROI
Cardinality	0–*n*
Attributes	Description string{abc} Value for completeness (0–*n*) € Value for accuracy (0–*n*) € Value for one-time costs (0–*n*) € Value for yearly costs (0–*n*) €
References	[28]

Table 16 Description of the ROI_{DAGO} element

Name of element	ROI_{DAGO} (return on information, data governance)
Definition	Describes data governance in plain language and assesses indicators as outlined in the assessment model description
Part of	Financial aspects
Related to	ROI
Cardinality	0–*n*
Attributes	TBD Description string{abc} Value for integrity (0–*n*) € Value for consistency (0–*n*) € Value for one-time costs (0–*n*) € Value for yearly costs (0–*n*) €
References	[28], own development

Table 17 Description of the ROI_{DAMA} element

Name of element	ROI_{DAMA} (return on information, data management)
Definition	Describes data management in plain language and assesses indicators as outlined in the assessment model description
Part of	Financial aspects
Related to	ROI
Cardinality	0–*n*
Attributes	TBD Description string{abc} Value for accessibility (0–*n*) € Value for credibility (0–*n*) € Value for interpretability (0–*n*) € Value for a correction factor Value for one-time costs (0–*n*) € Value for yearly costs (0–*n*) €
References	[28], own development

Table 18 Description of the ROI_{DANA} element

Name of element	ROI_{DANA} (return on information, data analysis)
Definition	Describes data analysis in plain language and assesses indicators as outlined in the assessment model description
Part of	Financial aspects
Related to	ROI
Cardinality	0–*n*
Attributes	Description string{abc} Value for value creation (0–*n*) € Value for implementation efforts (0–*n*) €
References	[11, 19], own development

Table 19 Description of the ROI_{DEC} element

Name of element	ROI_{DEC} (Return on information, decision)
Definition	Describes the decision in plain language and assesses indicators as outlined in the assessment model description
Part of	Financial aspects
Related to	ROI
Cardinality	0–*n*
Attributes	Description string{abc} Value for frequency of use (0–*n*) € Value for relevance (0–*n*) € Value for timeliness (0–*n*) € Value for a correction factor Value for one-time costs (0–*n*) € Value for yearly costs (0–*n*) €
References	[28], own development

The ROI_{DEC} is an indicator of detail and evaluates the decision value behind the DP (Table 19). The canvas models sometimes refer to insights as a basis for decisions [25].

Because some of the KPIs might use scaling, a correction factor must be used if the variables in the numerator and denominator are not of equal number.

6 Visualization

The Sankey diagram is a visualization tool for representing flows, which may include energy, resources, or costs. In this case, it can be used to represent the flow of the data value through organizational decision processes in the form of ROI. The main components of the Sankey diagram are flows (or data streams), whose thickness represents the amount or value of the flow, and nodes [20, 40].

Another application has been presented by [3], who compared the representation of Sankey diagrams with node-connectivity diagrams within cybersecurity projects. In this study, database administrators answered questions using both diagram types. The study found slightly faster answering of the questions using the node-connection diagram; however, the experts had a higher preference for the representation of the data in the form of the Sankey diagram [3]. It must be critically considered in this study that database administrators who were presumably familiar with the representation of the node-connection diagram were mainly included; thus, its general validity may be doubtful. [4] presents an example of how flow charts can be enriched with additional graphics to increase insights. A schematic draft of a possible visualization is shown in Fig. 6; the visualization also includes a sunburst diagram to add further indicators if needed and sample "lorem ipsum" text.

7 Discussion and Outlook

While a lot of work has been done in describing data value chains through canvas models, they still focus highly on certain viewpoints and thus are limited in their application. Unlike previous models that have focused on data infrastructure, data analysis, and interaction with decision-makers, the new model is broader and therefore more generally applicable. It consists of the essential elements of the identified DVCs and canvas models. These models have been merged, and evaluation criteria have been added to examine the results of the evaluation.

This chapter proposes an ontology and its visualization for establishing a data value chain (DVC) based on a literature review. First, a preliminary DVC ontology REV0 was generated inductively from the relevant literature. This concept was evaluated in workshops of the WiSA Big Data research project with industry partners, and improvements such as simplification were suggested. Based on these suggestions and a deductive literature review in relevant online databases, the DVC ontology REV1 was created as a result. The restriction of the keywords was also a challenge during the evaluation. Some databases allow ten keywords, while others

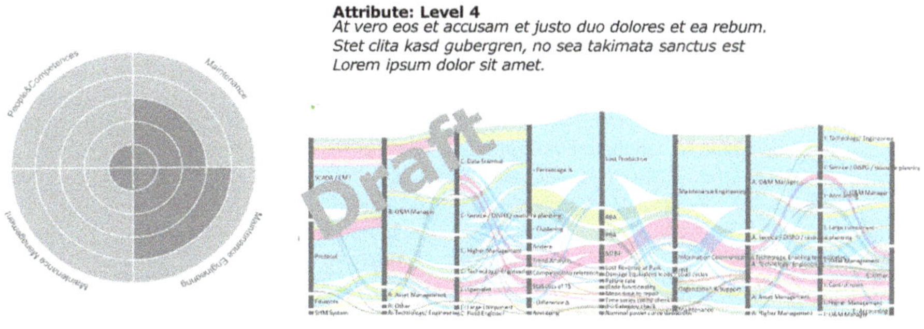

Fig. 6 Visualization mock-up for a DVC

limit the length of the search algorithm to three, making the direct comparison of search algorithms difficult; thus, adaptions needed to be implemented. Regardless of these limitations, the study of [31], in which only nine different canvas models were identified, and the study of [11], which added five additional models, have been substantially extended. The classification of [31] and [33] has also been reviewed and aligned to fit the scope of this work. The number of new canvas models found in this work suggests that the creation of value from data has gained momentum as a field of research. The existing canvas models lack an evaluation model which could be implemented in the suggested solution. By continuing and complementing existing studies, we have made a small contribution to increasing the understanding and research of DVCs.

There is still a long way to go to fully bridge the gap between decision-makers and data scientists. In this work, we have described a model which can be used to analyze, visualize, discuss, and optimize DVCs. Analysis of DVCs and understanding of ways in which organizations and industries can improve their processes for creating value from data will be a substantial element of further data-driven decision-making, future automatization, and other related improvements.

We hope that the model presented here can be used as a guide to further improve DVCs and create a general understanding of their qualities. This progression will enable researchers to evaluate DVCs in different industries or organizations or for different data value strategies.

References

1. Agrawal, A., Gans, J., Goldfarb, A.: A simple tool to start making decisions with the help of AI (2018). https://hbr.org/2018/04/a-simple-tool-to-start-making-decisions-with-the-help-of-ai. Accessed 17 Aug 2022
2. bi-survey: Global survey on data driven decision-making in businesses. 14 survey-based recommendations on how to improve data-driven decision-making (2016). https://bi-survey.com/data-driven-decision-making-business. Accessed 24 Sept 2021
3. Blinder, R., Biller, O., Even, A., Sofer, O., Tractinsky, N., Lanir, J., Bak, P.: Comparative evaluation of node-link and Sankey diagrams for the cyber security domain. In: Lamas, D., Loizides, F., Nacke, L., Petrie, H., Winckler, M., Zaphiris, P. (eds.) Human-Computer Interaction – INTERACT 2019. Springer International Publishing, Cham (2019)
4. Brath, R., Jonker, D.: Graph Analysis and Visualization. Discovering Business Opportunity in Linked Data. Wiley, Indianapolis (2015)
5. Breitfuss, G., Fruhwirth, M., Pammer-Schindler, V., Stern, H., Den-nerlein, S.: The data-driven business value matrix - a classification scheme for data-driven business models. In: Pucihar, A., Kljajić Borštnar, M., Vidmar, D., Baggia, A., Jereb, E., Kofjač, D., Lenart, G., Rajkovič, U., Rajkovič, V., Šmitek, B. (eds.) Humanizing Technology for a Sustainable Society. Conference Proceedings, 1st edn, pp. 803–820. University of Maribor Press, Maribor (2019)
6. Breitfuss, G., Fruhwirth, M., Wolf-Brenner, C., Riedl, A., de Reuver, M., Ginthoer, R., Pimas, O.: Data service cards - a supporting tool for data-driven business. In: Pucihar, A., Kljajić Borštnar, M., Bons, R., Cripps, H., Sheombar, A., Vidmar, D., Baggia, A., Jereb, E., Kofjač, D., Lenart, G., Marolt, M., Urh, M., Werber, B. (eds.) 33rd Bled eConference Enabling Technology for a Sustainable Society. June 28–29, 2020, Online: Conference Proceedings, 1st edn. University Press; Faculty of Organizational Sciences, Maribor (2020)

7. Cai, L., Zhu, Y.: The challenges of data quality and data quality assessment in the Big Data era. CODATA. **14**(0), 2 (2015). https://doi.org/10.5334/dsj-2015-002
8. Chandrasekaran, B., Josephson, J.R., Benjamins, V.R.: What are ontologies, and why do we need them? IEEE Intell. Syst. **14**(1), 20–26 (1999). https://doi.org/10.1109/5254.747902
9. Cohn, M.: User Stories Applied. For Agile Software Development/Mike Cohn. The Addison-Wesley Signature Series. Addison-Wesley, Boston/London (2004)
10. Davenport, T., Patil, D.J.: Data Scientist: the sexiest job of the 21st century. Meet the people who can coax treasure out of messy, un-structured data (2012). https://hbr.org/2012/10/data-scientist-the-sexiest-job-of-the-21st-century
11. Fruhwirth, M., Breitfuss, G., Pammer-Schindler, V.: The data product canvas: a visual collaborative tool for designing data-driven business models. In: Pucihar, A., Kljajić Borštnar, M., Bons, R., Cripps, H., Sheombar, A., Vidmar, D., Baggia, A., Jereb, E., Kofjač, D., Lenart, G., Marolt, M., Urh, M., Werber, B. (eds.) 33rd Bled eConference Enabling Technology for a Sustainable Society. June 28–29, 2020, Online: Conference Proceedings, 1st edn. University Press; Faculty of Organizational Sciences, Maribor (2020)
12. Gao, Y., Janssen, M.: The open data canvas–analyzing value creation from open data. Digit. Gov. Res. Pract. **3**(1), 1–15 (2022). https://doi.org/10.1145/3511102
13. Hadjoudj, Y., Pandit, R.: A review on data-centric decision tools for offshore wind operation and maintenance activities: challenges and opportunities. Energy Sci. Eng. **11**(4), 1501–1515 (2023). https://doi.org/10.1002/ese3.1376
14. Haneke, U., Trahasch, S., Zimmer, M., Felden, C. (eds.): Data Science. Grundlagen, Architekturen und Anwendungen, 1st edn. Edition TDWI, dpunkt.verlag, Heidelberg (2019)
15. Hay, L., Duffy, A.H.B., McTeague, C., Pidgeon, L.M., Vuletic, T., Grealy, M.: A systematic review of protocol studies on conceptual design cognition: design as search and exploration. Des. Sci. **3** (2017). https://doi.org/10.1017/dsj.2017.11
16. Hunke, F., Seebacher, S., Thomsen, H.: Please tell me what to do – towards a guided orchestration of key activities in Data-Rich Service systems. In: Hofmann, S., Müller, O., Rossi, M. (eds.) Designing for Digital Transformation. Co-Creating Services with Citizens and Industry: 15th International Conference on Design Science Research in Information Systems and Technology, DESRIST 2020, Kristiansand, Norway, December 2–4, 2020: Proceedings/ Sara Hofmann, Oliver Müller, Matti Rossi (eds.), vol. 12388. LNCS Sublibrary. SL 3, Information Systems and Applications, Incl. Internet/Web, and HCI, vol. 12388, pp. 426–437. Springer, Cham, Switzerland (2020)
17. Husáková, M., Bureš, V.: Formal ontologies in information systems development: a systematic review. Information. **11**(2), 66 (2020). https://doi.org/10.3390/info11020066
18. Kaufmann, M.: Big Data management canvas: a reference model for value creation from data. Big Data Cogn. Comput. **3**(1), 19 (2019). https://doi.org/10.3390/bdcc3010019
19. Kayser, L., Mueller, R., Kronsbein, T.: Data collection map: a canvas for shared data awareness in data-driven innovation projects. In: Proceedings of the 2019 Pre-ICIS SIGDSA Symposium (2019). https://aisel.aisnet.org/sigdsa2019/18/
20. Keimer, I., Egle, U.: Die Digitalisierung der Controlling-Funktion. Springer Fachmedien Wiesbaden, Wiesbaden (2020)
21. Kenett, R., Redman, T.C.: The Real Work of Data Science. Turning Data into Information, Better Decisions, and Stronger Organizations. Wiley, Hoboken (2019)
22. Koutsopoulos, G., Bider, I.: Business process canvas as a process model in a nutshell. In: Gulden, J., Reinhartz-Berger, I., Schmidt, R., Guerreiro, S., Guédria, W., Bera, P. (eds.) Enterprise, Business-Process and Information Systems Modeling. 19th International Conference, BPMDS 2018, 23rd International Conference, EMMSAD 2018, Held at CAiSE 2018, Tallinn, Estonia, June 11–12, 2018, Proceedings, vol. 318. Lecture Notes in Business Information Processing Ser, vol. 318, pp. 49–63. Springer, New York (2018)
23. Kronsbein, T., Mueller, R.: Data thinking: a canvas for data-driven ideation workshops. January 8-11, 2019, Maui, Hawaii. In: Proceedings of the 52nd Hawaii International Conference on System Sciences (HICSS-52), Grand Wailea, Hawaii, 8.01. - 11.01. AIS Electronic Library (AISeL) (2019). https://aisel.aisnet.org/hicss-52/. https://doi.org/10.24251/HICSS.2019.069

24. Kruse, F., Dmitriyev, V., Marx Gómez, J.: Building a connection between decision maker and data-driven decision process. Arch. Data Sci. Ser. A. **4**(1) (2018). https://doi.org/10.5445/KSP/1000085951/03
25. Kühn, A., Joppen, R., Reinhart, F., Röltgen, D., von Enzberg, S., Dumitrescu, R.: Analytics canvas – a framework for the design and specification of data analytics projects. Proc. CIRP. **70**, 162–167 (2018). https://doi.org/10.1016/j.procir.2018.02.031
26. Kuhrmann, M., Fernández, D.M., Daneva, M.: On the pragmatic design of literature studies in software engineering: an experience-based guideline. Empir. Softw. Eng. **22**(6), 2852–2891 (2017). https://doi.org/10.1007/s10664-016-9492-y
27. Lacruz, A.J., Oliveira Leite, M.C.: Research Project Canvas. v1.1 (2021). https://www.researchgate.net/publication/349771536_Research_Project_Canvas. Accessed 21 Aug 2022
28. Laney, D.B.: Infonomics. How to Monetize, Manage, and Measure Information as an Asset for Competitive Advantage. Bibliomotion, Oxon (2018)
29. Mayring, P., Fenzl, T.: Qualitative Inhaltsanalyse. In: Baur, N., Blasius, J. (eds.) Handbuch Methoden der empirischen Sozialforschung, pp. 633–648. Springer Fachmedien Wiesbaden, Wiesbaden (2019)
30. Nagle, T., Sammon, D.: The Data Value Map. A framework for developing shared understanding on data initiatives. In: European Conference on Informatoin Systems (ECIS). European Conference on Informatoin Systems (ECIS), Guimarães, Portugal, 05.06. - 10.06, pp. 1439–1452 (2017)
31. Neifer, T., Lawo, D., Esan, M.: Data science canvas: evaluation of a tool to manage data science projects. In: Proceedings of the 54th Hawaii International Conference on System Sciences. Hawaii International Conference on System Sciences (HICSS), Honululu, Hawaii (2021)
32. Noy, N.F., McGuinness, D.L.: Ontology development 101: a guide to creating your first ontology (2001). https://protege.stanford.edu/publications/ontology_development/ontology101.pdf. Accessed 18 July 2022
33. Osterwalder, A.: The Business Model ontology a proposition in a design science approach. PhD, 'Ecole des Hautes Etudes Commerciales de l'Université de Lausanne (2004). https://www.researchgate.net/publication/33681401_The_Business_Model_Ontology_-_A_Proposition_in_a_Design_Science_Approach. Accessed 26 March 2022
34. Peffers, K., Tuunanen, T., Rothenberger, M.A., Chatterjee, S.: A design science research methodology for information systems research. J. Manag. Inform. Syst. **24**(3), 45–77 (2007). https://doi.org/10.2753/MIS0742-1222240302
35. Provost, F., Fawcett, T.: Data science and its relationship to Big Data and data-driven decision making. Big Data. **1**(1), 51–59 (2013). https://doi.org/10.1089/big.2013.1508
36. Schmarzo, B.: Determining the economic value of data (2016). https://infocus.delltechnologies.com/william_schmarzo/determining-economic-value-data/. Accessed 12 June 2020
37. Schneider, L.: How to build a foundational ontology. In: Goos, G., Hartmanis, J., van Leeuwen, J., Günter, A., Kruse, R., Neumann, B. (eds.) KI 2003: Advances in Artificial Intelligence, vol. 2821. Lecture notes in computer science, pp. 120–134. Springer Berlin Heidelberg, Berlin, Heidelberg (2003)
38. Smith, B., Munn, K.: Applied Ontology. An Introduction. Metaphysical Research, vol. 9. De Gruyter, Berlin/Boston (2008)
39. Stuckenschmidt, H.: Ontologien. Konzepte, Technologien und Anwendungen. In: Informatik im Fokus, 2nd edn. Springer, Berlin, Heidelberg (2011)
40. Tomanek, D.P., Schröder, J.: Value Added Heat Map. Springer Fachmedien Wiesbaden, Wiesbaden (2018)
41. Vasandani, J.: A Data Science Workflow Canvas to Kickstart Your Projects. Use this guide to help you complete your data science projects (2019). https://towardsdatascience.com/a-data-science-workflow-canvas-to-kickstart-your-projects-db62556be4d0. Accessed 18 Dec 2022
42. Vollmuth, J.H., Zwettler, R.: Kennzahlen. In: Haufe TaschenGuide, vol. 186, 3rd edn. Haufe, Stuttgart (2015)
43. Wöltje, J. (ed.): Schnelleinstieg Unternehmensbewertung und Finanzkennzahlen. Haufe Lexware (2021)

Requirements for Machine Learning Methodology Software Tooling

Jochen L. Leidner and Michael Reiche

1 Introduction

Over the course of the last two decades, there has been an enormous growth in the importance of data-intensive projects, projects aiming at obtaining data-driven insights ("analytics"), and components that apply automatic induction, also known as machine learning, instead of traditional algorithms, to solve a problem at hand. This is because there is a human desire to push the "how" question onto the machine for solving.

This move from algorithms to "soft computing" models induced from data also means that new methodologies are needed that recommend news processes (e.g., to annotate datasets) and new best practices (e.g., to compute inter-annotator agreement between annotators) to build such models. These (we will take a closer look below) extend our toolbox of software engineering methodologies (waterfall, agile kanban,[1] etc.) with methodologies suitable for and indeed specifically designed for machine learning based working, which is typically quantitative, iterative, and experimental.

[1] A kanban board is a software tool to support the kanban (Japanese for "billboard") methodology, which relies on two primary practices: 1. to visualize work and 2. to limit work in progress.

J. L. Leidner (✉)
Coburg University of Applied Sciences, Coburg, Germany

KnowledgeSpaces UG (haftungsbeschränkt), Coburg, Germany
e-mail: leidner@acm.org

M. Reiche
Coburg University of Applied Sciences and Arts, Coburg, Germany
e-mail: michael.reiche@hs-coburg.de

M. Mora et al. (eds.), *Development Methodologies for Big Data Analytics Systems*, Transactions on Computational Science and Computational Intelligence, https://doi.org/10.1007/978-3-031-40956-1_4

As we transition to this new breed of methodology, naturally we would like to make use of state-of-the-art software tools that support our methodology of choice. Where traditional software engineering gave us *computer-aided software engineering* (*CASE*) tools [2], we hope for equivalent guidance in the new, data-centric world.

To this end, we present a collection of requirements for such a software stack: In this chapter, we will gather and organize requirements for software tooling to support machine learning/data science methodologies.[2] For the most part, we will be able to defer the choice of methodology, as it turns out the software tooling requirements can be separated from any particular methodology.

Software engineering—and in the era of soft computing this includes construction of machine learning models—does not happen in isolation: stakeholders need to be educated, influenced, convinced and kept informed, developers briefed about interfaces and non-formalized aspects of integration and maintenance; even holders of financial roles need to learn that model refresh is a recurring post-project activity that needs funding and staffing, which may be unwelcome news in projects motivated by the "saving through automation" promised that machine learning offers. While machine learning researchers typically zoom in on only the mathematical or experimental work surrounding parameter estimation and evaluation of their models, real-life projects require extraordinary amounts of interactions with the environment. This requires new methodologies that are now beginning to emerge, and these in turn require software tooling to capture, store, process, retrieve, etc. the project knowledge that needs to be managed.

The remainder of this chapter is structured as follows: Sect. 2 provides some background about requirements and the requirements capture process. Section 3 briefly recapitulates an exemplary selection of methodologies for machine learning projects. Section 4 describes our collected requirements for methodology software tooling, including requirements from the perspective of what stakeholders need from the system to support their work. Section 5 surveys some related work. In Sect. 6, we provide a critical discussion. Finally, Sect. 7 summarizes our findings and concludes with suggestions for future work.

2 Method: From Stakeholders to Requirements Capture

In order to develop software tools to support the application of (and compliance with) machine learning methodologies that assist project teams and other stakeholders with the functionalities required in the realization of machine learning projects, suitable requirements must be identified, formulated in high quality, and documented in a structured way [3–5]. Stakeholders are individuals or organizations with an interest in the planned system [4, p. 10]. Typical team members and other stakeholders' responsibilities of a machine learning project and their tasks can be found in Table 1.

[2] In the following, the term "machine learning" will be used, although it is also intended to refer to the entire field of data science.

Table 1 Team members and other stakeholders and their typical responsibilities

Team members (top) and other stakeholders (bottom)	Responsibilities
Data scientist	Model engineering including data preparation, algorithm selection, hyperparameters selection, and model evaluation [6]
Data engineer	Data processing including data collection, feature engineering, big data management, data pipeline management, and dataset building [6, 7]
Software engineer/ machine learning engineer	Turn the raw machine learning problem into a prototype or a well-engineered product, monitoring of drifts and adjustments of those [6, 8]
Technical lead	Take charge of all technical decisions
Project manager (PM)	Administer, facilitate, manage: setting, updating, and monitoring team activities, project, and time plans; communication with stakeholders; defines the business goal; review the achievement of objectives; mediate issues and fights in the team, enabling innovation [6, 8]
Machine learning solution architect	Integration of machine learning into the IT infrastructure [8]
User experience specialist	Ensures interacting with the system will be intuitive, pleasant, and successful (free from obstacles) [8]
Subject-matter expert	Bringing understanding of a subject area [8]
User (focus group)	Provides feedback and requirements. "User" refers to an end user of the system – product or service – that the machine learning models are going to be part of
Executive sponsor	Making budget decisions, establishing a vision and several main goals of the project [8]

The IEEE defines a requirement as follows [9, p. 62]: "(1) A condition or capability needed by a user to solve a problem or achieve an objective. (2) A condition or capability that must be met or possessed by a system or system component to satisfy a contract, standard, specification, or other formally imposed documents. (3) A documented representation of a condition or capability as in (1) or (2)" [9, p. 62].

Suitable requirements can be formulated generally at a higher level from a user perspective in natural language (user requirement) or in detail at a deeper level from a system perspective close to the software to be developed (system requirement) [5, 10]. Requirements do not provide information on how they are to be implemented. Rather, they provide information about what the system was intended to do [11, p. 230–231]. From the variants for classifying requirements, we use those that distinguish according to (1) priority; (2) necessity, i.e., must/should/will); and (3) functionality (nonfunctional/functional) [12, p. 80]. We use a template-based approach to formulate the requirements (see Fig. 1).

In principle, both analytical and empirical methods are suitable for the systematic eliciting of requirements. In Sect. 4, we elicit requirements for the needs of the typical stakeholders with the definition of characteristic user stories, which are derived from special properties of machine learning methodologies and from practical experience of the two authors of this chapter, respectively. From these, we extract a set of formal requirement templates (Fig. 1) and eventually entities that can form the basis for a class diagram (in an object-oriented analysis) or an ER (entity-relationship) diagram.

3 Machine Learning Process Models

In the following, we will recapitulate some selected methodologies starting with the oldest that have been proposed for projects using machine learning, data mining, and data science generally speaking. For detailed surveys, see the related work below.

3.1 KDD

The KDD process [14–16] resulted from the "Knowledge Discovery from Databases" community, which also created the conference series of the same name. Data mining, according to it, is a five-step process with iterations to get from raw data to knowledge. First, carrying out a data selection step results in target data, a preprocessing activity leads to preprocessed data, transformations lead to transformed data, and then the actual data mining turns it into patterns, which are interpreted by humans and/or evaluated by machine and/or humans (Fig. 2).

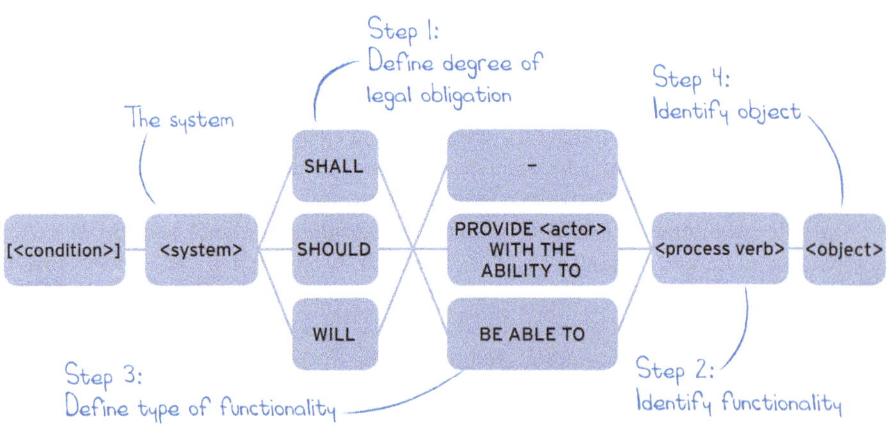

Fig. 1 Requirement template [13]

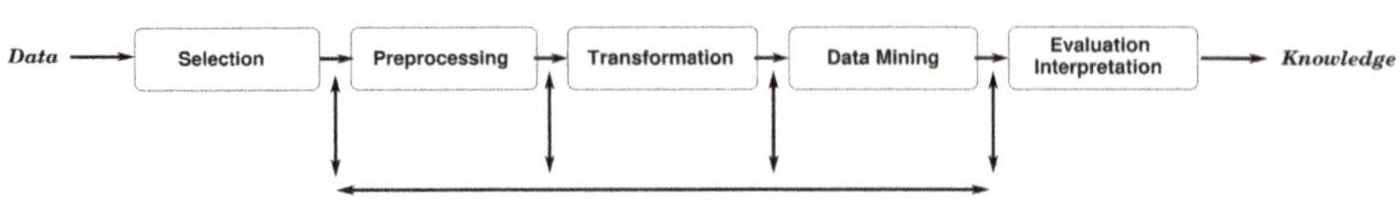

Fig. 2 The KDD process (simplified after Fig. 1 from [15, p. 29])

KDD starts in its first phase with the selecting a dataset (database, set of variables or data sample under consideration). Preprocessing work then removes noise/outliers, plugs gaps by imputation, and dealing with database schemas falls into this step. After that, the data is transformed (reduced and/or projected), which includes finding and extracting features that are useful for the task at hand and reducing its complexity. Next, the data mining algorithms/methods have to be selected (models, parameters). Framing or choosing the function of the data mining step then involves reflecting on the purpose of the model (e.g., summarization, classification, regression or clustering). The actual data mining phase applies classification, regression, some form of clustering, or sequence modeling suitable for the goals and dataset at hand. The final phase includes making sense of automatically discovered patterns, potentially using visualizations, weeding out superfluous or useless patterns, and translating useful patterns into language understandable by the project's stakeholders. Implicit, but accounted for in the methodology's prose descriptions, is also making use of the resulting knowledge gained, which often means integrating it into a software system, which the authors call "performance system"; this corresponds to the deployment phase in other methodologies. Documenting and reporting concludes a KDD process-based project.

3.2 SEMMA

The SEMMA methodology [17] was developed by SAS, Inc., a software company which also sells the SAS Enterprise Miner software, which is somewhat aligned with the SEMMA process. The process is divided into the following five phases:

- Sample: A subset of the appropriate data must first be selected. Identifying variables or factors (both dependent and independent) influencing the process is carried out in this phase, as is partitioning the data into training and test folds.
- Explore: In an exploratory stage, uni- or multivariate analysis is conducted to detect gaps in the data and to study interconnected relationships; this phase is expected to rely heavily on data visualization techniques.
- Modify: Data is cleaned and transformed in order to prepare it for the modeling, using insights from the previous exploration phase.
- Model: This phase is where the core modeling step applies, i.e., a variety of data mining techniques are applied with the intention of identifying the one most suitable for solving the business problem at hand.
- Assess: Evaluate the model (How useful and reliable is it? How well it solves the problem?). Computing quantitative evaluation metrics of the best model's quality is part of this last phase.

SEMMA has had limited impact to date, which is likely due to the fact that it was created (and is owned) by a single proprietary company.

3.3 CRISP-DM

In contrast to the KDD process, the Cross-Industry Standard Process for Data Mining (CRISP-DM for short) [18, 19] was developed in an industrial environment and emerged from the cooperation of the companies NCR System Engineering, SPSS Inc., and DaimlerChrysler AG. The iterative CRISP-DM starts with the Business Understanding phase before and ends with the Deployment phase after the cycle of the KDD process (Fig. 3). Because activities not previously considered are included here, the process is more comprehensive. The process is divided into the following six phases:

- Business Understanding: From the business perspective, project goals are determined and requirements and resources are defined in this initial phase. All findings are incorporated into a project plan.
- Data Understanding: Here, data is collected, described, and analyzed to explore it.
- Data Preparation: The goal of this phase is to create an adequate dataset for the following modeling. For this process, data is selected, cleaned, transformed, merged, and formatted.
- Modeling: In this phase, the modeling itself is performed. Therefore, modeling techniques are applied and parameters are calibrated.
- Evaluation: The obtained model is tested for its final use by evaluating how the defined business objectives from the Business Understanding phase have been achieved. In addition, past activities are reviewed, and next steps are determined.
- Deployment: The model and the knowledge gained through it are made usable.

During the use of the model, it is monitored and, if necessary, maintained. Besides that, the project is documented and a final report is prepared.

The recommended procedures of the CRISP-DM are described in a detailed handbook. It states that each phase has several generic activities, which in turn are associated with artifacts, mostly in the form of reports. A project that adapts the CRISP-DM in its pure form and is carried out entirely according to its manual will therefore result in a comprehensively documented project. However, the process is often adapted to the individual circumstances of a project, which leads to the omission of given elements or the addition of new ones [20]. It is still widely used today and can be considered the de facto standard in the field of data-intensive analytics projects [20–22].

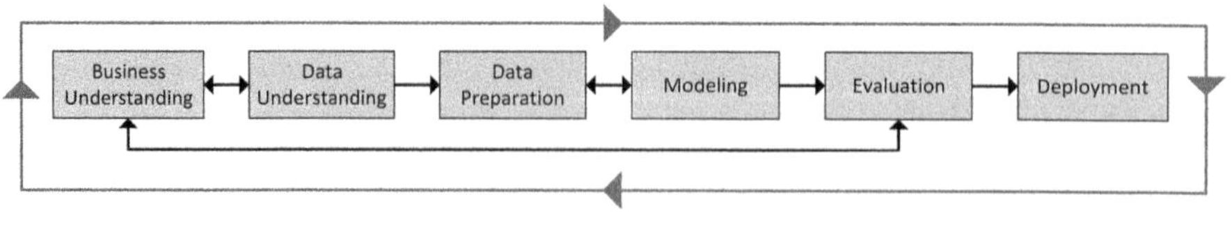

Fig. 3 The CRISP-DM methodology (simplified after Fig. 2 from [18, p. 10])

3.4 CRISP-ML(Q)

CRISP-ML(Q) (short for "Cross-Industry Standard Process for the development of Machine Learning applications with Quality assurance methodology") is an attempt (by a group that did not include the original CRISP-DM creators) to adjust CRISP-DM from data mining to machine learning work [23].[3] It takes into account the special characteristics of machine learning, such as monitoring and maintaining a machine learning application in a changing deployed environment. As the name already indicates, essential concepts have been taken over from the CRISP-DM. In CRISP-DM(Q), the two phases Business Understanding and Data Understanding are merged into the Business and Data Understanding phase. The term "Maintenance" has found its way into the name of the Monitoring and Maintenance phase. Quality assurance measures to mitigate risks are proposed for all six phases of the iterative methodology.

3.5 Data-to-Value (D2V)

The Data-to-Value methodology ("D2V" for short) is a development process model [24–26] for the construction of systems that use (mostly supervised) machine learning in at least some of its components; it was developed, tested, and taught during the teaching of university students at various universities (Essex, Zurich, Frankfurt, Sheffield, Coburg) over the course of a decade, and it is motivated by the first author's long-standing industry practice in research and development projects in natural language processing and information retrieval system construction of applications for professionals in the vertical domains of news/journalism, finance/insurance, legal, risk/security, and pharmacology. It offers several characteristics that are unique at the time of writing:

- It is an "evaluation first" methodology, which means that quantifying models is addressed before any models are actually built; evaluation scaffolding is constructed early, so ongoing quantitative evaluation can guide the development process, following the saying "what you can't measure, you can't improve."
- It has an intricate number of stages – over 30 – each of which is associated with some from a set of 100+ guidance questions to help more junior team members around standard pitfalls and to create more consistency for senior team members.
- In particular, acknowledging the importance of data quality, the gold data annotation process is spelled out in detail, which is surprisingly lacking from most natural language processing and machine learning text books published to date.

[3] https://ml-ops.org/content/crisp-ml

- Ethical and technology impact considerations are not treated as an optional afterthought; following [27, 28], they have been integrated into the process by design, in the form of various checkpoints.
- D2V features a "feasibility study" phase early in a project, which was motivated by real-life requests by managers for impossible projects, in particular predictive modeling of target variables for which no predictive signal is available in the available datasets. This step dramatically de-risks data-intensive projects and helps reduce sunk cost.

Figures 4 and 5 show the various stages of the D2V methodology.

Note that the boxes with double-line frames are indicative of subprocesses. The feasibility study is a subgraph that contains a copy of the whole graph (except feasibility study itself) in a way that permits many steps to be partially completed or skipped; its purpose is, ahead of committing to a full-blown project, to check whether there is enough signal in the data so that the predictive models conceived to be constructed later actually can be built. Due to space reasons, we must refer the interested reader to the open access technical report [25], which contains the most detailed description of the process model to date.

Now that we have sketched the space of common and recent methodologies, we can look into the list of core requirements for a software system that supports these and other similar methodologies.

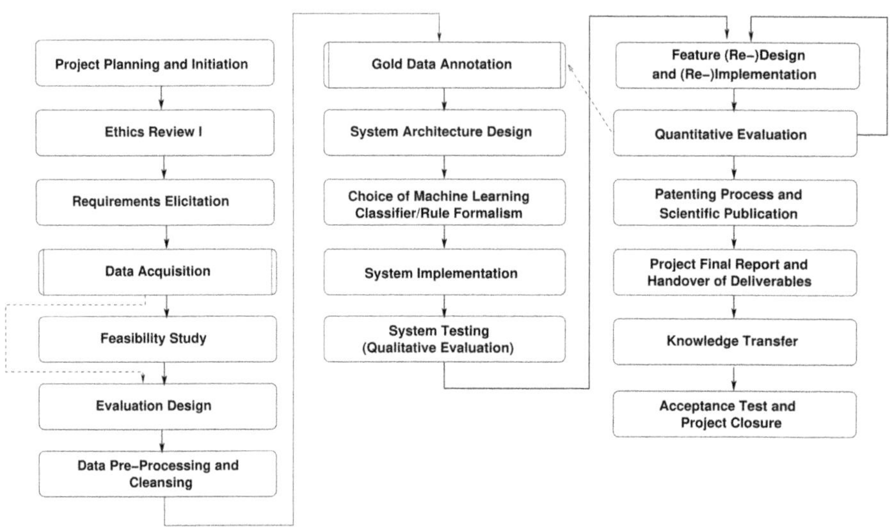

Fig. 4 The Data-to-Value phases: overall process (after [26])

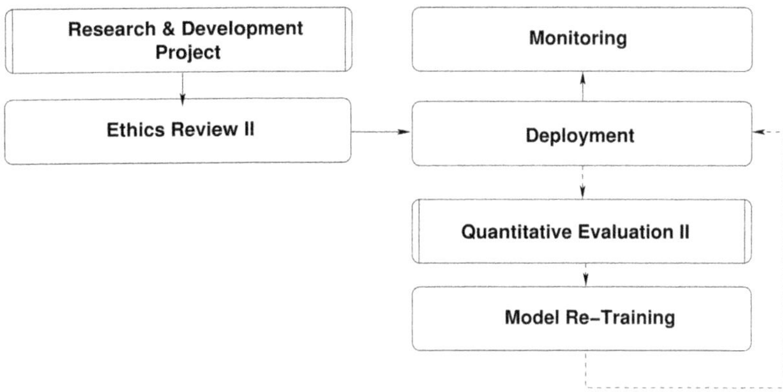

Fig. 5 Further Data-to-Value phases: (i) gold data annotation subprocess (left), (ii) system deployment (right; after [26])

4 Requirements for Software Support Tools

We sourced our requirements as follows: First, we derived many requirements directly or indirectly from properties of the methodologies that the software should support, especially where these overlap. Second, we supplemented these by requirements gained through introspection by extracting them from user stories by the two authors, who have a combined 30 years of experience.

We will describe our findings starting with user stories: A series of them put the users of the system (team, in particular PM, and other stakeholders) at the center. We then derive a longer list of requirements of instantiated requirement templates from these; for brevity, we show a selection of the most relevant requirements here, and we will exclusively focus on functional requirements. Finally, we derive a list of the key classes (in object-oriented thinking) or entities (in entity-relationship thinking), respectively.

4.1 Overall Vision

In addition to our presentation of the functional requirements, we begin with a vision statement to capture the overall spirit and scope of the system that we anticipate being a useful support tool for machine learning projects with the following particular methodologies:

> There is a need for a system that guides a project team to follow a chosen machine learning methodology and which manages the project's metadata centrally, persistently, and transparently.

This vision of a system that supports its users by providing process guidance and metadata management is like a meta-level requirement: All other requirements should at least not be in conflict with it. By saying "chosen" methodology, we imply that not a particular methodology is hardwired in the software we envisage but that it should be designed and implemented in a way that is separate from any particular methodology and thus be capable of supporting most, past, present, and future methodologies; in order words, we attempt to generalize across several or most approaches. This requirement seems realistic, as in the realm of workflow tools, such a generic approach has already been successful (see below).

4.2 User Stories and Requirement Templates

We now present several user stories, followed by the requirements extracted from them.

Jack is an experienced project manager. His company uses a new software to support a range of machine learning methodologies, as it improves his everyday professional life in many ways: Instead of having to gather specifics for a machine learning project (phases, paths, quality, etc.) from multiple systems, he can now use one system one-stop shop tool to capture, store, and communicate a project's status and progress to his superiors and team members as well as to control and monitor his projects.

Requirement 1 (Track Current Phase) (Must-Have) The system shall maintain the current state (active phase) of the project with respect to the methodology that it uses.

Requirement 2 (Monitor and Control Project) (Must-Have) The system shall offer possibilities to monitor and control the project.

Requirement 3 (Track Actions and Time per Phase) (Must-Have) The system shall capture "what happens when" in the project, including the duration (in calendar days) spent in each phase.

Requirement 4 (Import Methodologies) (Must-Have) The system shall let the user model import one (from a predefined library of given methodology) workflow.

Requirement 5 (Edit/Customize Methodologies) (Must-Have) The system shall let the user customize a methodology's workflow for a project.

At the beginning of a project, Jack must define which methodology (CRISP-DM, Data-to-Value, etc.) should be used. The phases and paths are then automatically recorded in the software as a default workflow of the methodology. This workflow, including phases and paths, can be adopted or customized. Jack is able to enrich the selected workflow with project specifics, including the planned duration of sections, the project participants (core team) and other stakeholders, and their roles and the RACI (responsible, accountable, consulted, informed) category assigned to them [29]. All changes to the project and their history are automatically stored and versioned. Target times specified by him can be compared with the actual numbers, or with those of past projects. If guidance questions are provided by the methodology or requested by the users, then Jack can answer them himself or pass them on to his team, thereby determining common processes of the machine learning project. All answers are stored in the system's database.

Requirement 6 (Support Multiple Methodologies) (Must-Have) The system shall associate one methodology with each project.

Requirement 7 (Track Actions and Time per Phase) (Must-Have) The system shall log project actions conducted in a permanent read-only activity protocol.

Requirement 8 (Capture and Categorize Stakeholders) (Must-Have) The system shall capture all project stakeholder names, roles, and their associated RACI categories.

Requirement 9 (Create Artifacts) (Must-Have) The system shall automatically create document-based and/or document-like artifacts for evaluating completed phases and/or activities. (Note: An artifact is a by-product created during the machine learning development process, such as datasets, models, or documents.)

Requirement 10 (Ask Guidance Questions and Record Answers) (Must-Have) The system shall ask the project manager the set of potential guidance questions associated with a phase (if the chosen methodology posits them) and capture their answers. If they do not know an answer, permit entry of a list of team members ("share/forward") that may be able to answer.

Jack can enter needed skills and skill levels for the project and then adds Sarah to the project as a team member in a data science role. The system is aware of her skills in data science, Python programming, and in particular her experience with clustering methods; however, as the system detects that Sarah has not yet got experience with Apache Hadoop and Apache Spark, two standard systems for distributed processing in big data projects, which Jack had indicated as needed for the project, the system alerts Jack to the skill gap, and because the formal start of the project is still a few weeks away, Jack proposes Sarah be sent to a training course as a means of upskilling to her line manager Joe, a suggestion to which he agrees.

Requirement 11 (Manage Team) (Must-Have) The system shall capture the initial team, the role(s) of each member, skills required, skills needed for the project, as well as ongoing team changes during the project.

All project participants can view the status of the project and the progress in the workflow at any time on the central view of a web-based dashboard, while stakeholders external to the team get to see a specific view. The team can view their personal task inbox as well as the global task list for current phrase with respect to the methodology. Important success indicators, activities already carried out, and activities still to be carried out are shown. RAG tags (red-amber-green) are used to convey high-level status information to senior stakeholders. On the one hand, the visualization of the dashboards at the moment of viewing can serve as a basis for discussion vis-à-vis stakeholders. On the other hand, standard reports can be sent at regular intervals, or ad hoc reports can be sent via email or other communication channels when defined events occur. A final report is semiautomatically generated by the system at the end of each phase and at the end of each project. Jack can view the current project and also past or concurrent projects he is entitled to. These are stored in the system directly or through links to a project database. The presentation of several projects side-by-side enables them to be compared with each other, and all data is persisted for long-term archival purposes.

Requirement 12 (Manage Tasks) (Must-Have) The system shall permit team members to view their assigned (or all) tasks that are associated with the given methodology's current phase. They can also change the status, using the ternary states "OPEN," "IN PROGRESS," and "COMPLETED"; tasks are automatically synchronized with external third-party software systems from the project management and the information management domains.

Requirement 13 (Manage Scope Change) (Must-Have) The system shall permit the PM to enter a scope change request received from the customer, and it collects sign-offs and comments from relevant stakeholders.

Requirement 14 (Regularly Update Stakeholders) (Must-Have) The system shall send regular project updates to all stakeholders honoring the RACI matrix and comprising RAG status for all milestones, potential delays accrued, best model quality information, and key performance indicators.

Requirement 15 (Create Final Report) (Must-Have) The system shall automatically generate a final report after a project is completed. The report is stored in the system and may be edited by the project manager and/or technical lead before sending it to relevant stakeholders. The customer of the project is asked to acknowledge initial receipt, and by answering to what extent the project has met the criteria for success, the project is formally closed.

Maria has a computing degree specializing in data engineering. In the team, it is her task to import different input datasets and enrich them with metadata. The software system supports her in comparing and preparing different input datasets. She can add useful comments to each dataset for her team colleagues. The system supports her by offering the possibility to integrate several systems, which she uses as data sources.

Requirement 16 (Plug-Ins) (Must-Have) The system shall be able to integrate different types of software via a plug-in mechanism, so that a team/organization can choose from a broad range of external tools.

Requirement 17 (Track Input Datasets Versions) (Must-Have) The system shall capture and store a short description from a business perspective, attributes, and metadata of each input dataset for machine learning. (Note: Metadata is data about data, which is the description and context of data.)

Requirement 18 (Support Team Activities) (Nice-to-Have) The system should provide team members with the ability to support and comment on their activities.

Requirement 19 (Track Code Versions) (Must-Have) The system shall capture and version code the data used for machine learning experiments.

Requirement 20 (Record Decisions and Rationales) (Must-Have) The system shall capture (and store in a database) any important decisions taken and the rationales behind them.

Wendy, a data scientist on the project, has an idea for a new feature that may make her company's classification model more accurate. As she is in the Feature Engineering phase of Data-to-Value methodology used in her project, she can remain in the current phase and initiate a "new feature" hypothesis action. She then codes up the feature extractor for her idea and triggers the evaluation action. It seems to help, as F1 goes up slightly, priming her to checks in the new version of the code. The system keeps track of the rationale behind her idea and links it to the change-set ID of the revision control system that has the code, the model version thus improved, and the database of the system records the new quality scores (precision, recall, F1) of her improved classifier. The system also automatically updates work statistics, adding 1 day to the Feature Engineering phase. Two days later, at the end of the month, stakeholders receive an email to the system's dashboard that shows the quality improvements achieved in this month's reporting period.

Requirement 21 (Track Output Datasets Versions) (Must-Have) The system shall capture and version the success metrics, attributes, and metadata of each output record for machine learning. Note: Success metrics are, e.g., accuracy, error rate, precision, and recall.

Bob is a junior data scientist that has recently joined Wendy's team from a local university. He is smart and hard-working, but naturally, he still lacks real-life project experience. While, officially, Wendy mentors Bob, in practice she has to spend much time in meetings as she receives more and more responsibilities. Bob is guided by the system's phases, each of which is associated with guidance questions that help Bob avoid some typical "beginners' pitfalls." As he is charged with the data annotation subproject, the system suggests some possible tools for annotation and for inter-annotator agreement, and it persists the quality of each round of gold data annotation. After working hard in phase Annotate More Data by Multiple Annotators, Bob initiates a re-training action of the latest model based on additional gold data that was just annotated, and the system calls the training and evaluation scripts automatically. Upon completion of the automatic evaluation, quality KPIs are stored in the system's database with a time stamp. Seeing the new learning curve of a baseline support vector machine regressor, the system suggests checking if it has flattened enough to suggest concluding the gold data annotation work, and in doing so, he concurs. Bob spends the next couple of weeks training and improving various model variants, and he likes that the system provides a central means for him to keep track of the hyperparameter setting and the evaluation results of each model. He also draws motivation from seeing the F1 score improve over time in the "best model to date" view, which also shows the remaining time available for experimenting in the current project. In the past, Bob kept track of experiments in "README" files, but project managers would typically keep asking for them to be sent by email to them, which cluttered both their email inboxes.

Requirement 22 (Track Model Provenance) (Must-Have) The system shall capture and store the short name, version number (release number), technical description, hyperparameter settings, training data used, and revision control identifier (pull request ID) of each machine learning model. Ideally, this should be done to minimize human effort (avoiding duplication of data by semiautomatic import/software integration).

Requirement 23 (Track Quality over Time) (Must-Have) The system shall capture and store in a database the absolute quality as well as relative progress or regress of all model variants with time stamps.

Alice is the Chief Technology Officer (CTO), to whom the Vice President of Research and Development reports, and to whom directs Wendy's data science group. As a key stakeholder, she receives monthly PDF reports that document the project's progress, and she has access to the dashboard for the project. Skimming over the PDF report, it occurred to her to compare the project with the most similar past project to see how the current performance holds up to that scrutiny. Click on the dashboard; she can view the actions, milestones, and KPIs of the current project; and by adding a second project to this view and by using the built-in search, she can make the dashboard show the most similar project side-by-side for easy comparison. The fact that the system's database keeps track of all past projects' history and outcomes permits to use an organization's past performance in order to predict future behavior: Alice finds that the most similar past project was delivered successfully, with a small-time delay of 5%, and she is relieved that this may suggest that this time, absent of unknown unknowns, perhaps a similar result may materialize.

Requirement 24 (Track Artifacts) (Must-Have) The system shall store all artifacts versioned and with project assignment.

Requirement 25 (Search) (Must-Have) The system shall be able to search for artifacts, stakeholders, skills, and project knowledge.

We should stress that we do not assume our catalog to be complete or even complete with respect to "must-have" requirements, but merely as a first attempt to spell out the needs for a set of systems that – once implemented – can support a range of methodologies. We encourage others to supplement our list.

4.3 From Requirements Toward OO Classes/ER Entities

The above requirements help us identify a set of classes (or relational entities) as follows:

- *Projects.* The system shall maintain a persistent list of projects, some of which are in progress, some of which are completed, and some of which have been put on hold.
- *Team Members.* A team member has a line manager, a set of skills and associated availability information, and a set of roles in the project.
- *Tasks.* Tasks are the ultimate constituent elements of work packages in a project plan.
- *Roles.* A role is the function (or set of functions) that a team member plays in a given project, e.g., "technical lead," "project manager," "data scientist," "software engineer," or "data quality specialist."
- *Skills and Skill Levels.* A team member can have a particular skill profile (set of skills at certain levels each) associated with them. A project can have a set of skills and levels needed associated with it.
- *Stakeholders.* The PM can create and maintain a list of stakeholders (typically, people outside the project team), who may be external or internal and have an interest in the project. The project's executive sponsor, who pays for it, is a prime example. The system shall maintain RACI information (a classification of how to communicate with the stakeholder; see [29]). A stakeholder can pose a question to the PM. A stakeholder can view one of several progress dashboards.
- *Models.* A project relies on one or more machine learning components or models, each of which exists in various iterations. Models have a technical short name and a description, as well as link to code and training data it was used to induce it from. Models can be run on datasets (experiments), resulting in experimental results. Model performance can be plotted on a timeline that documents progress regarding quality (time t shows all models available then, singling out the best one).
- *Experiments.* A project typically comprises several experiments per machine learning component. An experiment comes into life once code for a model type is run against one or more folds of one or more datasets. Experiments can conclude once time allocated to experimenting is used up, sufficient quality has been reached, or (most often) diminishing returns on efforts invested have been reached.
- *Datasets.* A project uses one or more datasets. Datasets can be human-annotated or not ("raw"). Datasets can either play the role of input or source of gold data for either training or evaluation.
- *Methodologies.* A methodology defines a set of phases and possible transitions between them in the form of a directed graph. Each phase may contain activity information and other elements such as questionnaires.
- *Textual Artifacts.* A textual artifact is a text document (regardless of file format) that is uploaded (mostly automatically) to serve as the documentation for the

project as a whole or serves as the result (deliverable) of a methodology's phase. One can view these as file attachments.
- *Phases.* A phase is a state (point in time associated with a set of activities and project subgoals) in a methodology; it can recursively contain other phases.
- *Activities.* An activity is something that a team member is expected to carry out, given the project is in a certain stage of a methodology.
- *Questionnaires.* A questionnaire is a set of predefined questions associated with some methodologies and their associated, project-specific answers. They support making assumptions and decisions in the project explicit, transparent, and shareable.
- *Code.* Model implementation program code is not contained in the system but referred to as a link to a source code change-set (e.g., in Git or a similar revision control system).

5 Related Work

We found five broad groups of prior art that are relevant to the work presented in this chapter: first, work on tool support for machine learning methodologies; second, relevant work on requirements capture, including requirement specifications for machine learning models; third, tools to construct machine learning pipelines; fourth, general work on tooling support for workflows, especially but not limited to the software engineering domain; and fifth, work on tool support for machine learning development, deployment, and operations.

5.1 *Work on Tooling for Machine Learning Process Methodology*

A useful starting point for abstracting over individual methodologies is comparisons: For instance, several papers offer survey and comparisons of CRISP-DM, SEMMA, KDD, and other methodologies [22, 30–34].

In a case study [35], 17 software engineers, data scientists, and others at a Dutch bank were interviewed to identify shortcoming of existing machine learning process models. They find that "existing development tools for machine learning are still not meeting the particularities of this field," and in particular, "feasibility assessments, documentation, model risk assessment, and model monitoring are stages that have been overlooked by existing lifecycle models."[4]

We are not aware of any previous work on actual requirements elicitation for software support tooling in the context of following a methodology.

[4] The interviews predate the major Data-to-Value publications, which address several aspects of all of these.

5.2 Requirements Capture

The computer science literature on requirements capture for designing software systems is vast, and it is covered in general software engineering textbooks [10, 11] as well as in dedicated monographs and papers [4, 12, 36–41], so we will just mention a few exemplary references important to our work here.

Mullery [42] describes an early method for controlled requirement specification. User stories have long been considered a useful tool in requirements elicitation [43].

Recently, design thinking has had a great influence on software engineering, in particular under the various agile methodologies. Canedo and Parente da Costa [44] provide a survey of the intersection of design thinking and agile software engineering. While we are not aware of previous work that derived requirement templates from user stories, in [45] goals are derived from a use-case based requirement specification. While we extracted the requirements from user stories manually, [46] propose an interesting idea, namely, the extraction using natural language processing methods (see [47] for a review).

5.3 Workbenches for Constructing Machine Learning Pipelines

There are a number of software systems that permit the technical experimentation with machine learning methods by defining workflows run by individuals, and without addressing any nontechnical issues such as project management, data and code provenance, etc. We will just name two popular and free example systems here, but there are many others. Weka (short for: Waikato Environment for Knowledge Analysis, [48]) is an open source project originating from the University of New Zealand at Waikato. It is implemented in Java, therefore cross-platform, and it lets users define data mining/machine learning workflows visually. These can then be executed end to end automatically at the click of a button from reading the data to showing an evaluation result table.[5] Orange [49] is a similar offering originating at the Bioinformatics Laboratory, Faculty of Computer and Information Science, University of Ljubljana, Slovenia. It is implemented in Python and C++ and likewise uses a graphical "no code" interface to define local data processing workflows that constitute machine learning experiments.[6] RapidMiner (formerly known as YALE) is a commercial offering originating from work done at the University of

[5] See http://old-www.cms.waikato.ac.nz/~ml/weka/ (accessed 2023-01-30) and https://en.wikipedia.org/wiki/Weka_(machine_learning) (accessed 2023-01-30).

[6] See https://orangedatamining.com/widget-catalog/ (accessed 2023-01-30) for a set of the available processing resources/components that can be used in a workflow and https://www.youtube.com/c/OrangeDataMining (accessed 2023-01-30) for a set of training videos.

Dortmund, Germany.[7] It is cross-platform and also permits the composition of data flows on screen via visual programming. SAS Inc., the largest privately owned software company, also has a range of offerings in this space (SAS Enterprise Miner, SAS Viya, etc.).[8]

These and similar systems are very useful for beginners or machine learning users without the skills or willingness to implement code, but contrary to our goal here they do not integrate documentation, team collaboration, and other nontechnical aspects that are essential for working in large and in particular distributed teams.

5.4 *Work on Support Tooling for Business Workflows*

There is a huge body of work in computer science and business informatics specifically on workflows[9] and computer-supported systems for workflow management (see [50] for a survey).

Besides the academic literature, there are software products available for BPM (business process modeling) and RPA ("robotic process automation"), in particular from the largest enterprise software providers: SAP SE's Signavio Process Manager and Oracle's BPM Suite.[10] The latter contains Oracle BPM Studio, a component that enables process developers to create process-based applications and for process analysts and developers to model business processes.

5.5 *Work on Support Tooling for Machine Learning Development, Deployment, and Operations*

The term "ML-DevOps" combines the areas of development, deployment, and operations of machine learning applications (in analogy to traditional software DevOps, cf. [51]) so that their cooperation is strengthened with a continuous pipeline. The objective is to reduce the release time of usable software products and to increase the quality of the software product in the production environment [52, 53].

[7] See https://rapidminer.com (accessed 2023-01-30).

[8] See https://www.sas.com (accessed 2023-01-30).

[9] In computer science, "workflow" is a highly ambiguous term, which may denote business processes (the word sense we are interested in here), allocation of work to workers in operating systems or high-performance compute clusters, and global distribution of work in grid computing (distributed scientific computing), among others.

[10] See https://www.oracle.com (accessed 2023-01-30) and https://www.sap.com (accessed 2023-01-30).

Available software products in this area are Domino's Enterprise MLOps platform, DataRobot's AI Cloud platform, and Weights & Biases's MLOps platform.[11] The differences to the software proposed in this chapter are the lack of integration of (and guidance by) different (customizable) methodologies and the lack of team functionality. This results in disadvantages: For example, a question catalog with guidance questions can be provided by a methodology to help in going through the phases. This question catalog could recommend certain workflows and exclude other workflows based on the response behavior of the team members.

6 Discussion

We have collected what we believe are the core requirements for a support tool that facilitates implementing (following, complying with) recently proposed machine learning methodologies, and we have shown a subset of higher-level ones in this chapter (some more granular ones were not presented for space reasons). The main lesson to be learned from this exercise is a reaffirmation that considerable value can be created through the combination of well-designed processes supported by well-designed support tooling; however, a strong symmetry can be observed in that the advantages benefit more the organization and management than the team members. All team members need to embrace the methodology as well as the tool, which requires a small overhead of time and effort from everyone; however, the immediate benefit is to have a single go-to location where project information can be accessed from. Project managers and less experienced team members benefit the most from the tooling as specified here, the former from the centralization of project-related information and the latter from the guidance that the tooling provides.

It is hard to check a set of requirements for completeness in isolation; the easiest way to find out that everything that is needed is covered is perhaps to implement the requirements in a running system, so that shortcomings will quickly become apparent. Ultimately, support tooling for methodologies ought to be evaluated in controlled experiments, as has been proposed for methodologies themselves [54], but the effort to carry out such experiments is substantial.

One of the hardest problems, in our view, will be to design any supporting software in ways that ensure it will be embraced or at least accepted by all team members. The past has shown that many systems (including useful and well-respected ones like Atlassian's JIRA issue tracking system) are rejected as "bureaucratic" or "overkill," whereas others (e.g., version control systems like Git and the online service offering it, Microsoft's github.com, or the team chat groupware Slack) have become more readily accepted. One key to success here might be offering

[11] See https://www.dominodatalab.com/product/domino-enterprise-mlops-platform, https://www.datarobot.com/platform/ and https://wandb.ai/site (accessed 2023-01-30).

alternative user interfaces (e.g., based on the command line) to accommodate varying preferences among software engineers.

Another serious practical challenge is the question of how to design the software tool that supports the methodology in a way so that the integration of existing software tools is possible from the project management domain (e.g., Microsoft Project, Omni Group's OmniPlan, Atlassian Trello, Atlassian JIRA, SAP Project System, etc.), the machine learning experimentation domain (e.g., RapidMiner, Python/Google Tensorflow, Weka, JetBrains PyCharm, JetBrains IntelliJ, Microsoft Code), team and communication domain (e.g., email, Slack, Zoom), and data and information management domain (e.g., PostgreSQL, Git, Amazon Web Services Simple Storage Service) (Fig. 6). Project plans need to be importable, and activities belonging to a methodology's phases must be able to trigger creation of the tasks assigned to team members, for instance, in JIRA, automatically.

The following four refined yet minimal requirements fall out of the above discussion and can be seen to specialize the original Requirement 16. They appear to be critical for the system's practical success. At the same time, it is challenging to implement them in a way that avoids the anticipated system to become more than loosely coupled with the various third-party systems supported. The plug-in mechanism, therefore, must be designed to be simple, generic, portable, and minimalistic.

Requirement 26 (Import Tasks) (Nice-to-Have) The system should provide an inbound interface for third-party plug-ins to import sets of tasks from external project management systems (e.g., Microsoft Project, Omni Group's OmniPlan).

Requirement 27 (Export Tasks) (Nice-to-Have) The system should provide an outbound interface for third-party plug-ins to export sets of tasks to external ticketing management systems (e.g., Atlassian JIRA or Trello).

Requirement 28 (Synchronize Contacts, Skills, and Comments) (Nice-to-Have) The system should provide an inbound and outbound interface for third-party plug-ins to import and export project-related communications (e.g., email, Slack), people names, roles, and contacts (e.g., Lightweight Directory Access

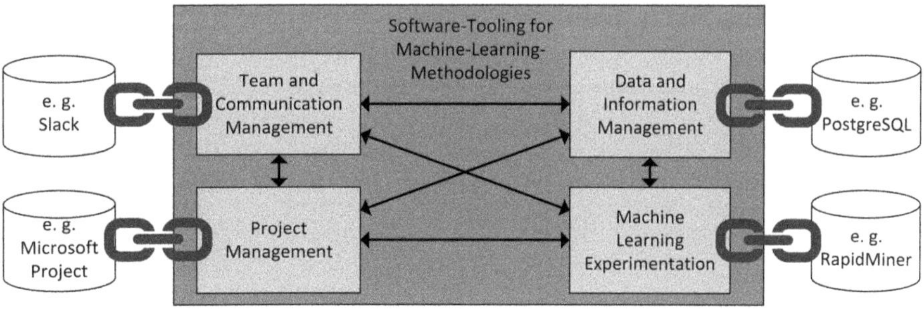

Fig. 6 Loosely coupled integration of external systems

Protocol, Internet Message Access Protocol, Microsoft Exc[]hange) from and to external communication as well as people skills, expertise, and experience via personnel management systems (e.g., SAP SuccessFactors, Kenjo, Workday).

Requirement 29 (Synchronize Development Activity) (Nice-to-Have) The system should provide an inbound and outbound interface for third-party plug-ins to import and export action or change notifications from and to external experimentation management systems (e.g., SAS, RapidMiner) and data (incl. model) stores (e.g., RNA Mapping Database, Amazon Web Services Simple Storage Service).

7 Summary, Conclusions, and Future Work

In this chapter, we have described a set of requirements for software tooling support for machine learning methodologies. We tried to do so without hard-wiring assumptions of any specific methodology, abstracting the narration to the level of following a predefined workflow.

Our result is a list of 29 presented functional requirements, with indications whether we rate each as essential or optional. In drawing up this list, we followed the proven template-based approach for requirements capture.

While there has been a lot of prior research on business process modeling, including software tooling, we are not aware of any previous requirements capture attempt for machine learning methodology software support.

In future work, our findings could be confirmed or challenged by holding interviews with experienced stakeholders to ensure our catalog of requirements is complete. High-level user requirements ought to be translated to specific system requirements that in turn inform a system architecture for the envisaged system. A system or a set of systems then ought to be implemented that embody subsets of these requirements as the logical next step. A prototype could then serve to affirm the consistency and completeness of the requirements.

While this is going to be a lot of work, it might not be hard work; in contrast, the adoption of such tools ought to be maximized, and the effectiveness of the support provided by such tools ought to be evaluated (e.g., using questionnaires similar to Lending and Chervany [55]), tasks we consider both very challenging. Furthermore, future work could study how to maximize adoption of tools by the various individuals that make up project teams and other stakeholders.

Acknowledgments The authors would like to thank Marco Zierl for discussions and two anonymous referees for helpful feedback that improved the quality of this chapter.

References

1. Weber, C., Hirmer, P.: P. Reimann. In: Abramowicz, W., Klein, G. (eds.) Business Information Systems, pp. 403–417. Springer International Publishing, Cham (2020)
2. Iivari, J.: Commun. ACM. **39**(10), 94 (1996)
3. Cheng, B.H., Atlee, J.M.: Future of Software Engineering (FOSE '07), pp. 285–303 (2007). https://doi.org/10.1109/FOSE.2007.17
4. Kotonya, G., Sommerville, I.: Requirements Engineering. Worldwide Series in Computer Science. Wiley, Nashville (1998)
5. Sommerville, I.: IEEE Softw. **22**(1), 16 (2005). https://doi.org/10.1109/MS.2005.13
6. Kreuzberger, D., Kühl, N., Hirschl, S.: Machine learning operations (mlops): overview, definition, and architecture (2022). https://doi.org/10.48550/ARXIV.2205.02302. https://arxiv.org/abs/2205.02302
7. Provost, F., Fawcett, T.: Data Science for Business. O'Reilly Media, Sebastopol (2013)
8. Taulli, T.: Implementing AI Systems. Apress, Berkeley (2021)
9. IEEE: IEEE standard glossary of software engineering terminology (1990). https://doi.org/10.1109/IEEESTD.1990.101064. IEEE Std 610.12-1990
10. Sommerville, I.: Software Engineering, 10th edn. Pearson Education, London (2015)
11. Braude, E.J., Bernstein, M.E.: Software Engineering, 2nd edn. Wiley, Nashville (2007)
12. Hull, E., Jackson, K., Dick, J.: Requirements Engineering. Springer-Verlag, London (2005)
13. The SOPHISTs: Requirements engineering: the sophists »a short RE primer« (2016). https://www.sophist.de/fileadmin/user_upload/Bilder_zu_Seiten/Publikationen/Wissen_for_free/RE-Broschuere_Englisch_-_Online.pdf. Accessed 2022-07-21
14. Fayyad, U., Piatetsky-Shapiro, G., Smyth, P.: AI Mag. **17**, 3 (1996). https://doi.org/10.1609/aimag.v17i3.1230
15. Fayyad, U., Piatetsky-Shapiro, G., Smyth, P.: Commun. ACM. **39**(11), 27 (1996). https://doi.org/10.1145/240455.240464
16. Fayyad, U.M., Piatetsky-Shapiro, G., P.: Smyth. In: Fayyad, U.M., et al. (eds.) Advances in Knowledge Discovery and Data Mining. MIT Press, Cambridge, MA (1996)
17. SAS Institute Inc.: Data mining using SAS Enterprise Miner: a case study approach (2013). https://support.sas.com/documentation/onlinedoc/miner/ casestudy_59123.pdf. Accessed 2022-07-21
18. Chapman, P., et al.: CRISP-DM 1.0 – step-by-step data mining guide. Tech. rep. The CRISP-DM Consortium (2000) https://www.kde.cs.uni-kassel.de/wp-content/uploads/lehre/ws2012-13/kdd/files/CRISPWP-0800.pdf. Accessed 2008-05-01
19. Shearer, C.: J. Data Warehous. **5**, 13–22 (2000)
20. Schröer, C., Kruse, F., Gómez, J.M.: Proc. Comput. Sci. **181**, 526 (2021). https://doi.org/10.1016/j.procs.2021.01.199
21. KDnuggets: What main methodology are you using for your analytics, data mining, or data science projects? poll (2014). https://www.kdnuggets.com/polls/2014/ analytics-data-mining-data-science-methodology.html. Accessed 2022-07-21
22. Saltz, J., Hotz, N.: 2020 IEEE International Conference on Big Data (Big Data), p. 2038–2042 (2020). https://doi.org/10.1109/BigData50022.2020.9377813
23. Studer, S., Bui, T.B., Drescher, C., Hanuschkin, A., Winkler, L., Peters, S., Müller, K.R.: Machine Learning and Knowledge Extraction. **3**(2), 392 (2021). https://doi.org/10.3390/make3020020
24. Leidner, J.L.: Project management for data science. Tutorial held at the IEEE international conference on data science and applications (DSAA 2018), Turin, Italy, 2018
25. Leidner, J.L.: Data to value: an 'evaluation-first' methodology for natural language projects. Tech. rep. Cornell University, New York, NY, USA (2022) https://arxiv.org/abs/2201.07725. ArXiv Pre-Print Server
26. Leidner, J.L.: In: Rosso, P., Basile, V., Martínez, R., Métais, E., Meziane, F. (eds.) Natural Language Processing and Information Systems: proceedings of the 27th International

Conference on Applications of Natural Language to Information Systems, 15–17 June, Valencia, Spain, pp. 517–523. Springer, Cham, Switzerland, NLDB 2022 (2022). https://doi.org/10.1007/978-3-031-08473-7_48
27. Hovy, D., Spruit, S.L.: Proceedings of the 54th Annual Meeting of the Association for Computational Linguistics, ACL 2016, pp. 591–598. ACL, Berlin, Germany (2016)
28. Leidner, J.L., Plachouras, V.: Proceedings of the First ACL Workshop on Ethics in Natural Language Processing Held at EACL, pp. 30–40. Association for Computational Linguistics, Valencia, Spain (2017). https://doi.org/10.18653/v1/W17-1604
29. Kerzner, H.: Project Management: a Systems Approach to Planning, Scheduling, and Controlling, 10th edn. Wiley, Hoboken (2009)
30. Azevedo, A., Santos, M.F.: Proceedings of the IADIS European Conference on Data Mining, 24–26 July 2000, pp. 182–185, Amsterdam (2008)
31. Haertel, C., Pohl, M., Nahhas, A., Staegemann, D., Turowski, K.: PACIS 2022 Proceedings (2022). https://aisel.aisnet.org/pacis2022/242
32. Kurgan, L.A., Musilek, P.: Knowl. Eng. Rev. **21**(1), 1 (2006). https://doi.org/10.1017/S0269888906000737
33. Mariscal, G., Óscar Marbán, C., Fernández: Knowl. Eng. Rev. **25**(2), 137 (2010). https://doi.org/10.1017/S0269888910000032
34. Martinez, I., Viles, E., Olaizola, I.G.: Big Data Res. **24**, 100183 (2021). https://doi.org/10.1016/j.bdr.2020.100183
35. Haakman, M., Cruz, L., Huigens, H., van Deursen, A.: Emp. Softw. Eng. **26**(5), 1 (2021)
36. Wiegers, K.: Software Requirements, 3rd edn. Microsoft Press, Redmond (2013)
37. Pohl, K.: Requirements Engineering: Grundlagen, Prinzipien, Techniken, 2nd edn. dpunkt, Heidelberg, Germany (2008)
38. Vessey, I., Conger, S.A.: Commun. ACM. **37**(5), 102 (1994). https://doi.org/10.1145/175290.175305
39. Herzwurm, G., Schockert, S., Mellis, W.: Joint Requirements Engineering: QFD for Rapid Customer-Focused Software and Internet-Development. Vieweg, Wiesbaden (2000)
40. Rupp, C.: Requirements-Engineering und -Management. Professionelle, iterative Anforderungsnanalyse für IT-Systeme. Hanser, Munich, Germany (2001)
41. Firesmith, D.: Engineering security requirements. J. Object Technol. **2**(1), 53–68 (2003)
42. Mullery, G.P.: Core -a method for controlled requirement specification (1979)
43. Lucassen, G., Dalpiaz, F., van der Werf, J.M.E., Brinkkemper, S.: International Working Conference on Requirements Engineering: foundation for Software Quality, pp. 205–222. Springer (2016)
44. Canedo, E.D., da Costa, R.P.: In: Marcus, A., Wang, W. (eds.) Design, User Experience, and Usability: theory and Practice, pp. 642–657. Springer International Publishing, Cham, Switzerland (2018)
45. Anton, A.I., Carter, R.A., Dagnino, A., Dempster, J.H., Siege, D.F.: Requir. Eng. **6**, 62 (2001)
46. Ghosh, S., Elenius, D., Li, W., Lincoln, P., Shankar, N., Steiner, W.: ARSENAL: automatic requirements specification extraction from natural language. Tech. rep. Cornell University, New York, NY, USA (2016)
47. Raharjana, I.K., Siahaan, D., Fatichah, C.: IEEE Access. **9**, 53811 (2021). https://doi.org/10.1109/ACCESS.2021.3070606
48. Witten, I.H., Frank, E., Hall, M.A., Pal, C.J.: Data Mining: practical Machine Learning Tools and Techniques, 4th edn. Morgan Kaufmann, Amsterdam (2016)
49. Demšar, J., Curk, T., Erjavec, A., Črt Gorup, T., Hočevar, M., Milutinovič, M., Možina, M., Polajnar, M., Toplak, A., Starič, M., Štajdohar, L., Umek, L., Žagar, J., Žbontar, M., Žitnik, B.Z.: J. Mach. Learn. Res. **14**, 2349 (2013)
50. La Rosa, M., Aalst, W.M.P.V.D., Dumas, M., Milani, F.P.: ACM Comput. Surv. **50**(1) (2017). https://doi.org/10.1145/3041957
51. Ebert, C., Gallardo, G., Hernantes, J., Serrano, N.: IEEE Softw. **33**(3), 94 (2016)

52. Bass, L., Weber, I., Zhu, L.: DevOps: a Software Architect's Perspective. Addison-Wesley Professional (2015)
53. Lwakatare, L.E., Kilamo, T., Karvonen, T., Sauvola, T., Heikkilä, V., Itkonen, J., Kuvaja, P., Mikkonen, T., Oivo, M., Lassenius, C.: Inf. Softw. Technol. **114**, 217 (2019). https://doi.org/10.1016/j.infsof.2019.06.010
54. Saltz, J., Shamshurin, I., Crowston, K.: Proceedings of the Hawaii International Conference on System Sciences, pp. 1013–1022. HICSS (2017)
55. Lending, D., Chervany, N.L.: Proceedings of the 1998 ACM SIGCPR Conference on Computer Personnel Research, pp. 49–58. ACM, New York, NY, USA (1998)

A Selective Conceptual Review of CRISP-DM and DDSL Development Methodologies for Big Data Analytics Systems

David Montoya-Murillo, Manuel Mora, Sergio Galvan-Cruz, and Angel Muñoz-Zavala

1 Introduction

Nowadays, many worldwide organizations are in the process of digital transformation that requires the development of useful, secure, and valuable software applications. Additionally, these software applications are expected to be available for short periods and generate quality services that respond to the organization's and its customers' needs [1]. In recent years, due to analytics techniques and the availability of massive data sources – both internal and external – in organizations, there has been an increase in big data analytics systems (BDAS), which demand modern BDAS development methodologies, i.e., lightweight and agile. Currently, the first of these kinds of modern methodologies has been proposed [2, 3].

Big data analytics systems (BDAS) are a particular category of software applications in the domain of data science-data analytics projects. Data science or data analytics is a recent discipline that combines statistics, artificial intelligence, and computer science to explore, predict, or prescribe decisional situations. However, only large business organizations are the usual customers and end users of data science-data analytics projects, and they focus on costly big data platforms [4]. Thus, small and medium-sized business organizations lose the benefits of using data science-data analytics projects. Nevertheless, data science-data analytics approaches can also be used for small data projects [5].

However, both small and big data science-data analytics projects are difficult projects to be successful [6]. Several international reports indicate that a large

D. Montoya-Murillo (✉) · M. Mora · A. Muñoz-Zavala
Autonomous University of Aguascalientes, Aguascalientes, Mexico
e-mail: david.montoya@edu.uaa.mx

S. Galvan-Cruz
Software Engineering Unit, CIMAT, Zacatecas, Mexico

M. Mora et al. (eds.), *Development Methodologies for Big Data Analytics Systems*, Transactions on Computational Science and Computational Intelligence, https://doi.org/10.1007/978-3-031-40956-1_5

percentage of data science-data analytics projects failed to be released with the budget, schedule, or functionality as planned. Agile methodologies in data science-data analytics have been proposed to cope with the problem of failed data science-data analytics projects [6], but agile methodologies have also been criticized for more stable software applications, and thus a more disciplined development approach must be used [7].

In this chapter, therefore, there are evidence [6, 7] that suggests that a lightweight development approach – neither agile nor rigorous – applied to the development of small data science-data analytics projects can generate benefits in the usability, security, and quality of the project while maintaining the established schedule and budgets foreseen for the project. Several studies report the benefits of using lightweight development methodologies [3, 7] while avoiding the limitations of agile and rigorous development approaches and attempting to leverage their advantages.

For very small entities (VSEs), a new family of software engineering standards, ISO/IEC 29110, has been developed to fit the lightweight approach [8]. Initial studies have reported multiple benefits on product quality, achieving budget and schedule as planned [9, 10]. This family of standards is composed of four profiles – entry, basic, intermediate, and advanced, and these do not imply any specific application domain. They are related to the number and types of projects and the size of the VSEs. In this chapter, we considered the Technical Report ISO/IEC TR 29110-5-1-2:11, Software Engineering-Life Cycle Profiles for Very Small Entities (VSEs) Part 5-1-2: Management and Engineering Guide – Generic Profile Group – Basic profile.

For the successful development of a project, it is relevant to characterize critical factors as reported in Fig. 1 adapted from [3]. This diagram represents five critical

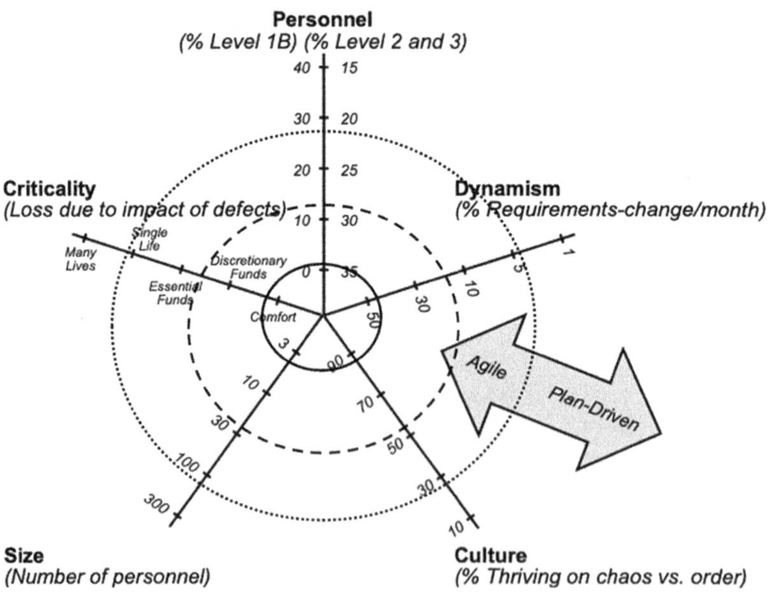

Fig. 1 Key discriminators of agile and plan-driven home grounds [3]

factors of a software development project. These five factors are culture, dynamism, personnel, (impact) criticality, and size. In general, those which are closer to the center are factors in favor to agility (internal circle) and their counterpart (external circle), those which are appropriated for rigorous projects in which time, technological complexity, or the number of people are considerably higher than in agile ones. This indicates that we should pay attention to people-related factors because the success of software projects is directly related to the people who develop them. Boehm and Turner [3] recommended looking for and taking care of how to balance technical and social skills. This recommendation suggests that a lightweight approach can offer advantages of both relative rigor for its compliance and relative agility for its sooner release (projects located between rigor and agile circle zones).

In the software development project context, the purpose of this chapter is to analyze the characteristics and favorable software process elements of a methodology for the project development of BDAS. Even though the BDAS area counts with many BDAS development tools, it lacks still well-tested methodologies specifically for data science-data analytics development projects [6, 7]. A comparison between the two main current proposed methodologies will be generated using the ISO/IEC 29110 standard – Basic profile – as an expected template of the software development process to identify the alignment of roles, phases-activities, and artifacts. The utilization of the ISO/IEC 29110 standard – Basic profile – can help as a factor of stability and success in the BDAS development projects of an organization [8]. This comparison is conceptual between two relevant methodologies against the indicated template of roles, phases-activities, and artifacts expected in the ISO/IEC 29110 standard – Basic profile.

The remainder of this paper continues as follows. In Sect. 2, the research methodology is reported. In Sect. 3, the theoretical background of BDAS and the ISO/IEC 29110 standard – Basic profile – are presented. In Sect. 4, the conceptual review of the two selected methodologies is presented, the first from a rigorous approach and the second from a lightweight approach. Finally, in Sect. 5, a discussion of the implications and conclusions of this research is presented.

2 Research Method

In this chapter, we use a selective conceptual review methodology with a descriptive and comparative dual goal [10]. According to [11], this review can be characterized by its goal, focus, perspective, coverage, organization, and expected audience. This conceptual review is focused on practices – i.e., empirical professional development methodologies for big data analytics systems (BDAS); it is realized from a non-neutral perspective – it aims to describe and compare two methodologies against a generic lightweight ISO/IEC 29110 basic profile development process; it uses a pivotal coverage – it describes and compares only the most relevant classic development methodology for BDAS vs. one of the most modern lightweight methodologies for BDAS; its organization is methodological – it is using the generic

Table 1 Selective conceptual review research steps

Step	Purpose	Outcomes	Outcomes in this research
(1) To formulate the research goal	To state the expected research goal indicating the theoretical or practical or both ones expected contributions	Research goal statement	To contribute to the literature with a conceptual descriptive-comparative review of two – one classic and one lightweight type – relevant development methodologies for BDAS and provide to the practice useful recommendations regarding both development methodologies.
(2) To define data sources and selective criteria	To identify and agree the set of data sources to collect the studies, as well as to define the selection criteria	List of data sources Selection criteria statements	The two development methodologies for BDAS were selected according to the next criteria: (1) to select the classic methodology most cited in the literature and (2) to select a modern and complete – i.e., it includes roles, phases, activities, and artifacts – lightweight development methodology reported from 2015 to 2022 period
(3) To collect studies	To get the studies	Set of selectedstudies	Two methodologies were identified, and theirs published references [13, 14] were obtained
(4) To review and synthetize the findings from the collected studies	To conduct the analysis and integration of finding	Structured schema of findings	We elaborated a generic lightweight development methodology using the ISO/ IEC 29110 – Basic profile – standard
(5) To elaborate report of findings	To produce visible results	Research results	This chapter was elaborated

development lightweight methodology as a template; and it is elaborated for a specialized audience – BDAS academics and professionals.

We applied the following steps adapted from [12] in this selective conceptual review: (1) to formulate the research goal, (2) to define data sources and selective criteria, (3) to collect studies, (4) to review and synthesize the findings from the collected studies, and (5) to elaborate report of findings. Table 1 summarizes the five selective review steps that were applied.

In the following section, the topics of BDAS and the ISO/IEC 29110 standard – Basic profile – are presented and introduced to understand the relationship between the development of such projects and the development guided by the standard.

3 Background

This section reviews the background of the fundamentals of big data analytics systems (BDAS) and the principles of the ISO/IEC 29110 standard – Basic profile – as a lightweight development process.

3.1 Foundations of Big Data Analytics Systems (BDAS)

Historically, two researchers – Michael Cox and David Ellsworth – from the National Aeronautics and Space Administration (NASA) [15; p. 5–5] were the first to refer to the term "big data" when they reported: "Visualization poses an interesting challenge for computer systems of computer systems: the data sets are often quite large, straining the capacity of main memory, local disk, and even remote disk. We call this the big data problem". They emphasized that even the supercomputers of this epoch could not process that amount of information, and thus they labeled this process as a "big data" handling problem. This refers to the problem of having volume of information that exceeds the capabilities of the available computers to handle them. Nowadays, the problem of big data is a factor that must be considered in any type of organization due to the increase in internal and external data generation. New sources of data include social media data, website clickstream data, mobile devices, sensors, and other machine-generated data. All these data sources must be managed in a consolidated and integrated way so that organizations obtain valuable inferences and knowledge [16].

With this development of big data, data warehouses, the cloud, and a variety of advanced software and hardware, data science-data analytics has evolved significantly. Data science-data analytics involve the investigation, discovery, and interpretation of patterns within data. Due to the growing enthusiasm around the use of data science-data analytics and many successful stories, more organizations are interested to exploit these approaches, as many companies in the industry offer similar products and use comparable technologies, making business processes a point of differentiation in projects or products [17]. This has generated business organizations that use data science-data analytics, for instance, to generate competitive advantages that allow them to better understand the situation of their organizations, the market, and the competition. These companies want to know more about their customers, what prices those customers will pay for it, how many items they will buy, and what triggers will make them buy more products. Similarly, they want to know when their inventories are running out of stock and need to anticipate demands and supply chain problems to achieve low-inventory rates and better order rates [17].

Managing the information captured from companies and their clients to obtain a competitive advantage has become a very expensive property when using traditional data analysis methods, which are based on structured relational databases [14]. This dilemma not only applies to large companies but also to small and medium-sized companies, research organizations, governments, and educational institutions, which need less expensive computing and storage power to analyze complex scenarios and models involving images, videos, and other types of data, such as textual data [18]. Big data is a business opportunity area for current international business due to the large amount of data that is generated and thus can be used for supporting data-based decision-making. Big data describes a holistic information management strategy that is constituted by a diversity of new types of data and the management of such data along with traditional data. Although many of the techniques for processing and analyzing these types of data have been around for some time, it has been the massive generation of data and lower-cost computational models that have fostered greater adoption [19]. According to [20], there are different ways to extract information from big data, and these can be divided into three types:

- *Traditional enterprise data*: Transactional ERP data includes customer information from CRM systems, general ledger data, and web store transactions.
- *Machine-generated/sensor data*: includes manufacturing sensors, call detail records, equipment logs, weblogs, trading systems data, and smart meters.
- *Social data*: social media platforms like Facebook, and micro-blogging sites like Twitter, including customer feedback streams.

With the growth of the study and development of big data and their data architecture, developments have grown exponentially. They have migrated their operation toward dynamic and flexible structures that leave behind the classic rigid structures, to give way to structures with the capacity to assimilate structured and unstructured data. Although big data in its origin only handled 3Vs, with the development of the area, the emergence of new technologies, and the increment in the generation of information, new Vs emerged. We reported the most agreed five characteristics that refer to volume, velocity, variety, veracity, and value. They are defined in Table 2.

Table 2 Big data features 5Vs

Attributes	Definition
Volume	The most recognized feature of big data is the presence of large datasets that allow us to analyze to extract valuable information. Organizations currently must learn to manage the large volume of data through new processes. Volume in big data can be defined as large volume of data that either consume huge storage or consist of large number of records [21]
Variety	The word "Variety" denotes the fact that big data originates from numerous sources that can be structured, semi-structured, or unstructured [22]. This is another critical attribute of big data as data is generated from a wide variety of sources and formats [21]
Velocity	Speed refers to the frequency of data generation and/or the frequency of data delivery [21]. The high speed of big data can allow analysts to make better decisions, generating commercial value [23]. To use the high speed of data, many companies now use sophisticated systems to capture, store, and analyze data to make real-time decisions and retain their competitive advantages [24]
Veracity	High data quality is an important big data requirement for better predictability in the trading environment [22]. Therefore, verification is necessary to generate authentic and relevant data and to have the ability to filter incorrect data [25]. This tells us that data verification is essential to the data management process since erroneous data will hinder decision-making or guide analysts down the wrong path. Similarly, incorrect data would have little relevance to add commercial value [24]
Value	It is the added value obtained by organizations; value is created only when data is analyzed and acted upon correctly. To do this, we must identify all the data that will help us in the best way to generate value. This can be interpreted as the extent to which big data generates economically worthy insights and/or benefits through extraction and transformation [24]

The architectural design of big data must be oriented to address these five characteristics recognized in big data known as the "5Vs."

Considering the importance of the aforementioned, the next logical step is to understand the basis of the technological architecture that supports such kind of projects. The main objective of a big data management architecture is the analysis and processing of large amounts of data that cannot be carried out conventionally since the capabilities of standard storage, management, and processing systems are exceeded [26]. A big data management architecture must be able to support the design of systems and models to process large volumes of data from a myriad of data sources quickly and economically, enabling better decision-making. In Fig. 2, a base structure for the potential big data management architecture is presented from Phillips-Wren et al. [27]. Before using big data, you must ensure that all components of the big data management architecture are in place. Without this proper configuration, it will be quite difficult to obtain valuable information and make correct inferences. Big data, analytics, data mining, or decision-making can be carried out in any type of company, whereas they can count with the adequate big data management architecture. Thus, large and medium-sized business companies are expected to count with costly IT infrastructures, but small companies will require less IT resources. Big data for small companies can provide some expected advantages of having small data sources, but more structured and with higher veracity of them for a more efficient processing.

In Table 3, we report a comparison between projects that comply with the theory of the 5Vs of big data and the so-called small data projects that do not have the exact

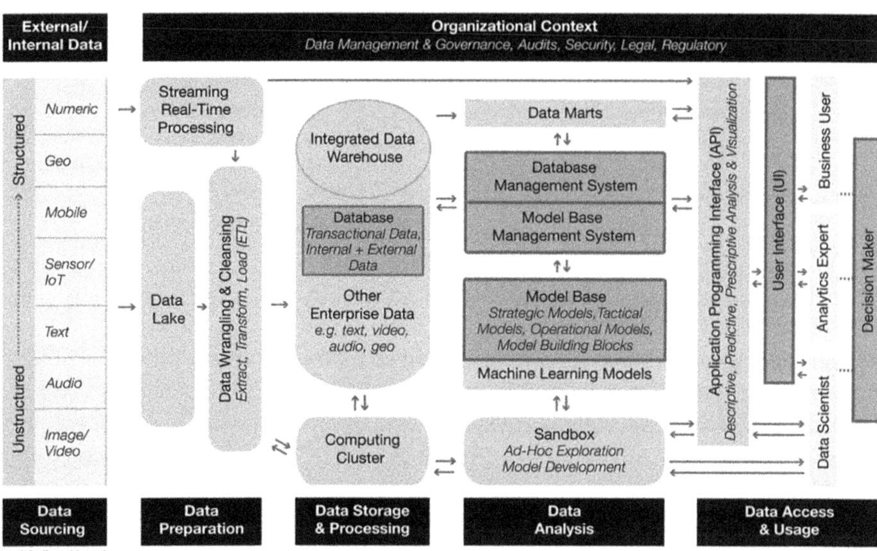

A. Muñoz-Zavala
Autonomous University of Aguascalientes, Aguascalientes, Mexico

Fig. 2 Architecture for BI&A and big data systems from [26]

Table 3 Comparison between small data and big data projects

Characteristics	Small data	Big data
Volume	In the range of GB to TB (10,000–100,000 records)	In the range of TB to ZB (1,000,000–1,000,000,000 records)
Velocity	Controlled and steady flow of data; accumulation of data is slow	Data arrives at very fast speeds; huge amount of data gets accumulated within a short period of time
Variety	Limited to wide (structured data)	Wide (huge variety of data)
Veracity	Contains less noise as data is collected in a controlled manner	The quality of data is not guaranteed. Rigorous validation of data is required prior its processing
Value	High	High
Data location	Data is located with an enterprise, local servers, and regional servers, among others	The data is present mainly in distributed storages in the cloud and in external unstructured databases of other owners and open data, combined with structured databases
Relationality data	Strong	Weak to strong
Flexibility and scalability	Low to middling	High
Example case	Facilitating maintenance decisions on the Dutch railways using big data: the ABA case study [28] This paper analyzes the applicability of big data techniques for facilitating maintenance decisions. However, the data is still not fully utilized due to the lack of adequate techniques to extract relevant events and crucial historical information because valuable information is hidden behind a huge number of terabytes coming from different sensors	Big Data Techniques for Public Health: A Case Study [29] The conclusions given are based on an exploratory big data case study conducted in San Diego County (California), where we analyzed thousands of health-related variables to gain interesting insights into the determinants of several health outcomes, such as life expectancy and anxiety disorders. For the purposes of this paper, the term "big data" refers to many variables (on the order of thousands)

fulfillment of all the Vs, but the efficiency of the results obtained is successful, proving that the possible benefits of this type of projects can be generated in any company or project. This table is proposed by us because although the term small data is commonly used nowadays, there is still no official table of the values or limitations of these. For instance, in [30] the differences are reported but no range of values of characteristics is presented. Similarly, a professional reporter [31] publishes "Top 10 Data and Analytics Trends for 2021" where Trend 4 states that changing from big to small data is supported to solve several problems for organizations facing increasingly complex questions about AI and challenges with sparse data use cases.

Big data leveraging "X-analytics" techniques enables the analysis and synergy of a variety of small and varied (big), unstructured, and structured data sources to improve contextual knowledge and decisions. Small data, as the name implies, can use data models that require less data but still provide useful insights [31].

However, these recent technical achievements do not go hand in hand with their application to real data science-data analytics projects. Studies from organizations and market research reports around the world [32, 33–36] indicate that problems such as BDAS implementation are widespread. Surveys applied [32] to data scientists over several years have shown that only a small percentage of respondents claim to use all their intended models. Other studies [33–36] have reported that DBAS implementation remains a major challenge for many companies. For instance, only 13% of data scientists claimed that their models are always used [33–36]. And that percentage is not improving in nearly a decade of surveys. When this question was first asked, almost identical results were observed which is an indication of little improvement in effectiveness in such projects.

These studies [32, 33–36] also indicated that, while the majority of companies state data science-data analytics as core expertise, only 15% claim to have deployed data science-data analytics projects in their organization. Furthermore, only 13% of respondents stated that their IT organizations put big data projects into production. These studies report similarly that the majority (80%) of early data science-data analytics projects in most US companies failed [33–36]; that 87% of data science projects never make it to production; and that in 77% of companies, the adoption of big data and artificial intelligence (AI) initiatives still represents a major challenge [32, 36].

3.2 *The ISO/IEC 29110 Standard – Basic Profile – as a Lightweight Development Process Template*

Many software development organizations are very small entities (VSE). According to the Organization for Economic Co-operation and Development (OECD) (2005), VSEs constitute many organizations in every country in the world, accounting for more than 95%, and in some countries up to 99%. This poses a challenge for the OECD by providing a business environment that supports the competitiveness of this large business population [8]. To help addressing the need for VSE-specific system and software life cycle profiles and guidelines, the International Organization for Standardization and the International Electrotechnical Commission have jointly published the four-phase ISO/IEC 29110 series of standards and guidelines. The ISO (International Organization for Standardization) and IEC (International Electrotechnical Commission) form the specialized system for worldwide standardization. The national member bodies of ISO and IEC participate in the development of International Standards through technical committees established by the respective organization to address some particular fields of technical activity. These publications enable VSE to select the appropriate process from software or systems engineering life cycle standards (such as ISO/IEC/IEEE 12207) from the outset and tailor it to the needs of the project [9]. ISO/IEC 29110 standard has been developed to improve the quality of products and/or quality of services and their performance.

Leading experts [9; p. 1] comment that: "The correct selection and application of the appropriate standards should increase the productivity of an organization and have a positive economic impact on that organization. In software engineering, a major challenge is the knowledge documented in the standards, this reaches an organization and is applied to its benefit." Now with these standards focused specifically on VSE, it is easier to get started with standards-managed projects for these types of organizations.

Many organizations, both in the public and private sectors, use standards and/or participate in their development. Some standardizations are closely related to their business strategy. Others approach the use of standards in an organized way and understand the direct impact of standards use on their activities and performance. Still, others may use standards from a more limited perspective, exclusively for a specific project, process, or activity. Most of them are aware that standards bring direct benefits to their organization. Some of the potential benefits include optimization of internal operations, innovative and scalable operations, and new market creation or entry [37]. Standards and associated technical documents could be considered a form of technology transfer and, if the right standards are selected and used correctly, should have an economic impact on an organization. In addition to the known benefits of standards, a French study has revealed five main lessons which were increased company value, innovation, transparency, ethics, international product, and service quality [38].

The core of ISO/IEC 29110 is a set of predesigned engineering and management guides that focus on project management and software or system development. ISO/IEC 29110 is designed for use with any life cycle, such as waterfall, iterative, incremental, evolutionary, or agile [39]. To understand the basis of this standard, we can observe in Fig. 3 from [9] the two phases (called processes) and activities of the software engineering ISO/IEC 29110 standard – Basic profile.

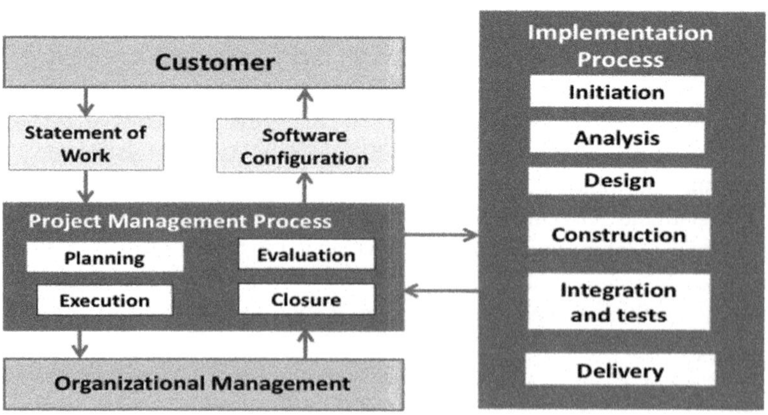

Fig. 3 ISO/IEC 29110 Basic profile processes and activities [9]

Table 4 Processes, tasks, work products, and functions of each software profile in ISO/IEC 29110

	Entry	Basic	Intermediate	Advanced
Number of processes	2	2	3 (+1 conditional)	3 (+3 conditional)
Number of tasks	40	67	107 (+8 conditional)	120 (+24 conditional)
Number of work products	14	22	39 (+3 conditional)	41 (+5 conditional)
Number of roles	3	7	8 (+1 conditional)	8 (+1 conditional)

The concept of "profile groups" is relevant to the ISO/IEC 29110 standard. This concept refers to a set of utilization profiles. The "Generic Profile Group" has been defined as applicable to VSEs that do not develop critical systems or critical software. The Generic Profile Group is a four-stage road map, called profiles, providing a progressive approach to satisfying a vast majority of VSEs. VSEs targeted by the "Entry profile" are VSEs working on small projects (projects that take no more than six person-months effort) and startups. The "Basic profile" targets VSEs developing a single application with a single work team. The "Intermediate profile" is targeted at VSEs developing more than one project in parallel with more than one work team. The "Advanced profile" is targeted to VSEs wanting to sustain and grow as an independent competitive system and/or software development business [8]. Table 4 reports the difference between these four profiles regarding the number of processes, tasks, products, and roles.

The ISO WG applied a survey to develop a set of requirements to produce a series of software and systems engineering standards and guides. Since 2011, hundreds of public and private organizations worldwide have implemented the ISO/IEC 29110 software series, as well as the systems engineering guides. For example, in Thailand, more than 450 public and private organizations have been certified as ISO/IEC 29110 compliant. Finally, trainers in more than 20 countries are teaching the ISO/IEC 29110 series.

Since ISO published the first standards and guides in 2011, more than 200 articles have been printed in peer-reviewed journals [39]. It should be understood that ISO/IEC 29110 is a set of documents that provide different levels of detail depending on the characteristics of each document. In this research, we will focus on the document called "Technical Reports ISO/IEC TR 29110-5-1-2:11, Software Engineering-Life Cycle Profiles for Very Small Entities (VSEs) Part 5-1-2: Management and Engineering guide – Generic Profile Group – Basic profile" where it specifies the standard from the Basic level profile, which has two main processes; the first one is Product Management. The second one is defined as software implementation. Figure 4 and Table 5 report these two processes – project management and software implementation – where roles, phases-activities, and artifacts – products – are also presented.

The explicit roles defined in the Basic profile are as follows:

- AN – analyst
- CUS – customer
- DES – Designer

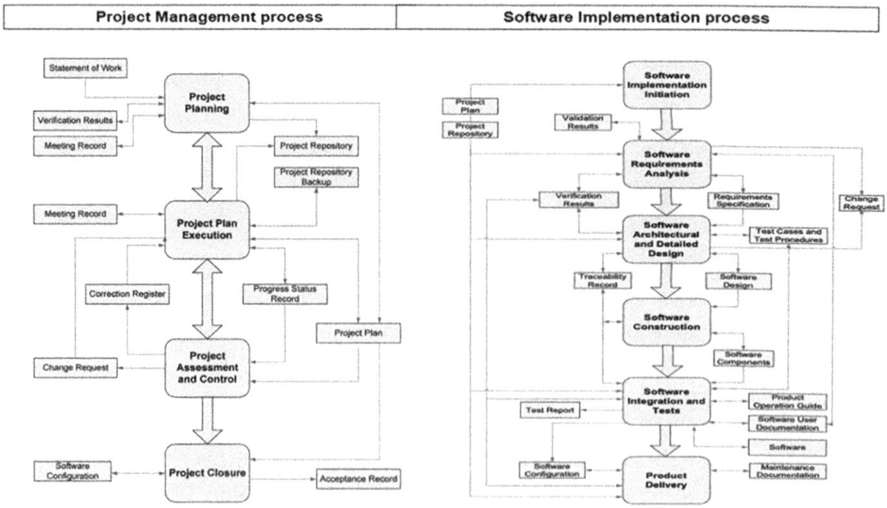

Fig. 4 Project management process and software implementation process ISO/IEC 29110 standard – Basic profile [40]

- PR – programmer
- PM – project manager
- TL – team leader
- WT – work team (TL, AN, DES, and/or PR)

Hence, the BDAS characteristics and the ISO/IEC 29110 standard – Basic profile – will be our theoretical basis for analyzing and identifying the critical factors of a lightweight development process.

4 Selective Comparative Analysis of BDAS Development Methodologies

In this section, we compare the two selected BDAS development methodologies: CRISP-DM (Cross-Industry Standard Process for Data Mining) [13] and DDSL (Domino Data Science Life Cycle) [14]. We review CRISP-DM v1.0 which is the free-access and most referenced methodology. There is a v2.0 but it is proprietary. DDSL is also a proprietary methodology, but a free 25-page document is available. Both methodologies will be compared against the ISO/IEC 29110 standard – Basic Profile – looking for how these BDAS development methodologies align against an expected template of roles, phases, activities, and artifacts for a lightweight development methodology.

Table 5 ISO/IEC 29110 standard - Basic profile

Process	Activities	Process	Activities	Description	Roles	Products (input and output)
Project management	Project planning	The project planning activity documents the planning details needed to manage the project	PM WT CUS	Statement of work Project plan Verification results Meeting record Version control strategy Project repository		
	Project plan execution			The project plan execution activity implements the documented plan on the project	PM WT CUS	Project Plan Progress status record Change request Correction register meeting record Meeting record Version control strategy Project repository backup Project repository
		Software implementation	Software implementation initiation	The software implementation initiation activity ensures that the project plan established in project planning activity is committed to by the work team.	PM WT	Project plan
			Software requirements analysis	The software requirements analysis activity analyzes the agreed customer's requirements and establishes the validated project requirements	WT CUS	Project plan Requirements specification Verification results Change request Validation results *Software user documentation Change request Software configuration

(continued)

Table 5 (continued)

Process	Activities	Process	Activities	Description	Roles	Products (input and output)
			Software architectural and detailed design	The software architectural and detailed design activity transforms the software requirements to the system software architecture and software detailed design	WT	Project plan Requirements specification Software design traceability Record Verification results Change request Test cases and test procedures Software configuration
	Project assessment and control		Software construction	The software construction activity develops the software code and data from the software design.	WT	Project plan Software design Traceability record Software components Software configuration
			Software integration and tests	The software integration and tests activity ensures that the integrated software components satisfy the software requirements	WT CUS	Project plan Test cases and test procedures Software components Traceability record Software Test report Product operation guide Verification results *Software user documentationSoftware configuration
			Product delivery	The product delivery activity provides the integrated software product to the customer	WT	Project plan Software configuration Maintenance documentationVerification results

				The project assessment and control activity evaluates the performance of the plan against documented commitments	PM WT CUS	Project plan Progress status record Correction register Change request
	Project closure	The project closure activity provides the project's documentation and products in accordance with contract requirements	PM CUS	Project plan Software configuration Acceptance record Project repository		

Adaptation from authors

4.1 Description of the Rigor-Oriented CRISP-DM Methodology

In response to common issues and needs in data mining projects in the mid-1990s, a group of organizations involved in data mining (Teradata, SPSS-ISL, Daimler-Chrysler, and OHRA) proposed a reference guide to developing data mining projects, named CRISP-DM [13]. CRISP-DM is considered the de facto standard for developing data mining and knowledge discovery projects. One important factor of CRISP-DM success is the fact that CRISP-DM is industry-, tool-, and application-neutral. The CRISP-DM process model for data mining provides an overview of the life cycle of a data mining project. It contains the corresponding phases of a project, their respective tasks, and relationships between these tasks.

The life cycle of a data mining project, according to CRISP-DM, consists of six phases, which can be observed in Fig. 5. It is always necessary to move back and forth between the different phases. Which phase or which task of a phase should be performed next depends on the outcome of each phase,. The arrows indicate the most important and frequent dependencies between the phases. The outer circle in the figure symbolizes the cyclic nature of data mining itself. Data mining does not end once a solution is deployed. Lessons learned during the process and from the deployed solution can trigger new, often more focused, business questions.

In the following statements, we briefly outline each phase

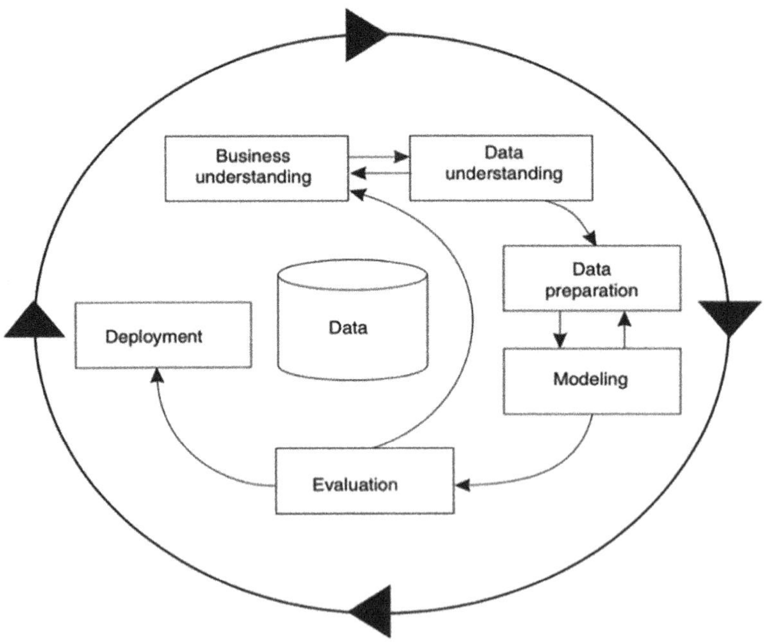

Fig. 5 The CRISP-DM methodology [13]

I. *Business understanding:* This initial phase focuses on understanding the project objectives and requirements from a business perspective and then converting this knowledge into a data mining problem definition and a preliminary plan designed to achieve the objectives.

II. *Data understanding:* This phase starts with an initial data collection and proceeds with activities that enable you to become familiar with the data, identify data quality problems, discover first insights into the data, and/or detect interesting subsets to form hypotheses regarding hidden information.

III. *Data preparation:* The data preparation phase covers all activities needed to construct the final dataset [data that will be fed into the modeling tool(s)] from the initial raw data. Data preparation tasks are likely to be performed multiple times and not in any prescribed order. Tasks include table, record, and attribute selection, as well as transformation and cleaning of data for modeling tools.

IV. *Modeling:* In this phase, various modeling techniques are selected and applied, and their parameters are calibrated to optimal values. Typically, there are several techniques for the same data mining problem type. Some techniques have specific requirements for the form of data. Therefore, going back to the data preparation phase is often necessary.

V. *Evaluation:* At this stage in the project, we have built a model (or models) that appears to have high quality from a data analysis perspective. Before proceeding to the final deployment of the model, it is important to thoroughly evaluate it and review the steps executed to create it, to be certain the model properly achieves the business objectives. A key objective is to determine if there is some important business issue that has not been sufficiently considered. At the end of this phase, a decision on the use of the data mining results should be reached.

VI. *Deployment:* The creation of the model is generally not the end of the project. Even if the purpose of the model is to increase knowledge of the data, the knowledge gained will need to be organized and presented in a way that the customer can use it. It often involves applying "live" models within an organization's decision-making processes, for example, real-time personalization of web pages or repeated scoring of marketing databases. Depending on the requirements, the deployment phase can be as simple as generating a report or as complex as implementing a repeatable data mining process across the enterprise. In many cases, it is the customer, not the data analyst, who carries out the deployment steps. However, even if the analyst will carry out the deployment effort, the customer needs to understand up front what actions need to be carried out to make use of the created models.

Figure 6 presents a scheme of phases accompanied by generic tasks (**in bold**) and results (*italic*), where the tasks and artifacts of this methodology are known. We can detect the focus of this methodology in the data, both in the tasks and the artifacts related to them; in the same way, we can detect why it is determined as a rigorous methodology when analyzing the number of artifacts required to be able to control the project according to the CRISP-DM metrics.

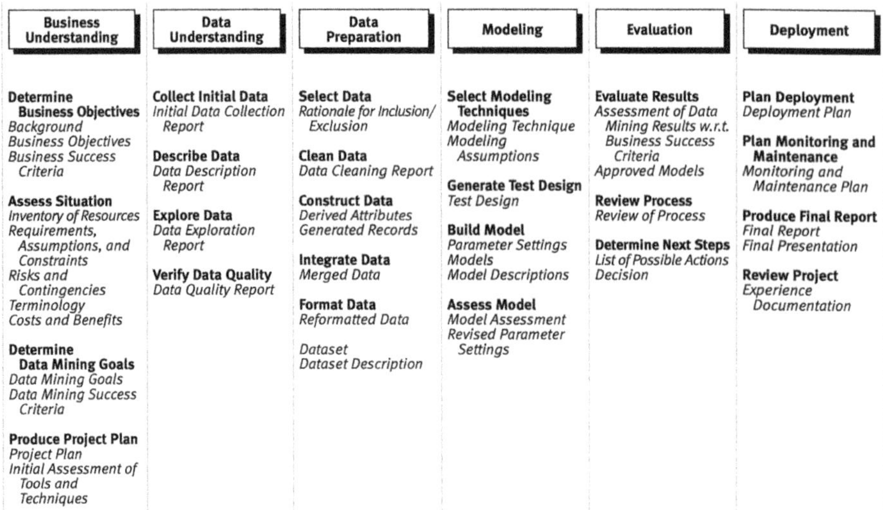

Fig. 6 Generic tasks and outputs of the CRISP-DM reference model [13]

Table 6 reports the relationship between each phase with its respective activities and artifacts generated. CRISP-DM does not report explicitly a set of roles [7, 13].

Software development, like many other engineering problems, has a structure that CRISP-DM resembles in many aspects (it starts with business needs and ends with the deployment and maintenance of the process outcome), but it would be equally inappropriate to use the same linear flow for all problems and circumstances. Similarities have suggested the application or adaptation of software development methodologies for data science (or big data) projects, but they must be carefully reviewed. We can also learn from some novel lightweight methodologies, which attempt to add flexibility to the process, allowing teams to develop software, from requirements to deployment, in a more efficient way [41]. The current theory also notes that future research should explore appropriate ways to integrate models into a productive environment [42].

Hence, CRISP-DM, even after more than 20 years of its creation, continues to play an important role as a common framework for the creation and management of data mining projects, even mentioned in different surveys as the most currently used in projects related to data science-data analytics [41].

4.2 Description of the Lightweight DDSL Methodology

Domino Data Lab, a Silicon Valley vendor that provides a data science platform, crafted its data science project life cycle framework in a 2017 whitepaper [14]. This organization based on over 4 years of working with data science organization

Table 6 The rigor-oriented CRISP-DM methodology

Phase.	Activities	Roles.	Products (output).
Business understanding	Determine business objectives Assess the situation Determine the objectives of data mining Create a plan for your data mining project	No explicit information is reported	Background Business objectives Business success criteria Inventory of resources Requirements, assumptions, and constraints Risks and contingencies Terminology Costs and benefits Data mining goals Data mining success criteria Project plan Initial assessment of tools and techniques
Data understanding	Collect initial data Describe the data Explore the data Check the quality of the data	No explicit information is reported	Initial data collection report Data description report Data exploration report Data quality report
Data preparation	Select data Data cleansing Construct data Integrate the data Format the data	No explicit information is reported	Rationale for inclusion/exclusion Data cleaning report Derived attributes Generated records Merged data Reformatted data Dataset Dataset description
Modeling	Select modeling technique Design the model tests Build the model Evaluate the model	No explicit information is reported	Modeling assumptions Test design Parameter settings Models Model descriptions Model assessment Revised parameter settings
Evaluation	Evaluate the result Process review Determine the next stages	No explicit information is reported	Assessment of data mining results w.r.t. business success criteria Approved models Review of process List of possible actions Decision
Deployment	Plan deployment Plan monitoring and maintenance Create a final report Project review	No information is reported	Deployment plan Monitoring and maintenance plan Final report Final presentation Experience documentation

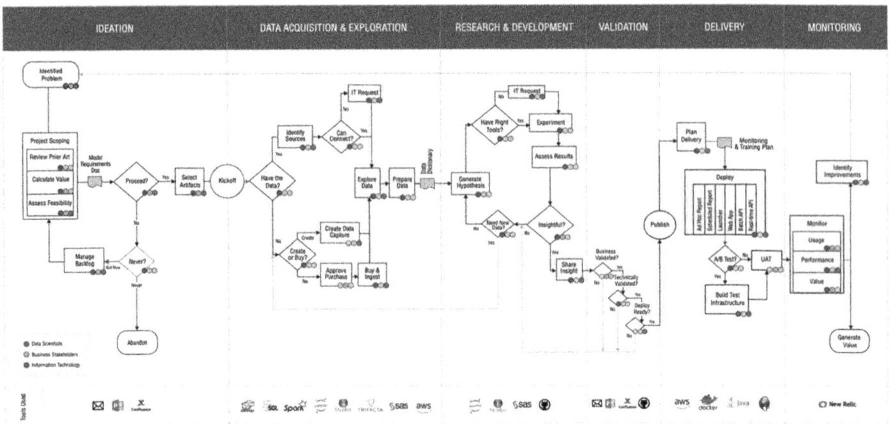

S. Galvan-Cruz
Autonomous University of Aguascalientes, Aguascalientes, Mexico

A. Muñoz-Zavala
Autonomous University of Aguascalientes, Aguascalientes, Mexico

Fig. 7 The lightweight DDSL methodology [14]

leaders, such as Allstate, Monsanto, and Moody's, observed that a plausible solution is a holistic approach to the entire project life cycle, from ideation through delivery and follow-up. Organizations that can develop a disciplined practice of iterative, self-measuring business value delivery, while using data science platform technology to support a hub-and-spoke organizational structure, can scale data science to a core capability and accelerate the delivery of a robust portfolio of models. While a complete transformation may take years, they suggest a "crawl, walk, and run" approach to building momentum toward the ultimate vision [14].

The Domino Data Science Life Cycle (DDSL) is a modern lightweight methodology [14] that integrates agile principles with data science projects, acknowledges the multiple team roles of data science projects, and extends the core data and modeling phases to first focus on the business problem and to finish with deployment or even production operations. Figure 7 portrays a flowchart that encompasses the roles, phases, activities, and artifacts that compose this methodology.

To get into this methodology, the organization Domino [14] recommends first understanding some of the concepts of the data science life cycle. Knowing which one to use and how to integrate it with a collaborative framework can be challenging. It should also be understood that DDSL is based on three guiding principles. *First*: Expect and embrace iteration, but avoid iterations that significantly delay projects or distract from the objective at hand. *Second*: Enable composite collaboration by creating components that are reusable in other projects *Third*: Anticipate auditability needs and preserve all relevant artifacts associated with the development and deployment of a model.

The core DDSL methodology itself divides a project into six iterative stages that reflect those of CRISP-DM, indicating that CRISP-DM provided a structural basis to generate DDSL. These stages are briefly described as follows:

I. *Ideation*: The initial phase puts the "problem first, not data first" by defining the underlying business problem and conducting business analysis activities such as current state process mapping, project ROI analysis, and upfront documentation. It also incorporates common agile practices including developing a stakeholder-driven backlog and creating deliverable mockups. IT and engineering are looped in early, and models might be baselined with synthetic data. The phase ends with a project "kickoff."
II. *Data Acquisition and Exploration*: Data science teams should identify data sources with help from stakeholders who can provide leads based on their intuition. Decisions are made to capture data or buy data from vendors. Exploratory data analysis is conducted, and the data is prepared for both the current project modeling and as reusable components for future projects.
III. *Research and Development*: Like the core modeling phase of CRISP-DM or any other data science process, this phase iterates through hypothesis generation, experimentation, and insight delivery. True to agile principles, Domino recommends starting with simple models, setting a cadence for insight deliveries, tracking business KPIs, and establishing standard hardware and software configurations.
IV. *Validation:* This phase focuses on both business and technical validations and loosely mirrors the evaluation phase from CRISP-DM. True to its principle to "enable compounding collaboration," Domino stresses the importance of ensuring the reproducibility of results, automated validation checks, and documentation. The main goal of this phase is to "ultimately receive sign-off from stakeholders."
V. *Delivery:* This is when models become products. Deployment, A/B testing, test infrastructure, and user acceptance testing, similar to those of any software project, are in this phase. Domino recommends additional considerations such as preserving links between deliverable artifacts, flagging dependencies, and developing a monitoring and training plan.
VI. *Monitoring*: Given the models' nondeterministic nature, Domino recommends monitoring techniques that extend beyond standard software monitoring practices. For example, consider using control groups in production models so that we can continually monitor model performance and value creation for the organization. Moreover, automatic monitoring of acceptable output ranges can help identify model issues before they become too pervasive.

The specific roles proposed in the DDSL methodology are data scientist, data infrastructure, data product manager, and on the customer side business stakeholder and a new role data storyteller [14]. To finish with this second methodology, we must remember that DDLS is a proprietary methodology, and the only official documentation that was free available is reported in [14]. This conceptual review was based on that document. Table 7 was generated as the summary for review.

Table 7 The lightweight DDSL methodology

Stage	Activities	Roles	Products
Ideation	Identified problem Project scoping Review prior art Calculate Assess feasibility Manage backlog Select artifacts	Data scientist Data product manager Business stakeholder Data storyteller*	Model requirements doc
Data acquisition and exploration	Getting the dataIdentify sources of the data Connect Create data (capture)Buy and ingest data Explore data Prepare data	Data scientist Data product manager Business stakeholder	Data dictionary
Research and development	Generate hypotheses Check that the appropriate tools are available Evaluate results Validate that no new data are needed	Data scientist Business stakeholder	Data model experiment
Validation	Validate the business Validate technically Validate ready to deploy Publish	Data scientist Data infrastructure Data product manager Business stakeholder	Validated data model
Delivery	Plan delivery Deploy Test	Data scientist Data infrastructure Data product manager Business stakeholder	Production data model.
Monitoring	Monitor Usage Performance Value Identify improvements Generate value	Data product manager Business stakeholder	Monitoring and training plan

4.3 Comparison of the Rigor-Oriented CRISP-DM and the Lightweight DDSL Methodologies

This section compares the two methodologies CRISP-DM and DDLS against the ISO/IEC 29110 standard – Basic profile. This is to detect if there is any circumstantial difference between the two approaches, the rigorous and the lightweight one, and, additionally, to determine if there is any area of improvement for this standard when applied to big data analytics systems projects.

To create the comparison, the activities of the two processes were placed – project management and software implementation – where these activities are related to the phases or stages of the two methodologies seeking to detect three possible cases which in the table are indicated with symbols: "≈" where there is a similarity between the standard and the methodology by commenting on some detail of importance in that similarity; "✘" for activities that failed to detect any similarity with the methodology or explicitly that it is some point that requires the standard but there is detail of this in the methodology compared; and "✔" where these indicate some areas in which the standard could be improved when applied to data-centric projects. This is because they already detected key points that may help in the effectiveness of projects or critical factors within big data analytics systems projects.

Therefore, in the column *Relationship* between the standard and the methodology is a brief description of the comparison and the case in which it coincides with the three previously mentioned, where the relation of the activity of the standard with that of the methodology is defined.

Only the activities of the processes are considered for the following comparisons in this study. In the **Appendix**, you can find tables of roles and artifacts for the methodologies compared to the standard, which can be used for future analyses.

The results of the comparison between the ISO/IEC 29110 standard – Basic profile – and CRISP-DM (Cross-Industry Standard Process for Data Mining) are depicted in Fig. 8. The figure illustrates that three of the activities prescribed by the ISO/IEC 29110 standard are conspicuously absent in the CRISP-DM methodology, denoted by the symbol "✘." This absence underscores a notable distinction between the traditional project management approach and the Data Mining approach. Conversely, there are four activities marked with "≈," indicating a substantial degree of similarity in this comparative analysis. In the "✔" the category, we observe that one of the activities outlined in the ISO/IEC 29110 standard is expounded upon more comprehensively within the CRISP-DM methodology, which is further subdivided into four segments: Data Understanding, Data Preparation, Modeling, and Evaluation. Consequently, CRISP-DM exhibits a more extensive approach concerning the delineation of specific data processes for a BDAS (Big Data Analytics and Solutions) project.

Figure 9 shows the comparison of ISO/IEC 29110 standard – Basic profile – and Domino's DDSL, where, in this case, it can be detected that only two of the standard activities are not mentioned in the methodology, so they are with "✘," indicating that they are not mentioned, which may show the difference between the traditional

ISO/IEC 29110 standard – Basic Profile -				Relationship between the ISO/IEC 29110 standard and the methodology		CRISP-DM Methodology
Process	Activities	Process	Activities			Phase
Project Management	Project Planning			≈	Understanding the project objectives and requirements. Generating a preliminary plan designed to achieve the project objectives.	Business Understanding
	Project Plan Execution	Software Implementation	Software Implementation Initiation	✗	Review of the project plan is omitted in crisp, in this methodology it jumps directly to implementation, without commenting on project management.	
			Software Requirements Analysis	≈	The analysis of the requirements given by the client and established for the project.	
			Software Architectural and Detailed Design	✓	This is a key point for data-centric projects, where besides analyzing the project architecture and the requirements of the client, it is the data-centric point where a more complex process of analysis (gathering, preparing, modeling, and testing the data) must be carried out.	Data Understanding Data Preparation Modeling Evaluation (Modeling)
	Project Assessment and Control		Software Construction	≈	Depending on the requirements, the deployment phase can be as simple as generating a report or as complex as implementing a repeatable data extraction process across the enterprise.	Deployment
			Software Integration and Tests	≈	CRISP-DM focuses more on testing at modeling time, it does not even talk about post-development testing, but it does talk about post-development integration.	Evaluation (software)
			Product Delivery	✗	Product Delivery is slightly mentioned in crisp, Project Evaluation Control is not detailed in crisp from the beginning of the project and Project Evaluation Control is not detailed in crisp from the beginning of the project.	Deployment (Monitoring and Maintenance)
	Project Closure			✗	Project closure is not described in detail in CRISP-DM.	

Fig. 8 Summary of the comparison between the ISO/IEC 29110 standard – Basic profile – and the CRISP-DM methodology

approach and the data science approach in mismanagement. There are five activities in "≈," denoting a broad similarity in this comparison. In the activity with "✓," we can see how one of the activities of the standard is described more extensively in the methodology. This is segmented into three: data acquisition and exploration, research and development, and validation. Similar to the previous scenario, it is assumed that a more detailed elaboration of the data processing procedure is given.

Now, with the knowledge gained from these two comparisons, between the BDAS development methodologies oriented to rigor and lightweight approach, we can appreciate both, similar characteristics and individual, starting with CRISP-DM. Being a rigorous methodology, many of its factors coincide with the standard in terms of processes or products related to a software project, showing that the standard handles project management more rigorously concerning the start and completion of each process. Highlighting the significance of the substantial role that these standards play as Project Manager, actively involved in overseeing the entire project from initiation to completion.

As an improvement factor for ISO/IEC 29110 standard – Basic Profile – a greater focus on data in the BDAS project is very important, because it is the key point of this type of project.

Following the individual characteristics of the methodology of Domino, you can appreciate a lighter approach that generally meets with great similarity in the activities or base processes. The standard can also detect that its approach makes little or no mention of the artifacts required or generated during the entire project, facilitating the processes from start to finish, but complicating the execution, due to lack of details. Even so, considering the great similarity, we can see an absolute approach to data in DDSL which shows the importance of such an approach in BDAS.

Giving way to the united analysis of methodologies to ISO/IEC 29110 standard – Basic profile – it is possible to observe how each one has its style (rigor or lightweight). The standard remains more meticulous in specifying the management throughout the project in each of the activities or processes when starting or ending the development, being a favorable point for the control of such projects. Taking into account the low level of success of such projects, control throughout the life cycle is an aspect that favors. Nested to the strengths of the standard, we find the characteristic of BDAS projects, which was detected in both methodologies, focusing the project on data. Although ISO/IEC 29110 standard – Basic profile – is rigorous enough to improve the success factor for data-centric projects, the core processes of both project management and software implementation must have a greater focus on data analysis and processing.

These differences were detected to understand how the standard helps when applying within a VSE to an analytics project, seeking with this to obtain a higher success rate and also to have a basis for companies that do not yet dare to enter into such projects applying the advantages of a standard and the specifications of a methodology specialized in a particular type of projects.

5 Discussion of Contributions and Conclusions

In this section, there will be a discussion of theoretical and practical implications, and the conclusions and recommendations for further research are going to be reported.

5.1 Discussion of Contributions

In the previous section, Figs. 8 and 9 showed the comparison of the activities of the methodologies against the ISO/IEC 29110 standard – Basic profile – showing a valuable summary description with a sufficient level of detail to contrast the principles of the ISO/IEC 29110 standard against these specialized methodologies for data science-data analytics projects. To space limitations, detailed comparison against roles and artifacts is not reported, but the detailed descriptions of the ISO/IEC 29110 standard – Basic profile – as well as of the two BDAS development methodologies reported are useful to estimate an overall perspective about roles and artifacts.

ISO/IEC 29110 standard – Basic Profile -				Relationship between the ISO/IEC 29110 standard and the methodology		DDSL Methodology
Process	Activities	Process	Activities			Stage
Project Management	Project Planning			≈	First identify the business problem after the analysis of the project and the data	Ideation
	Project Plan Execution	Software Implementation	Software Implementation Initiation	✗	Domino does not mention project planning or management. The methodology begins directly with developing and mentions nothing about planning	
			Software Requirements Analysis	≈	The requirements and limits of the project are analyzed	
			Software Architectural and Detailed Design	✓	Data science teams must identify data sources. A decision is made to capture the data or purchase it from vendors. Exploratory analysis of the data is conducted and prepared for project modeling. This phase iterates through hypothesis generation, experimentation and delivery of information. establish standard hardware and software configurations. Business and technical validations "finally receive stakeholder approval"	Data Acquisition and Exploration Research and Development Validation
	Project Assessment and Control		Software Construction	≈	The models are converted into products	Delivery
			Software Integration and Tests	≈	A/B testing, test infrastructure, and user acceptance testing	
			Product Delivery	≈	Domino recommends additional considerations, such as preserving linkages between deliverable artifacts, marking dependencies	
	Project Closure			✗	Domino recommends monitoring techniques, in order to continuously monitor the model's performance and value creation for the organization. But it does not comment on project closure	Monitoring

Fig. 9 Summary of the comparison between the ISO/IEC 29110 standard – Basic profile – and the DDSL methodology

Analyzing the results observed in Fig. 8, where the comparison between ISO/IEC 29110 and CRISPM-DM was made, we will start by taking as a reference the activities selected with the color "≈" which indicates the similarity between both; in these, we can detect that a key point is to start clearly with the objectives and clear requirements of the project to generate adequate planning of this from the beginning. In the part about user requirements and their analysis, both comment on the importance of the process, the review, and continuous contact with customers. For this, ending with the segments in common are the parts of developing or building the project where it is contemplated to start generating it and, in its completion, to test what has been developed. The activities that are required in the standard but are not mentioned in the theory of the methodology are the "✗" ones, which refer to activities that we must carry out in order not to lose the fulfillment of the standard even if they are not in CRISP-DM, mentioning that when we want to follow a standard, we must comply with it in its totality to be able to count on the benefits that this entails. Therefore, we cannot omit activities which in this case the ones missing in CRISP-DM are those related to project management such as initialization of the implementation, product delivery, controls during the implementation part, and project closure. For the software architectural and detailed design activity of the standard, the only one defined with the "✓" color was detected as the key point for the projects focused on data. This, we can observe just with the comparison against the methodology of which four of its six phases are focused on data. On the contrary, within the standard, only one artifact – "software design" – is focused on this, clearly showing that this can be a key factor to detect in the future for these areas of care for analytics projects; besides having more activities focused exclusively on data, CRISP-DM also has a remarkable amount of products that are focused exclusively on them.

The second comparative (Fig. 9) between ISO/IEC 29110 and DDSL, the methodology proposed by Domino (which is very similar to the previous one), says that the basis with which the company generated its methodology was CRISP-DM, but this new one has a lighter approach and focuses mainly on the process part and not so much on the artifacts for a data science project. In the case of DDSL, only two of the activities in "✘," we could not find any relationship with the documentation of the Domino methodology, specifically those related to project management, such as execution planning, evaluation, control, and project closure, suggesting that the standard has a stronger management approach. We continue with the "≈" color activities, which were related in both; these, as in the last case, coincide with the beginning of the project based on first identifying the problem, to begin with, the analysis of the project and its objectives, which must be clearly defined by the client and the development team and also validated. Having this clearly defined, we proceed to start the project. This part is the clear difference that is found in both, but once the design, scope, and architecture of the project are defined, both continue similarly in the initialization of generating the software, report, or base artifact to successfully achieve the objectives The generated must be tested and then delivered. The difference that DDSL has in comparison with CRISP-DM is that Domino did focus on the post-development with its stage called monitoring. Arriving at the activity of the standard "software architectural and detailed design" which is in "✔," analyzing DDSL, we found that this is an important point in the part of data science projects because there are three stages of Domino that focus exclusively on the data. These stages go from its acquisition, exploration, search, and development, ending even with the validation of them all. This specifically focused on data, which denotes that within the monitoring of our standard in this segment is where it is important to take into account the characteristics of a project, where the data is the main point of the whole development, that is to say, data science.

This highlights the similarities and differences found between methodologies and the standard. This gives way to generating certain specifications which can help the success of such projects based on a more detailed and clear approach to projects that focus on data. This strengthens the standard with the tasks, products, or activities that were highlighted within the specialized methodologies that give more focus to the way data is collected, selected, handled, and processed in projects when it comes to BDAs.

5.2 Conclusions

This research reviewed two development methodologies that focus on projects related to big data analytics systems; the first one was CRISP-DM defined as the traditional methodology most used nowadays fulfilling the characteristics of a rigor one, and the second one was DDSL which was generated by Domino company based on CRISP-DM and their experiences in projects, but looking for a lighter approach and applicable for new companies that start with this kind of projects. Both methodologies were compared against ISO/IEC 29110; such standard focused to be used by VSE, organizations, or development groups new to the area or simply small companies, which fits the research to find a way to improve the chances that such an organization can carry out projects related to analytics with a favorable success rate. This evaluation is worthwhile given the current growing interest and needs to implement the big data analytics systems project successfully due to the new business environment driven by the pressures of digital transformation.

Specifically in the comparison of ISO/IEC 29110 against the two methodologies, it is possible to detect the key point for a project that focuses purely on data, taking into account that if we analyze the documentation of the standard, there are very few times that it makes specific reference to data. To be clearer for the artifacts generated throughout the project of the standard, only one makes a direct mention of "data." We understand that the strong point of this standard is the management of the project as a whole, and it clearly defines at what point each part of the project starts, from the requirements gathering, its validation by both parties (team and client), and the development of the project to the final deliverables tests. But when analyzing the other two methodologies, we can see that this type of project is focused to a large extent on data, that is to say, unlike other developments, in these cases a large percentage of the entire project is focused exclusively on data, because it is important to detect what data we need, if we have it or we must get it, clean it, prepare it, and even validate it. Even test the data and the model before starting the development, so the testing part for this type of project is double, since first, the data must be tested before using them, and at the end, a test is also performed but already of the developed product. This can be seen in both CRISP-DM and DDSL to reach this conclusion, although each of them has its approach, as CRISP-DM being a traditional methodology is extensive and has a high number of artifacts, but during its development, it does not mention anything about roles. This is a big point against it, as it leaves this area without details.

In its counterpart, DDSL focuses entirely on the data process but without mentioning or making clear what artifacts or requirements are needed about the data. This is due to its light approach which is very useful but leaves some doubts regarding the outputs of the phases. Therefore, making use of the strengths of each of these, we can support the ISO/IEC 29110 standard to determine if we emphasize the data part and all the processes that must be performed with these for use in analytics. The success rate for the VSE in these projects will be increased.

Therefore, currently, no methodology complies in its entirety for any type of project, but it is required methodologies with specific particularities according to the type of project and even the type of organization, which also happens with the standards optimal that there is already one with the focus for VSE, but still missing more favorable characteristics for particular projects.

To retake the lessons learned, we conclude the following: With the detection that if the ISO/IEC 29110 standard supports VSE to face typical problems of software developers seeking to achieve a higher percentage of effectiveness in projects, simultaneously, if this standard is supported with further details or specifications of the entire data management based on the specific methodologies of such projects, it will be possible to further increase the effectiveness of the organization [8, 10], leaving aside the rigorous approach. When the comparison was made concerning activities, no matter how much detail is kept as in CRISP-DM or how lightweight the methodology is, focusing more on processes such as DDSL is noted as a more specialized method in a particular type of project that highlights critical or key points of this, for the case of BDAS makes clear the importance of focusing and detailing the approach to data treatments [7].

The study's limitations, as mentioned in Sect. 2, were that only two methodologies were taken through the analysis and the selective comparison was focused on phases-activities with the experts of the advisory group. Taking into account the compelling evidence presented in the methodology documents, the selection process considers various factors, addressing one aspect with a rigorous approach (CRISP-DM) and another with a lightweight touch (DDSL). Therefore, having only two methodologies selected against the standard was a limitation. This leads to future research being able to make a comparison with more methodologies, not just one of each approach as it was in this case and even the comparison is more focused on activities or roles. (**Appendix**). It also leaves future work where, in addition to making a comparison with more methodologies, they are also done in more detail, considering the roles and artifacts of both the methodologies and the standard. In this chapter, we only mention them.

Another branch of future work is the creation of a methodology following ISO/IEC 29110 but with more detail in the data management part, just as described in the specialized methodologies for this type of project where it could be suggested that within the standard, some of the activities, roles, or artifacts are added or further detailed to improve efficiency when used for BDAS. As a continuation of this line of inquiry, the identified enhancements to the standard can be subjected to a study aimed at assessing their usability, user-friendliness, compatibility, and overall value through pilot tests or expert surveys.

Appendix (Figs. 10, 11, 12 and 13)

<table>
<tr><th colspan="5">ISO 29110</th><th rowspan="2">Vs</th><th colspan="2">CRISP-DM</th></tr>
<tr><th>Process</th><th>Activities</th><th>Process</th><th>Activities</th><th>Roles</th><th>Process</th><th>Roles</th></tr>
<tr><td rowspan="13">Project Management</td><td colspan="3">Project Planning</td><td>Project Manager
Work Team
Customer</td><td rowspan="13"></td><td rowspan="4">Business Understanding</td><td rowspan="4">-</td></tr>
<tr><td rowspan="5">Project Plan Execution</td><td colspan="2"></td><td>Project Manager
Work Team
Customer</td></tr>
<tr><td rowspan="9">Software Implementation</td><td>Software Implementation Initiation</td><td>Project Manager
Work Team</td></tr>
<tr><td>Software Requirements Analysis</td><td>Work Team
Customer</td></tr>
<tr><td rowspan="4">Software Architectural and Detailed Design</td><td rowspan="4">Work Team</td><td>Data Understanding</td><td>-</td></tr>
<tr><td>Data Preparation</td><td>-</td></tr>
<tr><td rowspan="6">Project Assessment and Control</td><td>Modeling</td><td>-</td></tr>
<tr><td>Evaluation</td><td>-</td></tr>
<tr><td>Software Construction</td><td>Work Team</td><td rowspan="5">Deployment</td><td rowspan="5">-</td></tr>
<tr><td>Software Integration and Tests</td><td>Work Team
Customer</td></tr>
<tr><td>Product Delivery</td><td>Work Team</td></tr>
<tr><td colspan="2"></td><td>Project Manager
Work Team
Customer</td></tr>
<tr><td colspan="3">Project Closure</td><td>Project Manager
Customer</td></tr>
</table>

Fig. 10 Comparison of roles between ISO/IEC 29110 – Basic profile – and CRISP-DM

<table>
<tr><th colspan="5">ISO 29110</th><th rowspan="2">Vs</th><th colspan="2">CRISP-DM</th></tr>
<tr><th>Process</th><th>Activities</th><th>Process</th><th>Activities</th><th>Products (Output)</th><th>Process</th><th>Products (Output)</th></tr>
<tr><td rowspan="5">Project Management</td><td colspan="3">Project Planning</td><td>• Project Plan
• Verification Results
• Meeting Record
• Project Repository</td><td rowspan="5"></td><td rowspan="4">Business Understanding</td><td rowspan="4">• Background
• Business Objectives
• Business Success Criteria
• Inventory of Resources
• Requirements, Assumptions, and Constraints
• Risks and Contingencies
• Terminology
• Costs and Benefits
• Data Mining Goals
• Data Mining Success Criteria
• Project Plan
• Initial Assessment of Tools and Techniques</td></tr>
<tr><td rowspan="4">Project Plan Execution</td><td colspan="2"></td><td>• Project Plan
• Progress Status Record
• Change Request
• Meeting Record
• Project Repository Backup
• Project Repository</td></tr>
<tr><td rowspan="3">Software Implementation</td><td>Software Implementation Initiation</td><td>• Project Plan</td></tr>
<tr><td>Software Requirements Analysis</td><td>• Requirements Specification
• Verification Results
• Change Request
• Validation Results
• *Software User Documentation
• Change Request
• Software Configuration</td></tr>
<tr><td>Software Architectural and Detailed Design</td><td>• Software Design
• Traceability Record
• Verification Results
• Change Request
• Test Cases and Test Procedures</td><td>Data Understanding</td><td>• Initial Data Collection Report
• Data Description Report
• Data Exploration Report
• Data Quality Report</td></tr>
</table>

Fig. 11 Comparison of products between ISO/IEC 29110 – Basic profile – and CRISP-DM

				• Software Configuration
	Project Assessment and Control		Software Construction	• Traceability Record • Software Components • Software Configuration
			Software Integration and Tests	• Test Cases and Test Procedures • Traceability Record • Software • Test Report • Product Operation Guide • Verification Results • *Software User Documentation • Software Configuration

Data Preparation	• Rationale for Inclusion/ Exclusion • Data Cleaning Report • Derived Attributes • Generated Records • Merged Data • Reformatted Data • Dataset • Dataset Description
Modeling	• Modeling Assumptions • Test Design • Parameter Settings • Models • Model Descriptions • Model Assessment • Revised Parameter Settings
Evaluation	• Assessment of Data Mining Results w.r.t. Business Success Criteria • Approved Models • Review of Process • List of Possible Actions • Decision
Deployment	• Deployment Plan • Monitoring and Maintenance Plan • Final Report • Final Presentation • Experience Documentation

Fig. 5.11 (continued)

		Product Delivery	• Software Configuration • Maintenance Documentation • Verification Results		
			• Progress Status Record • Correction Register • Change Request		
	Project Closure		• Software Configuration • Acceptance Record • Project Repository		

Fig. 5.11 (continued)

ISO 29110				
Process	***Activities***	**Process**	***Activities***	Roles
Project Management	Project Planning			Project Manager Work Team Customer
Project Management	Project Plan Execution			Project Manager Work Team Customer
Project Management	Project Plan Execution	**Software Implementation**	Software Implementation Initiation	Project Manager Work Team
Project Management	Project Plan Execution	**Software Implementation**	Software Requirements Analysis	Work Team Customer
Project Management	Project Plan Execution	**Software Implementation**	Software Architectural and Detailed Design	Work Team
Project Management	Project Assessment and Control	**Software Implementation**	Software Construction	Work Team
Project Management	Project Assessment and Control	**Software Implementation**	Software Integration and Tests	Work Team Customer
Project Management	Project Assessment and Control	**Software Implementation**	Product Delivery	Work Team
Project Management	Project Assessment and Control			Project Manager Work Team Customer
Project Management	Project Closure			Project Manager Customer

Vs

DDSL	
Process	Roles
Ideation	Data Scientist Data Product Manager Business Stakeholder Data Storyteller*
Data Acquisition and Exploration	Data Scientist Data Product Manager Business Stakeholder
Research and Development	Data Scientist Business Stakeholder
Validation	Data Scientist Data Infrastructure Data Product Manager Business Stakeholder
Delivery	Data Scientist Data Infrastructure Data Product Manager Business Stakeholder
Monitoring	Data Product Manager Business Stakeholder

Fig. 12 Comparison of roles between ISO/IEC 29110 – Basic profile – and DDSL

<table>
<tr><th colspan="5">ISO 29110</th><th rowspan="2">Vs</th><th colspan="2">DDSL</th></tr>
<tr><th>Process</th><th>Activities</th><th>Process</th><th>Activities</th><th>Products (Output)</th><th>Process</th><th>Products (Output)</th></tr>
<tr><td rowspan="7">Project Management</td><td colspan="3">Project Planning</td><td>• Project Plan
• Verification Results
• Meeting Record
• Project Repository</td><td rowspan="7"></td><td rowspan="3">Ideation</td><td rowspan="3">• Model Requirements Doc.</td></tr>
<tr><td rowspan="6">Project Plan Execution</td><td colspan="2"></td><td>• Project Plan
• Progress Status Record
• Change Request
• Meeting Record
• Project Repository Backup
• Project Repository</td></tr>
<tr><td rowspan="5">Software Implementation</td><td>Software Implementation Initiation</td><td>• Project Plan</td></tr>
<tr><td rowspan="3">Software Requirements Analysis</td><td rowspan="3">• Requirements Specification
• Verification Results
• Change Request
• Validation Results
• *Software User Documentation
• Change Request
• Software Configuration</td><td>Data Acquisition and Exploration</td><td>• Data Dictionary.</td></tr>
<tr><td>Research and Development</td><td>• Data Model Experiment *</td></tr>
<tr><td>Validation</td><td>• Validated Data Model *</td></tr>
<tr><td>Software Architectural and Detailed Design</td><td>• Software Design
• Traceability Record
• Verification Results
• Change Request
• Test Cases and</td><td>Delivery</td><td>• Production Data Model*
• Monitoring & Training Plan.</td></tr>
</table>

Fig. 13 Comparison of products between ISO/IEC 29110 – Basic profile – and DDSL

			Test Procedures • Software Configuration"		
	Project Assessment and Control	Software Construction	• Traceability Record • Software Components • Software Configuration		
	Project Assessment and Control	Software Integration and Tests	• Test Cases and Test Procedures • Traceability Record • Software • Test Report • Product Operation Guide • Verification Results • *Software User Documentation • Software Configuration		
	Project Assessment and Control	Product Delivery	• Software Configuration • Maintenance Documentation • Verification Results		
	Project Assessment and Control		• Progress Status Record • Correction Register • Change Request		
	Project Closure		• Software Configuration • Acceptance Record • Project Repository	Monitoring	• Monitoring & Training.*

**Products that are not in the documentation but are required implicitly*

Fig. 5.13 (continued)

References

1. Kose, B.O.: Agile business analysis for digital transformation. In: Handbook of Research on Multidisciplinary Approaches to Entrepreneurship, Innovation, and ICTs, pp. 98–123. IGI Global (2021)
2. 15th Annual State of Agile Report. Digital.Ai. https://digital.ai/resource-center/analyst-reports/state-of-agile-report. Accessed 13 Dec 2022
3. Boehm, B., Turner, R.: Observations on balancing discipline and agility. In: Proceedings of the Agile Development Conference, 25–28 June 2003, pp. 32–39, Salt Lake City, UT, USA (2003, June)
4. Tsai, C.W., Lai, C.F., Chao, H.C., Vasilakos, A.V.: Big data analytics: a survey. J. Big Data. **2**(1), 1–32 (2015)
5. Estrin, D.: Small data, where n= me. Commun. ACM. **57**(4), 32–34 (2014)
6. Larson, D., Chang, V.: A review and future direction of agile, business intelligence, analytics and data science. Int. J. Inf. Manag. **36**(5), 700–710 (2016)
7. Martinez, I., Viles, E., Olaizola, I.G.: Data science methodologies: current challenges and future approaches. Big Data Res. **24**, 100183 (2021)
8. Laporte, C., O'Connor, R.: Software process improvement standards and guides for very small organization: an overview of eight implementations. CrossTalk, J. Def. Softw. Eng. **30**(3), 23–27 (2017)
9. Laporte, C.Y., Munoz, M., Miranda, J.M., O'Connor, R.V.: Applying software engineering standards in very small entities: from startups to grownups. IEEE Softw. **35**(1), 99–103 (2017)
10. Muñoz, M., Peña, A., Mejía, J., Gasca-Hurtado, G.P., Gómez-Alvarez, M.C., Laporte, C.Y.: Analysis of 13 implementations of the software engineering management and engineering basic profile guide of ISO/IEC 29110 in very small entities using different life cycles. J. Softw. Evol. Proc. **32**(11), e2300 (2020)
11. Cooper, H.M.: Organizing knowledge syntheses: a taxonomy of literature reviews. Knowl. Soc. **1**(1), 104–126 (1988)
12. Templier, M., Paré, G.: A framework for guiding and evaluating literature reviews. Commun. Assoc. Inf. Sys. **37**(1), 112–137 (2015)
13. Chapman, P., Clinton, J., Kerber, R., Khabaza, T., Reinartz, T., Shearer, C., Wirth, R.: CRISP-DM 1.0: Step-by-step data mining guide. SPSS Inc. **9**(13), 1–73 (2000)
14. Hotz, N.J., Saltz, J.: Domino Data Science Lifecycle – Data Science Project Management. The Practical Guide to Managing Data Science at Scale. Domino Data Lab (2018)
15. Cox, M., Ellsworth, D.: Managing big data for scientific visualization. In: ACM Siggraph, MRJ/NASA Ames Res. Center, vol. 97, No. 1, pp. 21–38 (1997)
16. Chang, H.C., Wang, C.Y., Hawamdeh, S.: Emerging trends in data analytics and knowledge management job market: extending KSA framework. J. Knowl. Manag. **23**(4), 664–686 (2018)
17. Davenport, T.H.: Competing on analytics. Harv. Bus. Rev. **84**(1), 98 (2006)
18. Sawant, N., Shah, H.: Big data application architecture. In: Big Data Application Architecture Q & A, pp. 9–28. Apress, Berkeley, CA (2013)
19. Heller, B., Röthlisberger, M.: Big data on trial: Researching syntactic alternations in GloWbE and ICE. Data to Evidence (d2e), Helsinki (2015, 19–22 Oct)
20. Dijcks, J. P.: Oracle: Big data for the enterprise. Oracle white paper, 16 (2012)
21. Russom, P.: Big data analytics. TDWI best practices report, fourth quarter. **19**(4), 1–34 (2011)
22. Schroeck, M., Shockley, R., Smart, J., Romero-Morales, D., Tufano, P.: Analytics: the real-world use of big data. IBM Global Bus. Serv. **12**, 1–20 (2012)
23. Gentile, B.: Top 5 myths about big data. Available at: http://mashable.com/2012/06/19/big-data-myths/#MwZnjjrOR8qV. Accessed 12 Jan 2023
24. Akter, S., Wamba, S.F., Gunasekaran, A., Dubey, R., Childe, S.J.: How to improve firm performance using big data analytics capability and business strategy alignment? Int. J. Prod. Econ. **182**, 113–131 (2016)
25. Beulke, D.: Big data impacts data management: The 5 vs of big data. Available from: Big Data Impacts Data Management: the 5Vs of Big Data, accessed (2011). https://davebeulke.com/big-data-impacts-data-management-the-five-vs-of-big-data/. Accessed 13 Jan 2023

26. Mikalef, P., Bourab, M., Lekakosb, G., Krogstiea, J.: Big data analytics and firm performance: findings from a mixed-method approach. J. Bus. Res. **98**, 261–276 (2019)
27. Phillips-Wren, G., Daly, M., Burstein, F.: Reconciling business intelligence, analytics and decision support systems: more data, deeper insight. Decis. Support. Syst. **146**, 113560 (2021)
28. Núñez, A., Hendriks, J., Li, Z., De Schutter, B., Dollevoet, R.: Facilitating maintenance decisions on the Dutch railways using big data: the ABA case study. In: IEEE International Conference on Big Data, pp. 48–53. IEEE, Washington, DC, USA (2014, Oct 27–30)
29. Katsis, Y., Balac, N., Chapman, D., Kapoor, M., Block, J., Griswold, W.G., et al.: Big data techniques for public health: a case study. In: 2017 IEEE/ACM International Conference on Connected Health: Applications, Systems and Engineering Technologies (CHASE), pp. 222–231. IEEE, Philadelphia, USA (2017, 17–19 July)
30. Kitchin, R., McArdle, G.: What makes Big Data, Big Data? Exploring the ontological characteristics of 26 datasets. Big Data Soc. **3**(1), 2053951716631130 (2016)
31. Panetta, K.: Gartner top 10 data and analytics trends for 2021. Gartner. https://www.gartner.com/smarterwithgartner/gartner-top-10-data-and-analytics-trends-for-2021, Accessed 13 Jan 2023
32. Davenport, T., Malone, K.: Deployment as a critical business data science discipline. Harvard Data Sci. Rev. (2021)
33. Demirkan, H., Dal, B.: "Why do so many analytics projects continue to fail? Key Considerations for Deep Analytics on. Big Data, Learning and Insights, INFORMS Analytics, 44–52 (2014)
34. Veeramachaneni, K.: Why you're not getting value from your data science. Harvard Bus. Rev. 12, 1–4 (2016)
35. Lohr, S., Singer, N.: How data failed us in calling an election. N. Y. Times, 10 (2016)
36. Reggio, G., Astesiano, E.: Big-data/analytics projects failure: a literature review. In: 46th Euromicro Conference on Software Engineering and Advanced Applications (SEAA), pp. 246–255. IEEE Portoroz, Slovenia (2020, 26–28 Aug)
37. Economic benefits of standards. ISO. https://www.iso.org/cms/render/live/en/sites/isoorg/contents/data/publication/10/04/PUB100403.html. Accessed 2 March 2023
38. Laporte, C.Y., Munoz, M.: Not teaching software engineering standards to future software engineers-malpractice? Computer. **54**(5), 81–88 (2021)
39. Laporte, C.Y., Miranda, J.M.: Delivering software-and systems-engineering standards for small teams. Computer. **53**(8), 79–83 (2020)
40. ISO/IEC TR 29110-5-1-2:2011 - Software engineering - lifecycle profiles for Very Small Entities (VSEs) - part 5-1-2: management and engineering guide – generic profile group: basic profile. International Organization for Standardization/International Electrotechnical Commission, Geneva, Switzerland
41. Martínez-Plumed, F., Contreras-Ochando, L., Ferri, C., Hernández-Orallo, J., Kull, M., Lachiche, N., et al.: CRISP-DM twenty years later: from data mining processes to data science trajectories. IEEE Trans. Knowl. Data Eng. **33**(8), 3048–3061 (2019)
42. Schröer, C., Kruse, F., Gómez, J.M.: A systematic literature review on applying CRISP-DM process model. Proc. Comput. Sci. **181**, 526–534 (2021)

A Selective Comparative Review of CRISP-DM and TDSP Development Methodologies for Big Data Analytics Systems

Gerardo Salazar-Salazar, Manuel Mora, Hector A. Duran-Limon, and Francisco Javier Álvarez Rodríguez

1 Introduction

In recent decades, the capacity of electronic devices and sensors, along with the use of social networks and the ability to store and exchange this data, has dramatically increased the opportunities to extract knowledge through data mining projects [1]. The diversity of data has increased in origin, format, and modes, as well as the variety of techniques from machine learning, data management, visualization, causal inference, and other areas [1]. In other words, not only the nature of the data has changed, but also the processes for extracting value from it.

Furthermore, the need for fast delivery of business intelligence has increased in the last 5 years due to the demand for real-time data analysis [2]. Added to the Internet of Things (IoT), where data collection is integrated into devices, it contributes to the demand for larger amounts of updated data. This generates many challenges for companies in processing such large amounts of data.

Generally, companies that implement big data analytics systems (BDAS) techniques have a great potential to improve their performance, enhance decision-making, improve their services, and create competitive advantages in their organizations. However, 85% of BDAS projects that are developed fail to deliver the expected value to organizations [3]. To guide these projects toward successful outcomes, the use of a process model or project methodology is recommended [4]. A well-defined methodology helps professionals to manage the tasks involved in executing these projects.

G. Salazar-Salazar (✉) · M. Mora · F. J. Á. Rodríguez
Autonomous University of Aguascalientes, Aguascalientes, Mexico
e-mail: al131651@edu.uaa.mx

H. A. Duran-Limon
University of Guadalajara, Guadalajara, Mexico

M. Mora et al. (eds.), *Development Methodologies for Big Data Analytics Systems*, Transactions on Computational Science and Computational Intelligence, https://doi.org/10.1007/978-3-031-40956-1_6

In 2019, VentureBeat revealed that 87% of data science projects never make it to production [5], due to the lack of methodologies for the development of this type of projects. In turn, a survey conducted in 2018 involving professionals from both the industry and nonprofit organizations found that 82% of those surveyed did not follow an explicit process methodology to develop data science projects. In addition, 85% of those surveyed believed that using an improved and more consistent process would produce more consistent and effective data science projects [6]. However, critics of process models and methodologies argue that they are too rigid and do not admit the iterative and open nature of most knowledge discovery (KD) projects [7].

This all indicates the lack of clear methodologies for developing BDAS projects, indicating a lack of research on the application of agile principles to these types of projects. However, current research suggests that agile would align well with the iterative discovery and validation supporting prescriptive and predictive analysis in BDAS projects [8], but it would have to be "short-cycle agile," suggesting a need for faster results [9].

This has led to the concern of implementing the agile software development (ASD) paradigm to BDAS-type projects. As it has been shown, agile methodologies align well with this type of projects, where little time is dedicated to defining requirements in advance and the emphasis is on the quick development of small projects. Therefore, more and more organizations are applying agile methods in data science projects to improve their success rate [10–12].

On the contrary, the life cycles of BDAS projects are currently in the same situation as software development before the introduction of agile methodologies, with issues in delivery time, early value generation, and reducing the risk of failure [13].

In the following sections, we will understand more about the origins of BDAS and its importance for different organizations, as well as the value and competitive advantages it generates for organizations. In turn, we will analyze the Agile Scrum methodology, which is the most popular agile methodology today, occupying 66% of the preference of developers and an additional 15% following Scrum derivatives (ScrumBan and Scrum-XP) [14].

In addition, we will analyze two of the most important and well-documented methodologies that exist for the development of BDAS projects, in which we can observe two distinct approaches: the rigorous methodologies with CRISP-DM, which has been considerably the most used methodology for analysis, data mining, and data science projects [15], and the agile methodologies with Team Data Science Process (TDSP) which in our analysis is evaluated as a well-documented and complete methodology with phases, roles, activities, and artifacts.

Finally, a small comparison of Scrum-XP with the two big data analytics methodologies are be made to compare phases, roles, activities, and artifacts of these methodologies and see a bit of the landscape of agile methodologies in big data analytics projects.

The purpose of this research is to compare two existing methodologies for the development of BDAS-type projects: one rigorous methodology and one agile methodology. The former has one of the most widely used agile methodologies worldwide, Scrum [14], in order to find similarities, differences, or areas for improvement in the development of BDAS projects with respect to the Scrum methodology, which is not focused on this type of project. Previous studies argue that the use of (elements of) the Scrum methodology improves the success rate of data science projects [16–18].

2 Research Methodology

In this chapter, we use a selective conceptual review methodology with a descriptive and comparative dual goal [19]. According to [19], this review can be characterized by its goal, focus, perspective, coverage, organization, and expected audience. This conceptual review focuses on practices – i.e., empirical professional development methodologies for BDAS; it is realized from a non-neutral perspective – it aims to describe and compare two methodologies against a generic agile Scrum-XP profile development process; it uses a pivotal coverage – it describes and compares only the most relevant classic development methodology for BDAS vs. one of the most modern agile methodologies for BDAS; its organization is methodological – it uses the generic development agile methodology as a template; and it is elaborated for a specialized audience – BDAS academics and professionals.

We applied the following steps adapted from [20]: (1) Formulate the research goal, (2) define data sources and selective criteria, (3) collect studies, (4) review and synthetize the findings from the collected studies, and (5) elaborate report of findings. Table 1 summarizes the five selective review steps that were applied.

3 Background

This section reviews the background on the fundamentals of BDAS and the fundamentals of the Scrum-XP workflow.

3.1 Foundations of Big Data Analytics Systems (BDAS)

NASA researchers Michael Cox and David Ellsworth [21] were the first to refer to the term "big data" when they report, "Visualization poses an interesting challenge for computer systems of computer systems: the data sets are often quite large, straining the capacity of main memory, local disk, and even remote disk, local disk, and even remote disk. We call this the big data problem." They emphasize that even the supercomputers of that time could not process that amount of information, which is why in the article they mention a process for handling "big data."

Business intelligence and analytics (BI&A) and the related field of big data analytics (BDAS) have become increasingly important for academic and business

Table 1 Selective conceptual review research methodology

Step	Purpose	Outcomes	Outcomes in this research
(1) Formulate the research goal	State the expected research goal indicating the theoretical or practical or both ones expected contributions	Research goal statement	We aim to contribute to the literature with a conceptual descriptive-comparative review of two – one classic and one agile type - relevant development methodologies for BDAS and provide useful recommendations regarding both development methodologies
(2) Define data sources and selective criteria	Identify and agree the set of data sources to collect the studies, as well as to define the selection criteria.	List of data sources Selection criteria statements	The two development methodologies for BDAS were selected according to the next criteria: (1) Select the classic methodology most cited in the literature; and (2) select a modern and complete - i.e., it includes roles, phases, activities, and artifacts -agile development methodology reported from 2015 to 2022 period
(3) Collect studies	Get the studies	Set of selected studies	Two methodologies were identified along with their published references [26, 29]
(4) Review and synthetize the findings from the collected studies	Conduct the analysis and integration of finding	Structured schema of findings	We elaborated a generic agile development methodology using the Scrum-XP profile standard
(5) Elaborate report of findings	Produce visible results	Research results	This chapter was elaborated

communities in recent decades. Companies and organizations in all industries have begun to obtain critical information from structured data collected through various business systems and analyzed by commercial relational database management systems. One of the main reasons for creating data warehouses in the 1990s was to store large amounts of data [22]. From an evolutionary perspective, big data is not new.

However, the current hype can be attributed to promotional initiatives of certain leading technology companies that invested in building the analytical market niche. Some academics and professionals have considered "big data" as data coming from various channels, including sensors, satellites, social media sources, photos, videos, cell phone, and GPS signals [23].

Figure 1 shows the rapid growth of data in recent years due to different factors such as the use of social networks, the increased use of mobile devices, and the implementation of sensors in various fields.

Business intelligence and analytics (BI&A) and the related field of big data analytics have evolved in three stages [25]. Through BI&A 1.0 initiatives, companies and organizations across all industries began to obtain critical information from structured data collected through various business systems and analyzed by

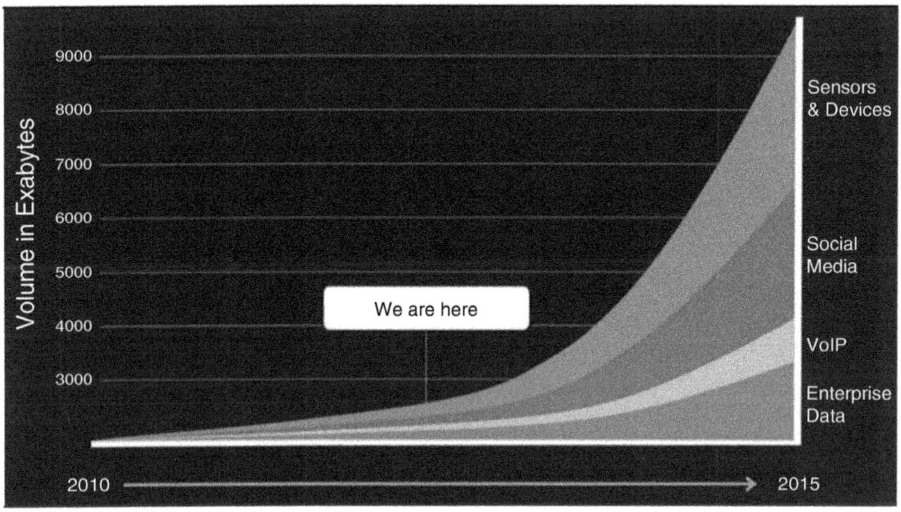

Fig. 1 The exponential growth of big data. (Source: Watson 2014) [24]

commercial database management systems [25]. In recent years, web intelligence, web analysis, web 2.0, and the ability to extract user-generated unstructured content have given rise to a new and exciting era of BI&A 2.0 research, which has led to unprecedented intelligence about consumer sentiment, customer needs, and the recognition of new business opportunities [25]. Now, in this era of big data, even if BI&A 2.0 is still maturing, we are on the verge of BI&A 3.0, with all the uncertainty that new and potentially revolutionary technologies entail [25].

Figure 2 shows the evolution of BI&A, emerging analytical research applications, and opportunities.

One of the most well-known characteristics of big data is undoubtedly the volume of data that can be stored; however, this is not the only characteristic of big data. Laney in 2001 – the creator of the concept of big data in business domain – suggested that volume, variety, and velocity (or the 3Vs) are the three dimensions of data management challenges. The 3Vs have emerged as a common framework to describe big data [25, 26]. In turn, today various authors mention the use of the 5Vs (volume, variety, velocity, veracity, and value), which are the main attributes of big data analytics systems. Table 2 describes each of these big data characteristics, the three mentioned initially, as well as the recently discovered characteristics.

However, the distinction between small and big data is recent. Before 2008, data was rarely considered in terms of "small" or "big." All data was, in effect, what is now sometimes referred to as "small data," regardless of volume. Due to factors such as cost, resources, and difficulties in generating, processing, analyzing, and storing data, limited volumes of high-quality data were produced through carefully designed studies using sampling frames designed to ensure representativeness [29].

Therefore, the term "big" is somewhat misleading, as big data is characterized by much more than volume. Some "small" datasets may be very large in size, such as

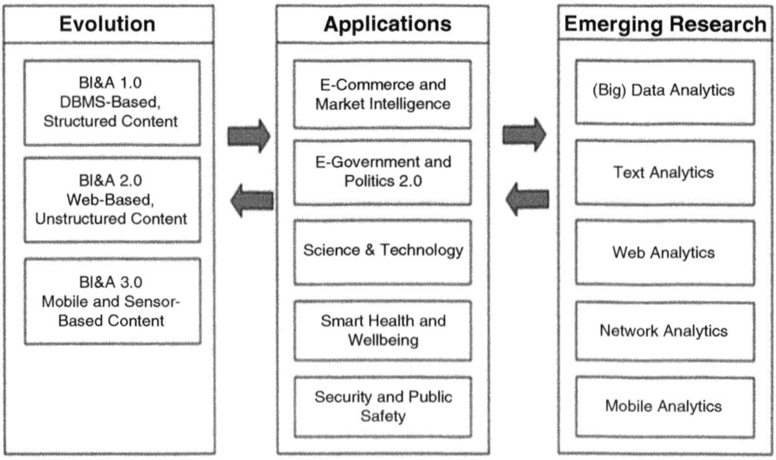

Fig. 2 BI&A overview: evolution, applications, and emerging research. (Source: Chen et al. 2012) [25]

Table 2 Big data features

Attributes	Definition
Volume	Volume in big data can be defined as: "Large volume of data that consume a large amount of storage or consist of a large number of records" [27]. Currently, organizations must learn to manage the large volume of data through new processes
Variety	"Variety" is another critical attribute of big data, as the data is generated from a wide range of sources and formats [28]. The word "variety" denotes the fact that big data originates from numerous sources that can be structured, semi-structured, or unstructured [27]
Velocity	Speed refers to the frequency of data generation and/or the frequency of data delivery [27]
Veracity	The high quality of data is an important requirement of big data for better predictability in the business environment [28]. This tells us that data verification is essential in the data management process, as erroneous data will make decision-making more difficult or lead analysts down the wrong path
Value	In the added value that organizations get, value is only created when data is analyzed and acted upon correctly. To do this, we must identify all the data that will help us in the best way to generate value

national censuses that also strive to be comprehensive. However, census datasets lack speed (usually conducted once every 10 years), variety (usually around 30 structured questions), and flexibility (once a census is established and administered, it is nearly impossible to modify questions or add new ones) [30].

Table 3 shows us the differences between small data and big data, where the main differences between them can clearly be seen in terms of volume, speed, and

Table 3 Comparison of SDAS vs. BDAS

Characteristic	Small data	Big data
Volume	In the range of GB to TB (10,000–100,000 records)	In the range of TB to ZB (1,000,000–1,000,000,000 records)
Velocity	Controlled and steady flow of data; accumulation of data is slow	Data arrives at very fast speeds; huge amount of data gets accumulated within a short period of time
Variety	Limited to wide (structured data)	Wide (huge variety of data)
Veracity	Contains less noise as data is collected in a controlled manner	The quality of data is not guaranteed. Rigorous validation of data is required prior its processing
Value	High	High
Case	Big data techniques for public health: a case study. In 2017 IEEE/ACM International Conference on Connected Health: Applications, Systems and Engineering Technologies (CHASE) (pp. 222–231) [31] Public health researchers increasingly recognize that to advance their field they must grapple with the availability of increasingly large (i.e., thousands of variables) traditional population-level datasets (e.g., electronic medical records) while at the same time integrating additional large datasets (e.g., data on genomics, the microbiome, environmental exposures, and socioeconomic factors)	Big data in big companies. International Institute for Analytics 3(1–31) [9] UPS is no stranger to big data, having begun to capture and track a variety of package movements and transactions as early as the 1980s. The company now tracks data on 16.3 million packages per day for 8.8 million customers, with an average of 39.5 million tracking requests from customers per day. The company stores over 16 petabytes of data

variety. However, despite these differences, we can see that the value of both is extremely important for companies, as we previously observed that they can generate value and competitive advantages with data of low volume, speed, and variety. Table 3 also shows a clear example of both, showing how there are organizations that generate value with both types of data.

This leads us to conclude that the use of BDAS techniques for much smaller data volumes is possible, as these techniques allow organizations to obtain value from the data and generate competitive advantages, as well as improve decision-making. That is, an organization can use BDAS techniques with small data volumes, a controlled data speed, limited data variety, and high accuracy of data with much less noise, all with the purpose of generating high value for the organization using the data.

3.2 The Scrum-XP Workflow: An Agile Framework of Practices

With the creation of the "Agile Manifesto" in 2001, the agile software development (ASD) was officially presented to the software engineering community through a set of 4 core values and 12 principles, established in the "Agile Manifesto" [32]. This manifesto sets out four main bases for agile software development:

1. Value people and their interactions more than processes.
2. Evaluate functional software over complete documentation.
3. Value collaboration with the client more than countercurrent negotiation.
4. Value the response to change rather than follow a plan.

The creation of these principles gave agile software development the momentum it needed to expand rapidly. The fundamentals and principles of the manifesto allowed for developing methods with a real-world approach, where the response to change became a success factor [33]. "Since its inception about two decades ago, ASD has quickly become a conventional software development model used today" [34] causing a dramatic impact on current software development, leading to the development of numerous manifestations, methods, frameworks, processes, and standards that comply with the foundations and principles of the agile manifesto. Two of the most widely used agile methodologies today are Scrum and Extreme Programing (XP), both of which are based on the agile manifest, but these methodologies advocate a significantly different set of agile practices.

Scrum is an agile method that focuses primarily on managing project team tasks through practices such as daily meetings, iteration planning, and short sprint delivery. In contrast, XP is an agile method that advocates practices focused on quality and software engineering techniques (peer programming, unit testing, etc.) [35]. Scrum is defined by the Scrum guide itself as: "A lightweight framework that helps people, teams and organizations generate value through adaptive solutions to complex problems" [36]. Figure 3 represents the Scrum life cycle with all the components that make up Scrum roles, events, and artifacts, as described by Ken Schwaber and Jeff Sutherland, who are the creators of the methodology.

Scrum is a management process that reduces the complexity of product development to meet customer needs. Scrum draws on the collective intelligence and experience of team members, rather than giving them detailed instructions for software development, allowing the team to use various processes, techniques, and methods within a single project. This framework consists of Scrum Teams and their roles, events, artifacts, and associated rules. Each component within the framework has a specific purpose and is essential to the success of Scrum [36]. An initial software engineering representation of the Scrum life cycle is the 1997 "Schwaber proposal" (Fig. 4), which consists of three phases: the pregame phase, the game phase, and the postgame phase.

These phases encompass all the roles, events, and artifacts that Scrum has, and they are seen as a more disciplined way of representing this methodology. We can

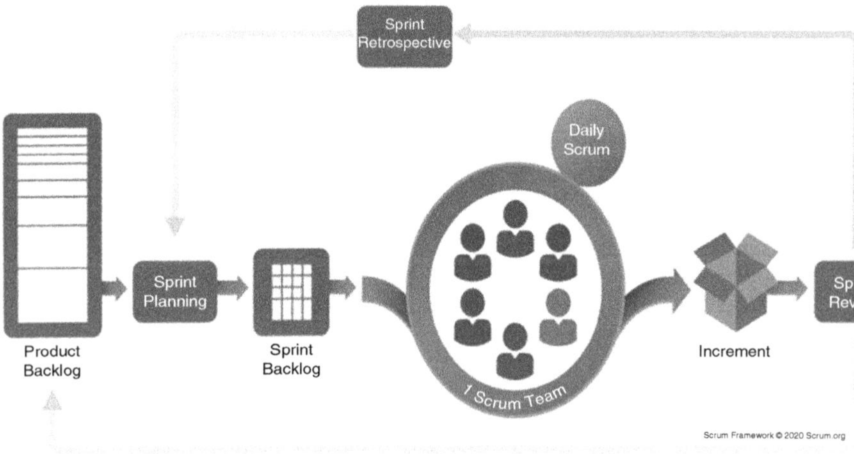

Fig. 3 Scrum life cycle. (Source: Sutherland and Schwaber 2020) [36]

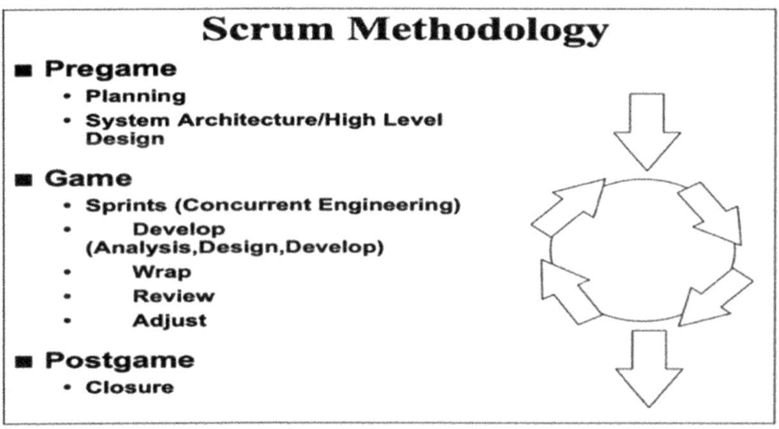

Fig. 4 Scrum methodology. (Source: Schwaber 1997) [37]

confirm this in Table 4, Scrum phases, where the objectives and functions of each of these phases are described. These phases help to establish the Scrum methodology in a more multidisciplinary context, since it shows us how is the implementation of the methodology from the planning, analysis, and design of the architecture, until the closure of the project. This is something that Scrum does not currently contemplate, due to the changes that the methodology has undergone since each work team that uses Scrum can adapt it according to how it works best or best fits your needs.

We can corroborate why Schwaber included some agile practices from the XP methodology, which establishes three very similar phases, consisting of different events and activities to perform to complete the launch of a product. Like Scrum, XP is divided into smaller mini projects that result in a functional boost known as

Table 4 Scrum phases [37]

Phases	Description
Pregame	This phase is the one in charge of making a schedule and cost estimate. For the development of a new system, this phase consists of planning and developing the architecture to a high-level design, while if it is an existing system, the analysis is much more limited
Game	This phase is the one in charge of making a schedule and cost estimate. For the development of a new system, this phase consists of planning and developing the architecture to a high-level design, while if it is an existing system, the analysis is much more limited
Postgame	Finally, the postgame phase prepares for release, including final documentation, prerelease staged testing, and launch

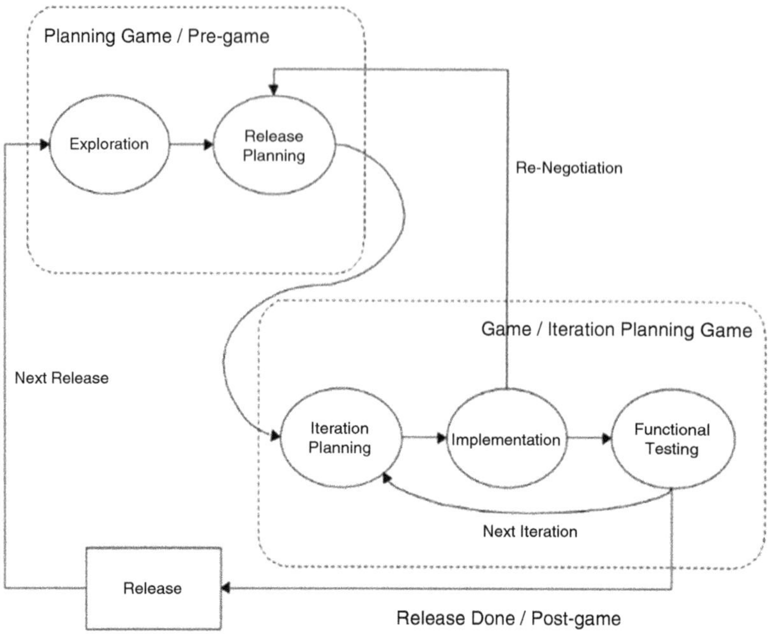

Fig. 5 Simplified process structure XP. (Source: Dudziak 1999) [38]

launch. An XP project creates frequent releases (every 2 weeks) for early and frequent feedback, gradually building iteration planned functionality [38].

The phases and events of XP are represented in Fig. 5. This figure shows a clear similarity to the Scrum methodology and even more so to the initial version proposed by Schwaber.

Given the established goals for both Scrum and XP, we can derive a Scrum-XP workflow reported in Table 5. This table shows us in a clearer way how both methods overlap, showing the events and roles involved in each of the phases established by Schwaber. Figure 6 shows a graphical representation of this combined Scrum-XP

Table 5 The Scrum-XP workflow of agile practices

Scrum phases	XP phases	Scrum event	Roles		Artifacts
			Principal	Secondary	
Pregame	Exploration	Create project Vision	Product owner	Scrum master	Project Vision Statement
		Develop epics	Product owner	Scrum master, Scrum Team	
		Create user Stories	Product owner	Scrum master, Scrum Team	User stories
	Release planning	Create prioritized product backlog	Product owner	Scrum master, Scrum Team	
		Conduct release planning	Product owner	Scrum master, Scrum Team	Product backlog
Game	Iteration planning + implementation + Functional testing	Create sprint backlog (sprint planning)	Scrum Team	Product owner, Scrum master	Spring backlog
		Conduct daily standup (daily Scrum)	Scrum Team	Product owner, Scrum master	Product and Sprint, Kanband Bord
		Increment development	Scrum Team	Product owner, Scrum master	Increment
		Review sprint	Scrum Team	Product owner, Scrum master	
		Retrospective sprint	Scrum Team	Product owner, Scrum master	Agreed Actionable improvements
Postgame	Release	Ship deliverables	Scrum Team	Product owner, Scrum master	Final release

workflow. Figure 6 presents the Scrum-XP workflow divided into the three phases, with the events and activities that take place in each phase. This diagram also shows what roles are responsible for carrying out each of these events and activities. Finally, it shows how the Scrum and XP methodologies overlap.

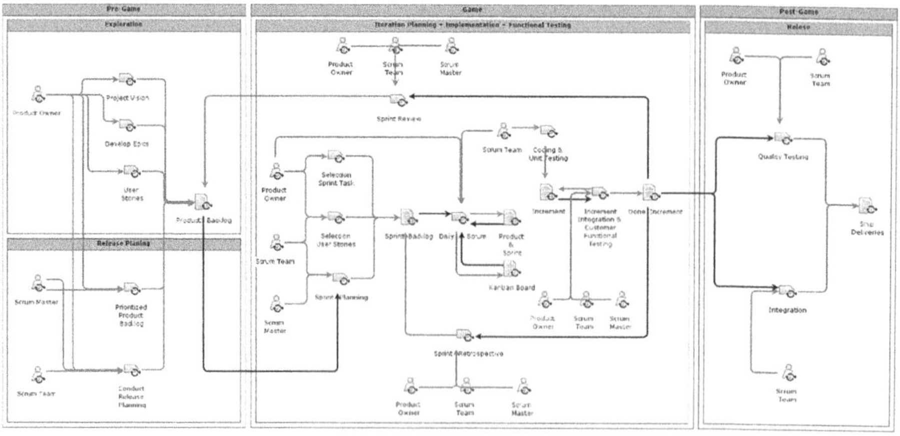

Fig. 6 The Scrum-XP workflow of agile practices

4 Selective Comparative Analysis

This section compares one of the most widely used rigor-oriented BDAS project methodologies (i.e., the CRISP-DM methodology [39]) and one of the best documented agile methodologies for BDAS projects (i.e., the TDSP methodology [41]).

4.1 Description of the CRISP-DM Methodology

In response to common problems and needs in data mining projects, in the mid-1990s, a group of organizations involved in data mining (Teradata, SPSS-ISL, Daimler-Chrysler, and OHRA) proposed a reference guide for developing data mining projects, called CRISP-DM (Cross-Industry Standard Process for Data Mining) [39]. CRISP-DM is considered the de facto standard for developing data mining and knowledge discovery projects.

The CRISP-DM data mining methodology is described in terms of a hierarchical process model, consisting of task sets described in four levels of abstraction (from general to specific) [4]. At the higher level, the data mining process is organized into several phases; each phase consists of several generic second-level tasks. This second level is called generic, because it aims to be general enough to cover all possible data mining situations. The third level, the specialized task level, is the place to describe how generic task actions should be carried out in certain specific situations. The fourth level, the process instance, is a record of actions, decisions, and results of a real commitment to data mining [4]. Figure 7 shows the life cycle of CRISP-DM.

The life cycle of a data mining project according to CRISP-DM consists of six phases; the sequence of phases is not strict. It is always necessary to move forward

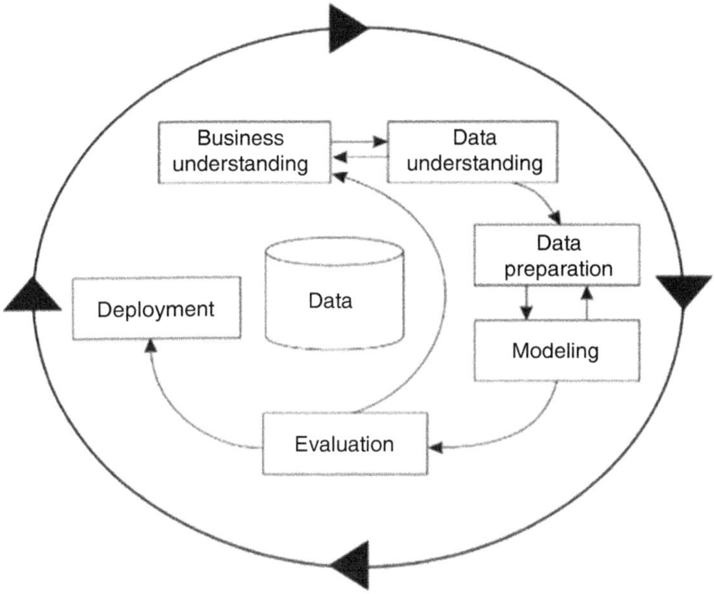

Fig. 7 The CRISP-DM methodology. (Source: Chapman et al. 2000) [39]

and back between the different phases. The arrows indicate the most important and frequent dependencies between phases.

In the following statements, we outline each phase briefly [39]:

1. *Business Understanding:* The business situation should be assessed to get an overview of the available and required resources. The determination of the data mining goal is one of the most important aspects in this phase. First the data mining type (e.g., classification) and the data mining success criteria (e.g., precision) should be explained. A compulsory project plan should be created
2. *Data Understanding:* Collecting data from data sources, exploring, and describing it and checking the data quality are essential tasks in this phase. To make it more concrete, the user guide describes the data description task by using statistical analysis and determining attributes and their collations.
3. *Data Preparation:* Data selection should be conducted by defining inclusion and exclusion criteria. Bad data quality can be handled by cleaning data. Dependent on the used model (defined in the first phase), derived attributes have to be constructed. For all these steps, different methods are possible and are model dependent.
4. *Modeling:* The data modeling phase consists of selecting the modeling technique, building the test case and the model. All data mining techniques can be used. In general, the choice depends on the business problem and the data. Another important aspect is defining how to explain the choice. For building the model, specific parameters must be set. For assessing the model, it is appropriate to evaluate the model against evaluation criteria and select the best ones.

5. *Evaluation:* In the evaluation phase, the results are checked against the defined business objectives. Therefore, the results must be interpreted, and further actions have to be defined. Another point is that the process should be reviewed in general.
6. *Deployment:* The deployment phase is described generally in the user guide. It could be a final report or a software component. The user guide describes that the deployment phase consists of planning the deployment, monitoring, and maintenance.

The reference model, Fig. 8, presents a quick overview of phases, tasks, and their artifacts and describes the steps to follow in a data mining project.

The user guide provides more detailed suggestions for each phase and each task within a phase and describes how to perform a data mining project.

Creating the model is usually not the end of the project. In general, the knowledge acquired must be organized and presented in such a way that the customer can use it [40]. Depending on the requirements, the implementation phase can be as simple as generating a report or as complex as implementing a repeatable data mining process. In many cases, it will be the user, not the data analyst, who will carry

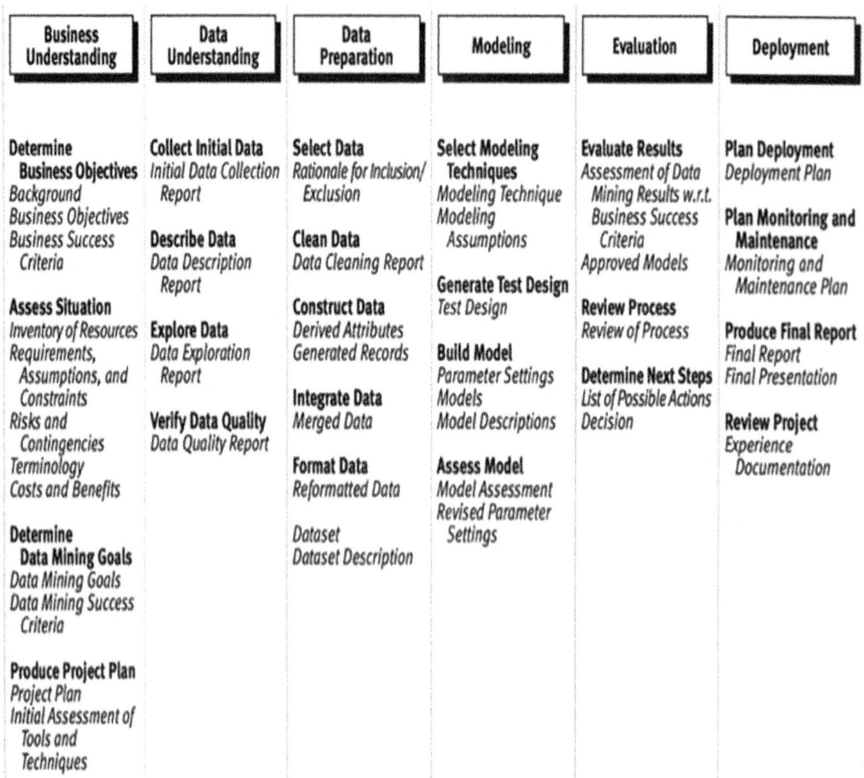

Fig. 8 The CRISP-DM methodology. (Source: Chapman et al. 2000) [39]

out the implementation steps. In any case, it is important to understand in advance what actions should be taken to make real use of the models created [40].

4.2 Description of the TDSP Methodology

Team Data Science Process (TDSP) [41] is an agile and iterative data science methodology to efficiently deliver predictive analytics solutions and smart applications. TDSP helps improve team collaboration and learning by suggesting how team roles work best together. Its main objective is to help companies make the most of the benefits of their analysis program. It is very well documented and provides several tools and utilities that facilitate its use.

TDSP provides a life cycle to structure the development of its projects; the TDSP project life cycle is like CRISP-DM and includes five iterative stages: commercial understanding, data acquisition and understanding, modeling, implementation, and customer acceptance; in fact, it is an iterative and cyclic process [41].

In the following statements, we outline each phase briefly [41]:

- *Business Understanding:* Initially, a question which describes the problem objectives should be defined clearly and explicitly. Relevant predictive model and required data source/s also must be identified in this step.
- *Data Acquisition and Understanding:* Data collection starts in this phase by transferring data into the target location to be utilized by analytic operations. The raw data needs to be cleaned. Also, either incomplete or incorrect values should be identified. Data summarization and visualization might help to find required cleaning procedures. Data visualization also could help to measure if data features and the collected amount of data are adequate over time. At the end of this stage, it might be necessary to go back to the first step for more data collection.
- *Modeling:* Feature engineering and model training are two elements of this phase. Feature engineering provides attributes and data features which are required for the machine learning algorithm. Algorithm selection, model creation, and predictive model evaluation are also subcomponents of this step. Collected data should be divided into training and testing datasets to train and evaluate machine learning model. It is important to employ different algorithms and parameters to find the best suitable solution to support the problem.
- *Deployment:* Predictive model and data pipeline are needed to be produced in this step. It could be either a real-time or a batch analysis model depending on the required application. The final data product should be accredited by the customer.
- *Customer Acceptance:* The final phase is customer acceptance which should be performed by confirming the data pipeline, predictive model, and product deployment.

Figure 9 shows the life cycle of the TDSP methodology, and we can better understand what activities are carried out in each of the phases, as well as some artifacts and the interaction that the phases have between them.

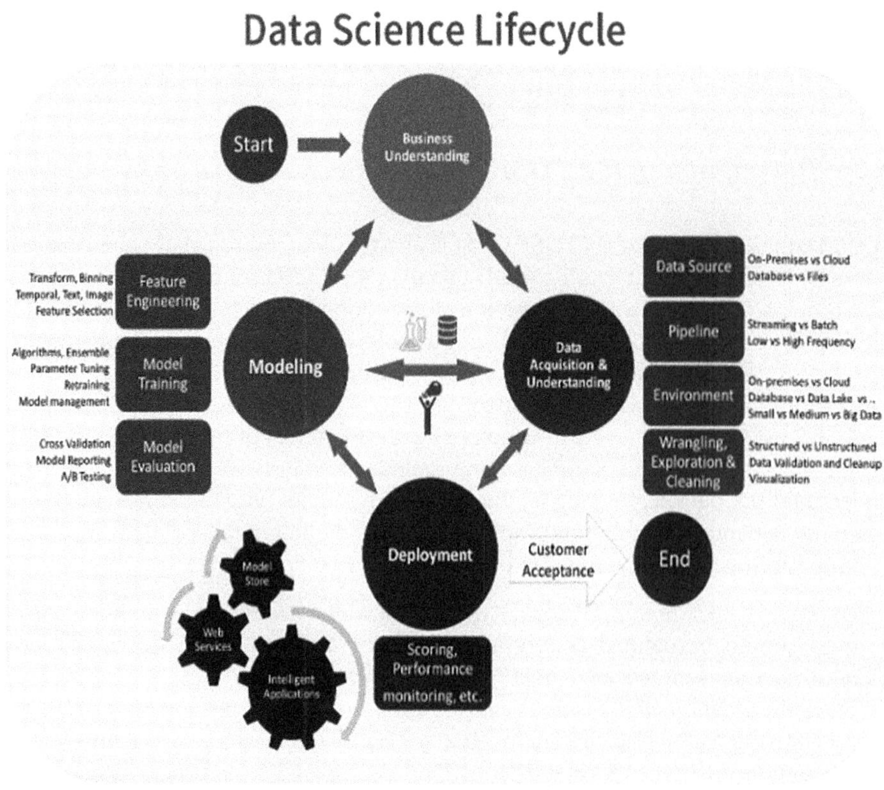

Fig. 9 The TDSP methodology. (Source: Microsoft Team 2017) [41]

TDSP addresses the weakness of CRISP-DM's lack of role definition by defining four distinct roles (solution architect, project manager, data scientist, and project leader) and their responsibilities during each phase of the project life cycle [41]. These roles are very well defined from the perspective of project management, and the team works under agile methodologies, which improve collaboration and coordination [41]. Its responsibilities for the creation, implementation, and development of the project are clear [41]. TDSP is one of the best documented methodologies available for BDAS projects, as it clearly specifies roles, tasks, and artifacts, as well as being a methodology easily combinable with other existing methodologies such as CRISP-DM or KDD. Table 6 reports the TDSP roles, activities, and artifacts.

4.3 *Comparison of the CRISP-DM and TDSP Methodologies*

When comparing the Scrum-XP workflow with the two methodologies for developing BDAS projects, we can observe similarities, differences, and areas of improvement for this methodology. At first glance, we can see that the Scrum-XP workflow

Table 6 TDSP roles, activities, and artifacts [41]

TDSP			
Phases	Activities	Roles	Artifacts
Business understanding	Define objectives Identify data source	Project lead, project manager	Charter document Data source Data dictionaries
Data acquisition and understanding	Ingest the data Explore the data Set up a data pipeline	Project lead, data scientist, solution architecture	Data quality report Solution architecture Checkpoint decision
Modeling	Feature engineering Model training Model evaluation	Data scientist, solution architecture, application developer, data engineer	Feature sets Model report Checkpoint decision
Deployment	Operationalize a model	Data scientist, solution architecture, application developer, data engineer	A status dashboard that displays the system health and key metrics A final modeling report with deployment details A final solution architecture document
Customer acceptance	System validation Project handoff	Project lead, project manager, data scientist	Exit report of the project for the customer

lacks data management activities (data acquisition, data cleaning, data preparation, data analysis, among others). These activities are indispensable in the methodologies for the development of BDAS-type projects.

The methodologies of BDAS analyzed are CRISP-DM (Cross-Industry Standard Process for Data Mining) which is the most widely used rigor methodology for more than 20 years. This methodology became soon the "de facto standard for the development of data mining and knowledge discovery projects" [42] and remains today as the most widely used analytical methodology according to many opinion polls [1]. On the other hand, TDSP (Team Data Science Process) is a modern agile methodology for the development of BDAS projects, which in our analysis is evaluated as a well-documented and complete methodology.

Key aspects such as phases, roles, activities, and artifacts of both methodologies were compared with Scrum-XP to analyze the similarities, differences, and areas of opportunity with this methodology.

Figures 10, 11, and 12 show the correspondence between the phases, activities, roles, and artifacts, respectively, of the Scrum-XP and CRISP-DM methodologies. The fourth column represents the relationship between the two methodologies and has sub-columns: one with symbols and the other one with a brief description of what those activities, roles, and artifacts do in the CRISP-DM methodology. The symbol (X) corresponds to the activities, roles, and artifacts that CRISP-DM has and that are not in the Scrum-XP methodology. In turn, the symbol $\approx$ indicates activities, roles, and artifacts that are similar in both methodologies. Finally, we denote with the symbol ✓ activities, roles, and artifacts that can be interpreted as an area of improvement for the Scrum-XP methodology.

Analyzing Figs. 10, 11, and 12, we can observe that the traditional CRISP-DM methodology is a very complete methodology, since it covers all the activities and artifacts expected in BDAS projects. On the other side, it is revealed that the methodology totally lacks roles. This indicates that despite being one of the best documented and complete rigorous methodologies, it has areas of improvement. Another area of improvement is the reduction of activities and artifacts that would allow us a more agile and iterative methodology. Another point to highlight is the indispensable role creation, since it is not officially established who is responsible for carrying out the activities and artifacts or who oversees controlling the project.

Figures 10, 11, and 12 show the lack of activities related to data acquisition and cleaning in the Scrum-XP methodology, which is fundamental in BDAS projects to generate AI models and the evaluation of these models. Another aspect that we can observe specifically in the artifact table is how broad is the CRISP-DM methodology is, since the methodology contemplates too many artifacts, and we can understand why it can be quite tedious to implement this methodology in BDAS projects.

Figures 13, 14, and 15 show a comparison of the phases, activities, roles, and artifacts of the Scrum-XP and TDSP methodologies. These tables show in their first columns the phases of the Scrum-XP and TDSP methodologies, continuing with the activities, roles, and artifacts of both methodologies. On the other hand, the relationship column between methodologies shows us two columns: one column with symbols and the other column with a brief description of what those activities, roles, and artifacts perform in the TDSP methodology.

Scrum Phases	XP Phases	CRISP-DM Phases	Relation Between Methodologies		CRISP-DM Roles	Scrum/XP Roles
Pre-game	Exploration	Business Understanding	X	Officially, CRISP-DM does not establish any role to carry out its tasks and its artifacts, which makes it an area of opportunity for new BDAS methodologies.	-	• Product Owner • Scrum Master • Scrum Team
		Data Understanding	X	Officially, CRISP-DM does not establish any role to carry out its tasks and its artifacts, which makes it an area of opportunity for new BDAS methodologies.	-	
	Release Planning	Data Preparation	X	Officially, CRISP-DM does not establish any role to carry out its tasks and its artifacts, which makes it an area of opportunity for new BDAS methodologies.	-	• Product Owner • Scrum Master • Scrum Team
Game	Iteration Planning + Implementation + Functional Testing	Modeling	X	Officially, CRISP-DM does not establish any role to carry out its tasks and its artifacts, which makes it an area of opportunity for new BDAS methodologies.	-	• Scrum Team • Scrum Master • Product Owner
		Evaluation	X	Officially, CRISP-DM does not establish any role to carry out its tasks and its artifacts, which makes it an area of opportunity for new BDAS methodologies.	-	
Post-game	Release	Deployment	X	Officially, CRISP-DM does not establish any role to carry out its tasks and its artifacts, which makes it an area of opportunity for new BDAS methodologies.	-	• Product Owner • Scrum Team • Scrum Master

Fig. 10 CRISP-DM vs. Scrum-X roles

Scrum Phases	XP Phases	CRISP-DM Phases	Relation Between Methodologies		CRISP-DM Task	Scrum/XP Event
Pre-game	Exploration	Business Understanding	≈	Focuses on understanding project objectives. Turns this knowledge of the data into the definition of a data mining problem and a preliminary plan designed to achieve the objectives.	• Determine business objectives. • Assess the situation. • Create a plan for your data mining project.	• Create Project Vision • Develop Epics • Create User Stories
			X	A data mining goal states project objectives in technical terms.	• Determine the objectives of data mining.	
		Data Understanding	X	The data understanding phase starts with an initial data collection and proceeds with activities to get familiar with the data, to identify data quality problems, to discover first insights into the data.	• Collect initial data. • Describe the data • Explore the data. • Check the quality of the data.	
	Release Planning	Data Preparation	X	The data preparation phase covers all activities to construct the final dataset (data that will be fed into the modeling tool(s)) from the initial raw data.	• Select data. • Data cleansing. • Data construction. • Integrate the data. • Format the data.	• Create Prioritized Product Backlog • Conduct Release Planning
Game	Iteration Planning + Implementation + Functional Testing	Modeling	X	Select the actual modeling technique that is to be used. Before we build a model, we need to generate a procedure or mechanism to test the model's quality and validity	• Select modeling technique. • Design the model tests.	• Create Sprint Backlog (Sprint Planning) • Conduct Daily Standup (Daily Scrum) • Increment Development • Review Sprint • Retrospective Sprint
			≈	Run the modeling tool on the prepared dataset to create one or more models. The data mining engineer interprets the models according to his domain knowledge, the data mining success criteria, and the desired test design. This task interferes with the subsequent evaluation phase.	• Build the model. • Evaluate the model.	
		Evaluation	≈	Before proceeding to final deployment of the model, it is important to evaluate the model and review the steps executed to construct the model to be certain it properly achieves the business objectives more thoroughly.	• Evaluate the results. • Process review. • Determine the next stages.	
Post-game	Release	Deployment	≈	Creation of the model is generally not the end of the project. In order to deploy the data mining result(s) into the business, this task takes the evaluation results and concludes a strategy for deployment.	• Plan deployment. • Plan monitoring and maintenance. • Create a final report. • Project review.	• Ship Deliverables

Fig. 11 CRISP-DM vs. Scrum-XP activities

Scrum Phases	XP Phases	CRISP-DM Phases	Relation Between Methodologies		CRISP-DM Artifacts	Scrum/XP Artifacts
Pre-game	Exploration	Business Understanding	≈	Record the information that is known about the organization's business situation. Describe the customer's primary objective. Describe the criteria for a successful or useful outcome to the project. List the resources available to the project. List all requirements of the project. List the risks or events that might occur. List the stages to be executed in the project	• Background • Business objectives • Business success criteria • Resource inventory • Requirements, Assumptions, and Constraints • Risks and contingency plans • Glossary with relevant terminology • Project Plan • Initial evaluation of techniques and tools • Costs and Benefits	• Project Vision Statement
			X	Describe the intended outputs of the project that enables the achievement of the business objectives. Define the criteria for a successful outcome to the project in technical terms.	• Data mining goals • Define a success criterion for the mining project	
		Data Understanding	X	Describe and list the sets of acquired data, the methods used to acquire them, the locations. List the results of the quality check	• Initial data collection report • Data description report • Initial data exploration report • Data quality report	
	Release Planning	Data Preparation	X	This is the dataset (or datasets) produced by the data preparation phase. Describe the dataset (or datasets) that will be used for the modeling. List the data to be included/excluded. Describe what decisions and actions were taken to address the data quality problems.	• Rationale for inclusion / exclusion • Data cleaning report • Derived attributes • Records created • Merged data • Reformatted data set • Dataset • Dataset Description	• User Stories • Product Backlog
Game	Iteration Planning + Implementation + Functional Testing	Modeling	✓	Document the actual modeling technique that is to be used. Many modeling techniques make specific assumptions on the data. Describe the intended plan for training, testing and evaluating the models	• Modeling technique • Model assumptions • Test design • Models produced by mining tools	• Spring Backlog • Product & Sprint • Kanban Board • Increment • Agreed Actionable Improvements
			≈	These are the actual models produced by the modeling tool, not a report. Describe the resultant model. revise parameter settings and tune them for the next run in the Build Model task	• Selected parameters • Description of the models • Evaluation of the models • Parameter evaluation	
		Evaluation	≈	This step assesses the degree to which the model meets the business objectives. Summarize the process review and highlight activities that have been missed and/or should be repeated.	• Models evaluated and approved • Process review • List of possible actions • Description of the decision made	
			X	Summarize assessment results in terms of business success criteria	• Valuation of data mining results	
Post-game	Release	Deployment	≈	Summarize deployment strategy. Summarize monitoring and maintenance strategy. This is the final written report of the data mining engagement. There will also often be a meeting at the conclusion of the project. Summarize important experiences made during the project.	• Implementation plan • Maintenance and monitoring • Final report of the project • Final presentation to the client • Documentation of the experience acquired during the development of the project	• Final Release

Fig. 12 CRISP-DM vs. Scrum-XP artifacts

In the symbol column, the symbol X indicates the activities, roles, and artifacts that are not included in the Scrum-XP methodology, which are indispensable in any BDAS project. In turn, it shows us with symbol ≈ the activities, roles, and artifacts that are similar in both methodologies. Finally, the symbol ✓ shows us activities, roles, and artifacts that can be improved in the Scrum-XP methodology.

Scrum Phases	XP Phases	TDSP Phases	Relation Between Methodologies		TDSP Roles	Scrum/XP Roles
Pre-game	Exploration	Business Understanding	≈	The roles fulfill the same functions, which are to manage a team and manage daily activities.	• Project Lead • Project Manager • Team Lead	• Product Owner • Scrum Master • Scrum Team
	Release Planning	Data Acquisition and Understanding	✓	For this section there is a fundamental role that is the Data Engineer who oversees collecting and cleaning the data. The other roles are very similar.	• Project Lead • Team Lead • Project Individual Contributions (Data Engineer)	• Product Owner • Scrum Master • Scrum Team
Game	Iteration Planning + Implementation + Functional Testing	Modeling	X	In this section, the roles of Data Scientist and Data Engineer are fundamental since they oversee designing and training the BDAS model.	• Data scientist • Project Manager • Project Individual Contributions (Application developer, Data Engineer, Data Scientist)	• Scrum Team • Scrum Master • Product Owner
		Deployment	X	The Data Scientist and the Solution Architecture are essential to monitor and establish the architecture design where the BDAS project will run.	• Project Manages • Project Individual Contributions (Data Scientist, Application developer, Data Engineer, Solution Architecture)	
Post-game	Release	Customer acceptance	≈	The Project manager together with the Data Scientist oversee checking that the objectives are met, delivering the necessary documentation, and implementing the project.	• Project Lead • Project Manager • Project Individual Contributions (Data Scientist)	• Product Owner • Scrum Team • Scrum Master

Fig. 13 TDSP vs. Scrum-XP roles

Scrum Phases	XP Phases	TDSP Phases	Relation Between Methodologies		TDSP Activities	Scrum/XP Event
Pre-game	Exploration	Business Understanding	≈	Works with the client to define objectives.	• Define a central Objective.	• Create Project Vision • Develop Epics • Create User Stories
			X	Identify relevant data sources.	• Identify data sources.	
	Release Planning	Data Acquisition and Understanding	X	Produce a clean, high-quality dataset.	• Ingest the data • Explore the data • Set up a data pipeline	• Create Prioritized Product Backlog • Conduct Release Planning
Game	Iteration Planning + Implementation + Functional Testing	Modeling	✓	Create data features and find the model that answers the question most accurately by comparing their success metrics.	• Feature engineering • Model training	• Create Sprint Backlog (Sprint Planning) • Conduct Daily Standup (Daily Scrum) • Increment Development • Review Sprint • Retrospective Sprint
			≈	Determine if your model is suitable for production.	• Model evaluation	
		Deployment	≈	Deploy the model and pipeline to a production or production-like environment for application consumption.	• Operationalize model	
Post-game	Release	Customer acceptance	≈	Confirm that the pipeline, the model, and their deployment in a production	• System validation • Project hand-off	• Ship Deliverables

Fig. 14 TDSP vs. Scrum-XP activities

Regarding the TDSP methodology, we can observe that it is a very complete and well-documented methodology, since it shows in an appropriate way the phases, roles, activities, and artifacts, in an agile and iterative way, compared it with CRISP-DM.

Scrum Phases	XP Phases	TDSP Phases	Relation Between Methodologies		TDSP Artefacts	Scrum/XP Artefacts
Pre-game	Exploration	Business Understanding	≈	A standard template is provided in the TDSP project structure definition.	• Charter Document	• Project Vision Statement
			X	This document provides descriptions of the data that's provided by the client.	• Data Source • Data Dictionaries	
	Release Planning	Data Acquisition and Understanding	X	Includes data summaries, the relationships between each attribute and target, variable ranking, and the solution architecture. The Project can be evaluated before development begins.	• Data Quality Report • Solution Architecture • Checkpoint Decision	• User Stories • Product Backlog
Game	Iteration Planning + Implementation + Functional Testing	Modeling	✓	• It contains pointers to the code to generate the features and a description of how the feature was generated. • For each model that's tried, a standard, template-based report that provides details on each experiment is produced. • Evaluate whether the model performs sufficiently for production.	• Feature Sets • Model Report • Checkpoint Decision	• Spring Backlog • Product & Sprint • Kanban Board • Increment • Agreed Actionable Improvements
		Deployment	≈	• A status dashboard that displays the system health and key metrics. • A final modeling report with deployment details. • A final solution architecture document.	• A status dashboard that displays the system health and key metrics • A final modeling report with deployment details • A final solution architecture document	
Post-game	Release	Customer acceptance	≈	The main artifact produced in this final stage is the Exit report of the project for the customer.	• Exit report of the project for the customer	• Final Release

Fig. 15 TDSP vs. Scrum-XP artifacts

In turn, it retains similarities in phases, activities, and artifacts with those of CRISP-DM. These similarities are based on CRISP-DM but focus on agility and complete the aspect of roles that are an area of improvement in the CRISP-DM methodology. On the other hand, we can observe that Scrum-XP lacks data acquisition and cleaning, which is fundamental in the TDSP methodology, where we observe phases dedicated specifically to this process.

By comparing both methodologies, we can see how agile TDSP is, since it shows us that it has a much smaller number of activities and artifacts than the CRISP-DM methodology and is much more like the number of activities and artifacts that Scrum-XP has but retains the foundation for BDAS project development.

The roles that TDSP handles are very similar to those of Scrum-XP such as the project lead or project manager. In addition, TDSP considers specific roles such as data scientist, data engineer, and solution architecture, among others. These are specific roles in BDAS projects that specifically allow for acquisition phases, data cleanup, and AI modeling. The artifact table shows us how there is a clear difference between TDSP and CRISP-DM where we can see a considerable reduction of artifacts, where TDSP is closer to the Scrum-XP methodology in terms of agility.

5 Discussion of Contributions and Conclusions

In this section, there is a discussion of the theoretical and practical implications, and the conclusions and recommendations for further research are reported.

5.1 Discussion of Contributions

Analyzing two of the most important methodologies in the BDAS area, we realize that both methodologies have areas of opportunity that can be exploited. This fact has led to an increase in the development of agile methodologies for BDAS projects that are useful, easy to use, compatible, and valuable to organizations.

Specifically, the CRISP-DM methodology establishes very well the phases, activities, and artifacts that must be fulfilled in the methodology, but the lack of roles in the methodology gives us an area of improvement for new methodologies. Despite not having these roles, CRISP-DM is the most used methodology for the creation of BDAS projects, as it is a very complete and well-documented methodology that has more than two decades in use. It is the basis for many other modern methodologies. Despite this, it is a methodology that has been created for more than two decades, focused on large companies or large BDAS projects and having a rigorous approach that does not consider the flexibility that most organizations demand today. This is why many organizations prefer to opt for other agile methods or even to develop without any model or methodology.

On the other hand, the methodology of TDSP is a very well-defined and well-documented methodology but unfortunately depends to a large extent on the services and policies of its owner company. This makes it difficult for small and medium-sized enterprises to adopt it on a larger scale, since all the documentation and tools suggested by this methodology are linked to this company.

Another key point of this methodology is that most of its roles and activities are focused on large companies that have many resources and a large staff for this type of project. The clearest example of this can be seen with the variety of roles requested by TDSP, since it organizes the roles according to the management of the team in data science, through different projects.

Finally, the Scrum-XP methodology is a complete methodology that shows us to be one of the most flexible and fast methodologies that exist in the current situation for software engineering projects and even in other types of projects. However, it lacks activities, roles, and artifacts that are indispensable in the development of BDAS projects like everything related to data acquisition, the cleaning of these, and the creation of models, among others. Similarly, we can see it reflected in the roles where Scrum-XP does not have a data engineer, data scientist, and more. This tells us why studies of this methodology have been conducted for BDAS projects but have not been widely used due to the lack of analysis of these techniques for data scientist.

5.2 Conclusions

There is a considerable growth in BDAS-type projects. Several experts agree that very few of these projects are carried out under a methodology; as a consequence, many projects do not reach production. Therefore, there is the lack of clear and well-defined methodologies for the development of this type of projects. As we saw above, there are very few methodologies, and the most used methodology, created more than two decades ago, does not adapt to the changing and flexible needs of today's organizations. However, there are large companies where it is better to use a rigorous methodology (e.g., in very large projects, large organizations, critical projects that depend on human lives, among others).

Given this, we can conclude that the use of a methodology to develop projects of the BDAS type is essential, but there is still a lack of clear, well-defined and well-documented methodologies for the development of such projects. Also, the use of agile methodologies focused on BDAS are essential to meet the needs of certain organizations that have unclear or undefined limitations in the use of some proprietary tools and do not provide the flexibility that developers need.

This research is limited to one agile Scrum-XP software engineering methodology and two BDAS methodologies, and their analysis was based on interpretation of the BDAS research team but looking at objective evidence from official documents of these methodologies.

This gives us guidance for further research regarding the comparison of more BDAS methodologies or the implementation of different agile methodologies to develop this type of projects and not only Scrum-XP. Likewise, it gives us a guideline to identify the need to generate a methodology aimed at free BDAS projects, which are focused on small and medium-sized enterprises, where such methodology will be agile, useful, easy to use, compatible, and not only for a specific project or organization but also for a generic solution for these projects. This can make developers and organizations more efficient in the key points necessary to optimize the development of BDAS projects. Recalling that there is no universal methodology for all types of projects, enterprises can generate competitive advantages, and as the data landscape can change so rapidly, BDAS projects are also changing the methodologies that are already used.

References

1. Martínez-Plumed, F., Contreras-Ochando, L., Ferri, C., Orallo, J.H., Kull, M., Lachiche, N., et al.: CRISP-DM twenty years later: from data mining processes to data science trajectories. IEEE Trans. Knowl. Data Eng. **33**(8), 3048–3061 (2019)
2. Halper, F.: Next-generation analytics and platforms for business success. TDWI Research Report. https://tdwi.org/webcasts/2015/01/next-generation-analytics-and-platforms-for-business-success.aspx. Accessed 13 Dec 2022

3. Walker, J.: Big data strategies disappoint with 85 percent failure rate. Digital J. https://www.digitaljournal.com/tech-science/big-data-strategies-disappoint-with-85-percent-failure-rate/article/508325. Accessed 13 December 2022
4. Mariscal, G., Marban, O., Fernandez, C.: A survey of data mining and knowledge discovery process models and methodologies. Knowl. Eng. Rev. **25**(2), 137–166 (2010)
5. Why do 87% of data science projects never make it into production?, VentureBeat. https://venturebeat.com/ai/why-do-87-of-data-science-projects-never-make-it-into-production/. Accessed 14 Dec 2022
6. Saltz, J., Hotz, N., Wild, D., Stirling, K.: Exploring project management methodologies used within data science teams. In: Paper presented at 24th Americas Conference on Information Systems 2018: digital Disruption, AMCIS 2018. Association for Information Systems (2018, 16–18 Aug)
7. Saltz, J.S.: The need for new processes, methodologies and tools to support big data teams and improve big data project effectiveness. In: Paper presented at 2015 IEEE International Conference on Big Data (Big Data). IEEE (2015, 29 Oct–01 Nov)
8. Ambler, S.W., Lines, M.: The disciplined agile process decision framework. In: Software Quality. The Future of Systems-and Software Development: paper Presented at 8th International Conference, SWQD 2016, Vienna, Austria, Proceedings. Springer International Publishing (2016, Jan 18–21)
9. Davenport, T. H., Dyché, J.: Big data in big companies. International Institute for Analytics, 3(1–31). https://www.iqpc.com/media/7863/11710.pdf. Accessed 14 Dec 2022
10. Larson, D., Chang, V.: A review and future direction of agile, business intelligence, analytics and data science. Int. J. Inf. Manag. **36**(5), 700–710 (2016)
11. Dremel, C., Wulf, J., Herterich, M.M., Waizmann, J.C., Brenner, W.: How AUDI AG established big data analytics in its digital transformation. MISQE. **16**(2), 81 (2017)
12. Baijens, J., Helms, R.W.: Developments in knowledge discovery processes and methodologies: anything new? In: Paper presented at Twenty-fifth Americas Conference on Information Systems (2019, 15–19 Aug)
13. Grady, N.W., Payne, J.A., Parker, H.: Agile big data analytics: AnalyticsOps for data science. In: 2017 IEEE International Conference on Big Data (Big Data), pp. 2331–2339. IEEE (2017, 11–14 Dec)
14. 15th Annual State of Agile Report. Digital.Ai. https://digital.ai/catalyst-blog/15th-state-of-agile-report-agile-leads-the-way-through-the-pandemic-and-digital/. Accessed 15 Dec 2022
15. Piatetsky, G.: CRISP-DM, still the top methodology for analytics, data mining, or data science projects. KDD News. https://www.kdnuggets.com/2014/10/crisp-dm-top-methodology-analytics-data-mining-data-science-projects.html. Accessed 15 Dec 2022
16. Schmidt, C., Sun, W.N.: Synthesizing agile and knowledge discovery: case study results. J. Comput. Inf. Syst. **58**(2), 142–150 (2018)
17. do Nascimento, G.S., de Oliveira, A.A.: An agile knowledge discovery in databases software process. In: Paper presented at Data and Knowledge Engineering: third International Conference, ICDKE 2012, Wuyishan, Fujian, China. Proceedings. Springer, Berlin Heidelberg (2012, 21–23 Nov)
18. Grady, N.W., Payne, J.A., Parker, H.: Agile big data analytics: AnalyticsOps for data science. In: Paper presented at 2017 IEEE international conference on big data (big data). IEEE (2017, 11–14 Dec)
19. Cooper, H.M.: Organizing knowledge syntheses: a taxonomy of literature reviews. Knowl. Soc. **1**(1), 104–126 (1988)
20. Templier, M., Paré, G.: A framework for guiding and evaluating literature reviews. Commun. Assoc. Inf. Syst. **37**(1), 112–137 (2015)
21. Cox, M., Ellsworth, D.: Managing big data for scientific visualization. In: ACM siggraph, vol. 97, pp. 21–38. MRJ/NASA Ames Research Center (1997)
22. Gandomi, A., Haider, M.: Beyond the hype: Big data concepts, methods, and analytics. Int. J. Inf. Manag. **35**(2), 137–144 (2015)

23. Rich S.: Big Data is a "New Natural Resource" http://www.govtech.com/policy-management/Big-Data-Is-a-New-Natural-Resource-IBM-Says.html. Accessed 20 Dec 2022
24. Watson, H.J.: Tutorial: Big data analytics: concepts, technologies, and applications. Commun. Assoc. Inf. Syst. **34**(1), 65 (2014)
25. Chen, H., Chiang, R.H., Storey, V.C.: Business intelligence and analytics: from big data to big impact. MISQ. **36**, 1165–1188 (2012)
26. Lee, T., Lee, H., Rhee, K.H., Shin, U.S.: The efficient implementation of distributed indexing with Hadoop for digital investigations on Big Data. Comput. Sci. Inf. Syst. **11**(3), 1037–1054 (2014)
27. Russom, P.: Big data analytics. TDWI best practices report, fourth quarter. **19**(4), 1–34 (2022) https://tdwi.org/research/2011/09/best-practices-report-q4-big-data-analytics.aspx?tc=page0&tc=assetpg&tc=page0&tc=assetpg&m=1. Accessed 21 Dec 2022
28. Schroeck, M., Shockley, R., Smart, J., Romero-Morales, D., Tufano, P.: Analytics: the real-world use of big data. IBM Global Business Services. Accessed https://www.ibm.com/downloads/cas/VXOJQW1L. 21 Dec 2022
29. Kitchin, R., Lauriault, T.P.: Small data in the era of big data. GeoJournal. **80**(4), 463–475 (2015)
30. Kitchin, R.: The Data Revolution: Big Data, Open Data, Data Infrastructures and Their Consequences. Sage (2014)
31. Katsis, Y., Balac, N., Chapman, D., Kapoor, M., Block, J., Griswold, W.G., et al.: Big data techniques for public health: a case study. In: Paper presented at 2017 IEEE/ACM International Conference on Connected Health: applications, Systems and Engineering Technologies (CHASE), pp. 222–231. IEEE (2017, 17–19 July)
32. Fowler, M., Highsmith, J.: Manifesto for Agile Software Development. https://agilemanifesto.org/. Accessed 22 Dec 2022
33. Campanelli, A.S., Parreiras, F.S.: Agile methods tailoring–a systematic literature review. J. Syst. Softw. **110**, 85–100 (2015)
34. Stavru, S.: A critical examination of recent industrial surveys on agile method usage. J. Syst. Softw. **94**, 87–97 (2014)
35. Tripp, F., Armstrong, D.J.: Agile methodologies: organizational adoption motives, tailoring, and performance. J. Comput. Inf. Syst. **58**(2), 170–179 (2018)
36. Sutherland, J., Schwaber, K.: The scrum guide. The definitive guide to scrum: the rules of the game. Scrum.org. https://scrumguides.org/docs/scrumguide/v2020/2020-Scrum-Guide-US.pdf. Accessed 10 Jan 2023
37. Schwaber, K.: Scrum development process. In: Business Object Design and Implementation, pp. 117–134. Springer, London (1997)
38. Dudziak, T.: Extreme programming an overview. Methoden and Werkzeuge der Software: produktion WS. pp. 1–28 (1999)
39. Chapman, P., Clinton, J., Kerber, R., Khabaza, T., Reinartz, T., Shearer, C., Wirth, R.: CRISP-DM 1.0: step-by-step data mining guide. SPSS inc. **9**(13), 1–73 (2000)
40. Wirth, R., Hipp, J.: CRISP-DM: towards a standard process model for data mining. Paper presented at Proceedings of the 4th international conference on the practical applications of knowledge discovery and data mining. 11–13 April 2000
41. Data Science Process Documentation.: Microsoft Team https://learn.microsoft.com/en-us/azure/architecture/data-science-process/overview. Accessed 23 Jan 2023
42. Marbán, O., Segovia, J., Menasalvas, E., Fernández-Baizán, C.: Toward data mining engineering: a software engineering approach. Inf. Syst. **34**(1), 87–107 (2009)

BDAS-EPM: An Integrated Evolution Process Model for Big Data Analytics Systems

Fen Wang, Tiko Iyamu, Gloria Phillips-Wren, and Jeffrey Yi-Lin Forrest

1 Introduction

Due to the overwhelming amount of web-based, mobile, and sensor-generated data arriving at a terabyte and even exabyte scale, BDA has become a significant research area and has played a critical role in many decision-makings and forecasting domains around the globe [1, 2]. Decision support information, insights, and values can be obtained and derived from the highly detailed, contextualized, and rich contents of relevance to any business with the appropriate BDA tools, platforms, and systems [3–5]. Indeed, organizations have become more competitive through the proper use of BDA systems in this big data era [5, 6].

BDA systems generally refer to software systems designed and developed to gather from multiple sources, manage, analyze, and transform massive volumes of data into valuable insights to support decision-making in large organizations [7]. Although successful BDA systems have been observed in diverse industrial domains, they have been mostly developed by and utilized in large business organizations due to the complex nature of the systems and extensive resources required for such deployment [8]. Furthermore, a lack of cohesive and systematic development of as

F. Wang (✉)
Central Washington University, Ellensburg, WA, USA
e-mail: fen.wang@cwu.edu

T. Iyamu
Cape Peninsula University of Technology, Cape Town, South Africa

G. Phillips-Wren
Loyola University Maryland, Baltimore, MD, USA

J. Y.-L. Forrest
Slippery Rock University, Slippery Rock, PA, USA

M. Mora et al. (eds.), *Development Methodologies for Big Data Analytics Systems*, Transactions on Computational Science and Computational Intelligence, https://doi.org/10.1007/978-3-031-40956-1_7

well as a high failure rate of BDA systems projects in practice has been reported recently [9].

Motivated by the flourishing opportunities as well as the lack of practical transference of deploying BDA systems in the big data era, we conduct a selective review [10, 11] on the BDA systems evolution, from theory to practice. This includes assessing the BDA systems techniques, frameworks, and emerging trends with the aim of providing a summary of core concepts, a succinct but valuable description, and an account for addressing the big data challenges and enhancing its opportunities. Accordingly, this chapter presents a big data analytics systems evolution process model (BDAS-EPM), which is an integrated and organized view of the BDA systems and techniques. The framework can be adopted by and guide organizations in achieving their goals and objectives with the use of appropriate BDA systems and solutions. Centered on the BDAS-EPM, the chapter offers a set of practical recommendations for data analysts, data scientists, and data architects including the executives and leaders in organizations, in their strategic and operational pursuits of innovative advancement and competitive edges.

The remainder of this chapter is structured as follows: In Sect. 2, a background overview of the key concepts, terms, and techniques of BDAS is depicted from the descriptive, predictive, and prescriptive landscape; the specification of the selective review research method as well as the research questions is explained in Sect. 3; results derived from the selective review are reported in Sect. 4, including the BDAS-EPM and the key characteristics from core studies as well as a discussion of implications and insights for applying the model in practice; and finally, in Sect. 5, we conclude with the research limitations, recommendations, and conclusions.

2 Background Overview

The convergence of near-real-time network speed, inexpensive massive storage of digital data, cloud computing technologies, and advances in artificial intelligence in the early twenty-first century enabled the creation, acquisition, storage, and analysis of "big data." The term "big data" was coined by Roger Magoulas from O'Reilly media in 2005 [19] to differentiate the new and increasingly large and complex datasets that could not be effectively managed with traditional technologies such as relational databases. For instance, healthcare data that include clinical tests, patient records, images, physician notes, genome sequencing, and medications can be combined and analyzed to study population response to treatments and to enable breakthroughs in medicine. The development of a successful COVID-19 vaccine in an astounding 9 months during 2020, compared to an average of 10–15 years for traditional vaccine development [20, 21], is an example of the use of big data such as genome sequencing to address a "wicked" problem [22] at speed.

There is no agreed-upon definition of big data (for various definitions, see, e.g., [23]). To provide a working definition, we accept the one offered by SAS [24] for the purposes of this chapter: "Big data is a term that describes large, hard-to-manage volumes of data – both structured and unstructured – that inundate businesses on a day-to-day basis." The concept of big data is better described in terms of the data's primary characteristics, sometimes referred to as the 3Vs: *volume*, *velocity*, and *variety* [18, 28]. Volume refers to the massive and growing amount of data with its large number of variables and voluminous number of observations. Velocity refers to the speed of data creation, change, and acquisition. Variety refers to the many types of data. A fourth V, Value, was recently proposed and added to the concept that refers to the low value density in contrast to the huge volume of the large dataset. Figure 1 summarizes and depicts the 4Vs of big data. Data types range from structured numeric data that can be effectively coded into relational databases to unstructured imagery. In between these two types are semi-structured data that have features of both; for example, textual data has some structure, and yet one needs context to understand the meaning. Some researchers and organizations add additional Vs such as variability and veracity (see, e.g., [24]) to draw attention to specific concerns such as data changeability and data quality within a specific industry and across the entire business world. In the current chapter, we focus on volume, velocity, variety, and value as the foundational description of big data.

Of course, big data by itself is not sufficient to drive insight. While the traditional term "analysis" is used to describe the mathematical and statistical methods for data investigation, the terms "data science," "business intelligence," and "analytics" have been associated with the analysis of big data for the purpose of deriving insights and supporting decision-making [25]. Although there are subtle differences in the meaning of the terms, they encompass a specialized set of methods needed to analyze and

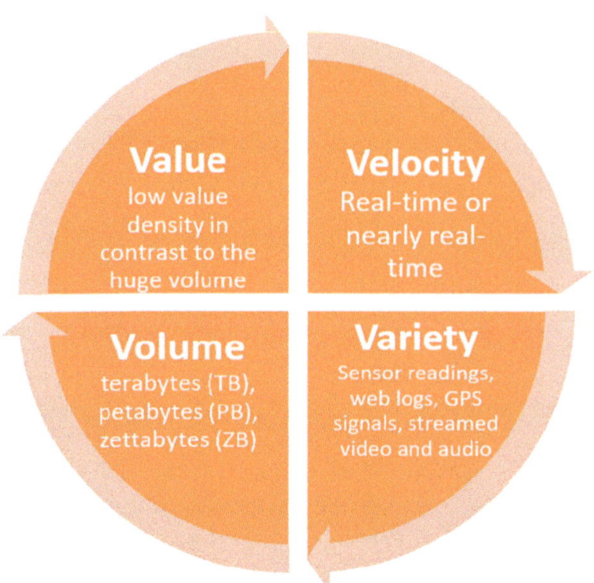

Fig. 1 The 4Vs characteristics of big data

use big data. For example, the extraction and query of big data demands more powerful tools and languages such as GraphQL used by Facebook, the visualization of big data requires new methods and software such as Tableau, and machine learning is key to discerning patterns and developing predictions from big data. Our context in this chapter is business; thus, we utilize the definition of business analytics offered by Delen and Ram [26, p. 3] as "big data analytics": "Business analytics is the art and science of discovering insight – by using sophisticated mathematical, statistical, machine learning, and network science methods along with a variety of data and expert knowledge – to support better and faster/timely decision making."

Analytics can be utilized to uncover patterns and deliver insights from big data, particularly using machine learning and artificial intelligence techniques [27]. Users employ analytics with a purpose of developing descriptive, predictive, or prescriptive models that provide insight for decision-making [18]. Descriptive analytics uses statistical techniques such as clustering or association to describe the current data and find linkages between variables. Predictive analytics are commonly used to predict a future state based on analysis of past patterns and behaviors. Techniques in predictive analytics include linear and logistic regression, decision trees, and neural networks. Prescriptive analytics attempt to delineate optimal outcomes using mathematical and computer science techniques such as optimization or genetic algorithms. Many companies today focus on descriptive analytics in executive dashboards and predictive analytics to gain an estimate of a future state based on past behavior. However, "institutional support remains the biggest barrier to AI adoption" [27] and the concomitant use of big data analytics. We aim to help overcome organizational barriers by describing a process to incorporate big data analytics into decision-making through the BDAS-EPM described in this chapter.

3 Selective Review Research Method

The aim of this research is to develop an integrative selective review and synthesis of the BDA systems techniques, frameworks, and applications to assess the BDA systems evolution, from theory to practice. For this aim, we establish the following specific research questions: RQ.1 What are the main concepts and evolution of BDA systems? RQ.2 What are the most relevant BDA systems frameworks? RQ.3 What are the main domains of applications reported for BDA systems? RQ.4 What are the main trends and challenges for effective decisional support with BDA systems?

To address these four research questions, we conducted a selective review of the scientific literature on the core topics of the BDA systems. A selective literature review is a descriptive and literature analysis research method [10] that addresses only a small sample of the most relevant studies on a topic of interest [12]. This methodology is appropriate in that the focus and scope of the current study is on examining the top 20 most recent and relevant studies with established high quality in the field to address the specified four research questions on the theoretical and practical BDA systems evolution. To achieve this selective literature review, we consider a worthy inclusion strategy to include relevant academic-oriented studies and high-quality professional literature for the 2000–2021 period. Big data phenomenon started at the dawn of the twenty-first century and became relevant ever since, while analytics and big data clearly emerged in the recent 2010–2022 period. The inclusion criteria for academic-oriented documents were established as follows:

C.1) Type of document (journal article, research-oriented book, or conference paper)
C.2) Status of authors (well-recognized in the topics of BDA and/or BDA systems)
C.3) Quality of publisher (document is published by a well-recognized international scientific editorial company)
C.4) Citations (the document is highly cited or the document was considered relevant despite a low number of citations; a very recently published paper may not have a very high citation count due to the limited time span of publication)

Accordingly, the inclusion criteria for high-quality professional literature were the following ones:

C.1) Journal impact factor (IF > = 3)
C.2) Type of source (the publishing organization is well-recognized in the topics of BDA and/or BDA systems)
C.3) Value of document (the document has been previously cited in relevant academic papers)

Tables 1 reports the list of the 20 selected papers in the period 2010–2022 on BDA and/or BDA systems, respectively, from the academic and professional literature.

4 Results and Synthesis

The systematized synthesis and analysis from the study are organized and elaborated into the following subsections, to answer the four research questions RQ.1 through RQ.4 as posed earlier in the chapter.

4.1 The Main Concepts and Evolution of BDA

Rapidly, the concept of BDA has, in all sectors, evolved over the years. The concept began as legacy systems, relational database management, and database warehousing using queries and reporting tools. In its evolution, the concept evolves from the timid and conservative forms of data collection and use to unprecedented datasets that require more sophisticated solutions ranging from web analytics and mining to cloud-based computing. This end entails new forms of dimensionality in data collection, processes, and analyses, which necessitate an array of analytics, such as descriptive, diagnostic, predictive, and prescriptive [13, 14]. Through its

Table 1 List of the 20 selected studies on BDA systems

ID	Relevance	Selected article citations	C.1 latest if >5?	C.2 index	C.3 citations >50?
1	RQ.1	Chen, H., Chiang, R. H., & Storey, V. C. (2012). Business intelligence and analytics: From big data to big impact. *MIS Quarterly*, 1165–1188	8.513	JCR	6075
2	RQ.1	Holsapple, C., Lee-Post, A., & Pakath, R. (2014). A unified foundation for business analytics. *Decision support systems*, *64*, 130–141	6.969	JCR	362
3	RQ.1	Abbasi, A., Sarker, S., & Chiang, R. H. (2016). Big data research in information systems: Toward an inclusive research agenda. *Journal of the association for information systems*, *17*(2), 3	5.346	JCR	610
4	RQ.1	Liang, T. P., & Liu, Y. H. (2018). Research landscape of business intelligence and big data analytics: A bibliometrics study. *Expert systems with applications*, *111*, 2–10	8.665	JCR	98
5	RQ.1	Niu, Y., Ying, L., Yang, J., Bao, M., & Sivaparthipan, C. B. (2021). Organizational business intelligence and decision making using big data analytics. *Information Processing & Management*, 58(6), 102,725	6.222	SCOPUS	25
6	RQ.2	Grover, V., Chiang, R. H., Liang, T. P., & Zhang, D. (2018). Creating strategic business value from big data analytics: A research framework. *Journal of management information systems*, 35(2), 388–423	7.582	SCOPUS	601
7	RQ.2	Osman, A. M. S. (2019). A novel big data analytics framework for smart cities. *Future generation computer systems*, 91, 620–633	7.187	SCOPUS	184
8	RQ.2	Ullah, F., & Babar, M.A. (2019). Architectural tactics for big data cybersecurity analytics systems: A review. *Journal of systems and software*, 151, 81–118	3.514	SCOPUS	75
9	RQ.2	Mohamed, A., Najafabadi, M.K., Wah, Y.B., Zaman, E., & Maskat, R. (2020). The state of the art and taxonomy of big data analytics: View from new big data framework. *Artificial intelligence review*, 53, 989–1037	9.588	SCOPUS	84
10	RQ.2	Ahmed, I., Ahmad, M., Jeon, G., & Piccialli, F. (2021). A framework for pandemic prediction using big data analytics. *Big data research*, 25	3.739	SCOPUS	39
11	RQ.3	Choi, T. M., Wallace, S. W., & Wang, Y. (2018). Big data analytics in operations management. *Production and operations management*, 27(10), 1868–1883	11.251	SCOPUS	439

(continued)

Table 1 (continued)

ID	Relevance	Selected article citations	C.1 latest if >5?	C.2 index	C.3 citations >50?
12	RQ.3	Jiang, D., Wang, Y., Lv, Z., Qi, S., & Singh, S. (2019). Big data analysis based network behavior insight of cellular networks for industry 4.0 applications. *IEEE transactions on industrial informatics*, 16(2), 1310–1320	11.648	SCOPUS	157
13	RQ.3	Galetsi, P., Katsaliaki, K., & Kumar, S. (2020). Big data analytics in health sector: Theoretical framework, techniques and prospects. *International journal of information management*, 50, 206–216	18.958	SCOPUS	115
14	RQ.3	Kaffash, S., Nguyen, A. T., & Zhu, J. (2021). Big data algorithms and applications in intelligent transportation system: A review and bibliometric analysis. *International journal of production economics*, 231, 107,868.	11.251	SCOPUS	86
15	RQ.3	Wang, J., Xu, C., Zhang, J., & Zhong, R. (2021) Big data analytics for intelligent manufacturing systems: A review. *Journal of manufacturing systems*	9.498	SCOPUS	53
16	RQ.4	Demirkan, H., & Delen, D. (2013). Leveraging the capabilities of service-oriented decision support systems: Putting analytics and big data in cloud. *Decision support systems*, *55*(1), 412–421	6.969	JCR	670
17	RQ.4	Maass, W., Parsons, J., Purao, S., Storey, V. C., & Woo, C. (2018). Data-driven meets theory-driven research in the era of big data: Opportunities and challenges for information systems research. *Journal of the Association for Information Systems*, 19(12), 1	5.346	SCOPUS	73
18	RQ.4	Dai, H. N., Wong, R. C. W., Wang, H., Zheng, Z., & Vasilakos, A. V. (2019). Big data analytics for large-scale wireless networks: Challenges and opportunities. *ACM computing surveys (CSUR)*, 52(5), 1–36.	14.324	JCR	95
19	RQ.4	Escobar, C. A., McGovern, M. E., & Morales-Menendez, R. (2021). Quality 4.0: a review of big data challenges in manufacturing. *Journal of intelligent manufacturing*, 32(8), 2319–2334	7.136	SCOPUS	29
20	RQ.4	Jagatheesaperumal, S. K., Rahouti, M., Ahmad, K., Al-Fuqaha, A., & Guizani, M. (2021). The duo of artificial intelligence and big data for industry 4.0: Applications, techniques, challenges, and future research directions. *IEEE internet of things journal*	10.238	SCOPUS	14

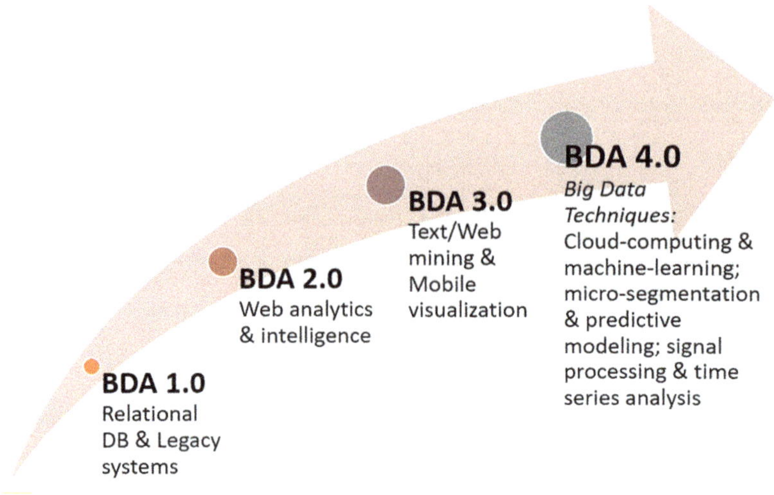

Fig. 2 BDA evolution phases and main techniques. (Adapted from Ref. [15])

sophistication, the evolution of BDA is changing the computing landscape and advancing analysis techniques along the phases 1.0 through 4.0 [15–17]. Figure 2 illustrates the BDA evolution process with main concepts and techniques highlighted to address the RQ.1 What are the main concepts and evolution of BDA systems.

The evolution of BDA is attributed to the availability of rich sources of data enabled by cloud storage and processing, virtualization, fast Internet speeds, and new methods of analysis such as visualization [18, 29]. This progression has made distributed computing and more sophisticated hardware a reality in many companies [30, 31]. Although BDA has evolved significantly, applying available technologies differs in both the business and technology perspectives from one organization to another. Escobar, McGovern, and Morales-Menendez [32] suggest that quality is one of the most relevant challenges of BDA evolution. Although some organizations have been able to demonstrate added value through BDA solutions, in many organizations the challenges posed by BDA remain tough, critical, and disruptive. Methodologies needed to manage overall BDA projects are still evolving including interdependent management of the data science team, the project, and data/information [29]. The challenges are often underpinned by the inability to select the most appropriate BDA frameworks and techniques for organization-specific purposes. To help data scientists resolve these and other challenges, Martinez et al. [29] propose a conceptual framework for practitioners of BDA to choose an appropriate methodology to manage data projects with a holistic point of view. In short, there is a persistent need to gain insights into the evolution of the BDA problem, examine why and how derived and data-based solutions can be deployed, and understand the contribution and value to the organization.

4.2 The Most Relevant BDA Frameworks

As the interest in the BDA concept and practice increases, the development of frameworks grows. The developed frameworks provide critical guide for the adoption, use, and management of BDA. This includes structure, capability, and criteria from both technology and nontechnical (human and process) perspectives. Thus, the requirements underlying each particular framework are critical for an organization to determine which framework is relevant and most appropriate for materializing its goal and objectives. Currently, there are five most popular BDA frameworks, which are Hadoop, Spark, Flink, Storm, and Samza [13, 14, 33]. Niu et al. [16] argue that, although BDA reveals more organizational insights and strengthens decision-making frameworks, it exposes potential risks. Figure 3 depicts the comparison of the five basic BDA processing frameworks.

Primarily, BDA frameworks are intended to holistically structure and integrate functions and measurable values and ensure flexibility while making produced products independent. Some of the common strengths among the frameworks include rapid collection of data, visualization of information, detection of deficiency, and effective assessment [34]. Thus, each of the frameworks should consist of the business and technical requirements, strategies, technical tools, and capabilities, to embed a successful BDA practice through the selection of most appropriate tools and techniques in an organization. However, the challenges remain in the selection, deployment, and practice of the frameworks, including the integration and management systems in a unified platform [35].

4.3 Applications of BDA

BDA are highly applied and can leverage challenges and opportunities presented by big data and domain-specific analytics needed in many high-impact application areas [3, 36, 37]. Based on the selective review, several domains were identified to examine the current and potential values BDA systems and applications can derive in the big data era, including operations management, telecommunications, healthcare, transportation management, and manufacturing systems. Figure 4 summarizes the promising and high-impact BDA applications in the five selected domains.

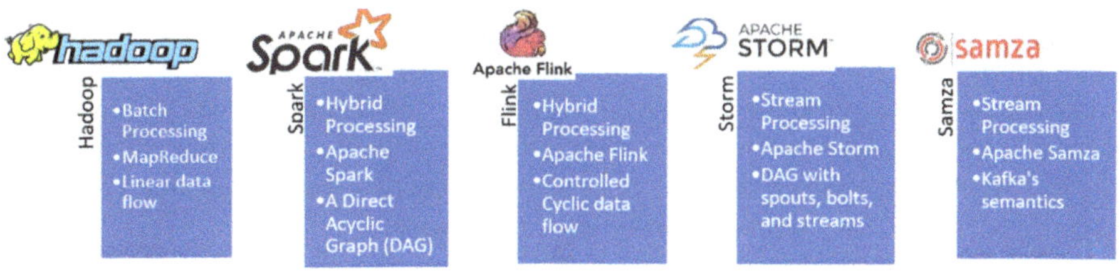

Fig. 3 Comparison of BDA system frameworks

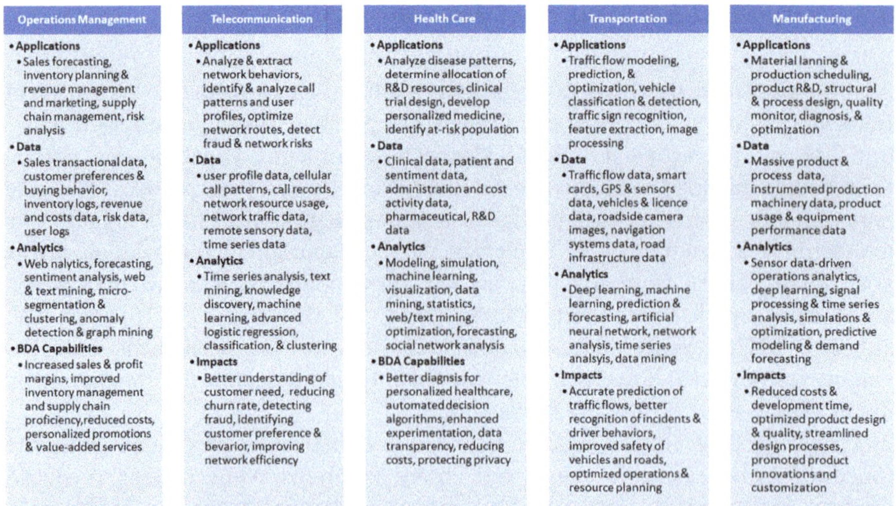

Fig. 4 BDA applications in five domains

The main application elements of BDA systems include a dataset, analytics integration, and architecture [31]. The connectedness of the domains is based on process. Sequentially, each process guides the connection and relationship between the domains. First, organizations employ big data analytics systems to formulate data-driven decisions that can improve sustainability and competitiveness. Second, the element "integration" enables effective complementary use of BDA systems, which offers operational efficiency and transformational advances. Third, the element "architecture" facilitates and ensures coexistence, flexibility, and compatibility of the adopted BDA systems. Dai et al. [38] suggest that despite the understanding of the domains, challenges remain and sometimes derail BDA deployment. To increase level of success in the deployment of BDA, complexities must be reduced through a process-driven approach [32] to ensure that the most appropriate frameworks and techniques are selected.

Each BDA application often consists of sets of data collected both internally and externally to an organizational system, enacted by requirements. This makes the data sources central, integration inevitable, and the architecture critical. Kaffash, Nguyen, and Zhu [36] reveal integration as a crucial aspect of BDA applications and suggest that models can assist to bridge this gap. Also, the sources could include both quantitative and qualitative data, which increases the quality of data, to enhance decision-making processes in an organization. Each application of BDA frameworks helps to provide an in-depth analysis that is useful for sustainability, competitiveness, and management purposes [34]. However, some organizations attempt to deploy BDA frameworks without developing necessary processes that uniquely interconnect domains, which causes many challenges.

4.4 BDA Challenges and Trends

Despite the premise, BDA is compounded with challenges, and the trends are viewed from both positives and negatives perspectives. Some of the challenges include the quantity and multiplicity of data sources and applications [33]. High quantity overwhelms and makes it challenging to holistically demystify the datasets, which can have negative impacts on the eventual decision-making. Huge amounts of data usually come from multiple sources and are often disjointed. Such multiplicity can lead to incomplete or inaccurate analysis of data and affect speedy decision-making. A relevant advancement is the state-of-the-art in handling the data extraction, data cleaning, data integration, data versioning, and metadata management with the introduction of data lakes, viewed mostly as intermediate repositories for big data [39]. According to Choi, Wallace, and Wang [40], the current trend includes machine learning, deep learning, data mining, and optimization. These trends are determined by both business and technology requirements. The trends focus on data-driven decisional support, which increases the amount and speed at which big data is collected, cleaned, and integrated.

BDA is a process-oriented approach that focuses on extracting relevant and useful insights from data for organizations to potentially materialize their objectives. In many organizations, it is used to guide managers to acquire deeper insights about processes and decision-making toward the efficiency of business operations and the development and adoption of strategies. In building BDA capabilities, datasets present three critical challenges in the forms of data quality, data integration, and data security [14]. These challenges can only be addressed through a process-oriented model that is based on governance. In addressing the challenging nature of BDA, solutions should provide platform scalability and logical flexibility [41] and must be process-oriented.

4.5 Illustration and Discussions on the BDAS-EPM

Based on the results presented above, a big data analytics systems evolution process model (BDAS-EPM) is developed, as shown in Fig. 5. The model (BDAS-EPM) is process-oriented and consists of five main phases, P^1 … P^5. The relationship and interactions between the phases are through processes. Also, each component of the phases can only be executed through processes.

It is well documented that many BDA frameworks exist, as comprehensively discussed in the earlier sections. With respect to structured and unstructured data, Papadopoulos et al. [42] developed a framework for organizational resilience. Lopez-Cuevas et al. [43] proposed a framework to analytically reveal disruptive events. A framework proposed by Grover et al. [14] focuses on the linkage between capabilities and value of BDA. From the analysis presented above, EPM for BDAS

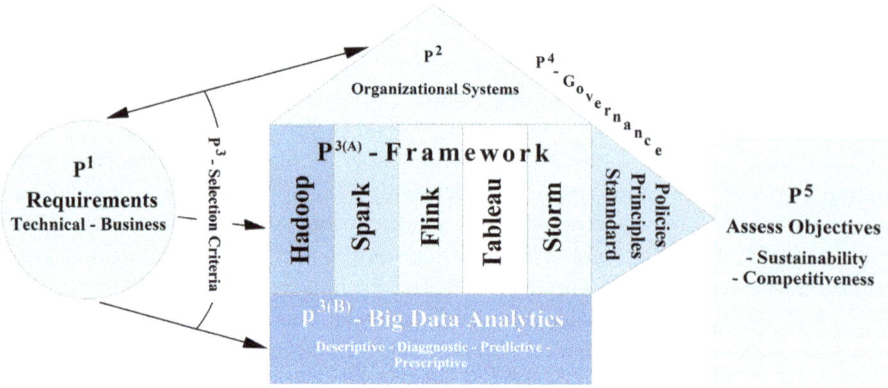

Fig. 5 Big data analytics systems evolution process model (BDAS-EPM)

is developed. The discussion that follows should be read with the BDAS-EPM in mind in order to gain a better understanding.

Requirements: The requirements dictate the adoption of strategies, which include the selection of appropriate BDA framework and analytics technique. From the frameworks' perspectives, many developed methods are not flexible and scalable enough to adapt to the requirements of BDA [40]. Thus, it is critical and necessary to develop a process model for implementation and practice purposes.

Organizational systems: Organizations host many computer systems, which bring about unprecedented complexity. Usually, the complexity emanates from various aspects, such as heterogeneity of data and diversity in the systems. Processes facilitate the classification features and functions including the relationships between applications [32].

Criteria for selection: The list of BDA frameworks and products is growing over time. This includes Hadoop, Apache Mahout, and Storm [14] and machine learning (ML) libraries, Spark ML and Flink ML [33]. Thus, criteria are essential. In the third phase, criteria are formulated based on the requirements. The set of criteria is used to select the most appropriate framework (P3, A) and BDA (P3, B). Framework: Wang et al. [33] argue that each application requires a decision-making mechanism to facilitate the deployment and practice of BDA for an organization. The process-oriented nature of this model does not only facilitate practice, but it also enables complementarity of multiple frameworks on the same platform to enhance a strong analytics capability. BDA: The efficiency of BDA application focuses on the transformation and improvement of the adopted strategy and implemented operations of an organization [16]. Thus, the selection of applications is fundamental to the success of BDA practice. Chen, Chiang, and Storey [3] argue that the selection of appropriate applications helps with the optimization of employed techniques.

Governance: The governance provides policies, principles, and standards, through which the BDA evolution can be implemented in practices, to support data-driven decision-making processes in achieving the objectives of an organization. Dai et al. [38] explain how an integration of diverse types of data and their heterogeneity are cumbersome for frameworks and techniques. Critically, this challenge necessitates the appearance and existence of governance to guide the deployment and use and management of BDA for organizational purposes.

Assess objectives: BDA projects, like all organizational projects, need to be continually assessed as to whether they are evolving and improving to meet the objectives and needs of the organization [44]. Organizations should determine if the project is offering business value such as organizational agility, sustainable competitive advantage, and knowledge creation that can be realized as financial or strategic performance and adjust their approach over time [45]. Specifically, assessing the objectives is guided by the governance criteria. It provides ability to learn more about the process, which inevitably causes evaluations of integral part of an organization, which include value, strategy, and quality [46]. Based on the governance activities, as shown in the BDAS-EPM (Fig. 5), the objectives are assessed from two main fronts. First, the assessment is aimed at gaining better understanding of how the objectives can be used to improve competitiveness. This is done by observing the conditions that are triggered and the measurement of actions that are executed [47]. Second, the assessment of the objectives provides insights on the levels of sustainability [48]. This enables an organization to initiate change effort, which includes adoption of appropriate tools and selection (upskilling) of skill set, for managing events and activities toward sustainability. Some studies in the manufacturing sector have shown that the adoption of BDA has a positive impact on knowledge management, green purchasing, and operational capabilities [49], demonstrating that organizational objectives such as sustainability can also be influenced.

The BDAS-EPM, in its process-oriented approach, draws the model from theoretical to a more practical solution by focusing on the steps, sequentially. According to Galetsi, Katsaliaki, and Kumar [50], such sequential approach is systematic and focuses on practice. A process orientation is exhaustive through detailing the relevant steps. The BDAS-EPM helps to consistently and uniformly implement and practice BDA as it continues to evolve from various perspectives, such as distributed computing, machine learning, and artificial intelligence. Centered on the proposed BDAS-EPM, this chapter presents the following recommendations to executives and leaders interested in such productive yet challenging investments:

- Defining *BDA requirements and selection criteria* is the most important step in the process and should involve multiple user groups and use cases. In addition, the assessment of an enterprise's culture and its ability to change or to become more data-driven in decision-making is needed before beginning to develop requirements.
- *BDA analytics frameworks and tools* embed their individual processes that ingest and transform data into information and knowledge. Those embedded processes may be similar to or quite different from those used in the enterprise. Thus, needed are an assessment of the match or mismatch and a realistic determination of the ability and willingness of the enterprise to modify current practices. In addition, each use of BDA tools necessitates data cleansing and robust datasets that must be accomplished before deployment.

- BDA systems require *clear governance and continual management* to define access, scope, security, and legal obligations, among others. Thus, there is an ecosystem that should be created and that will require oversight.
- The deployment of BDA systems assumes that value can be derived for the enterprise. *Continual assessment* of the quality of results and the ultimate transformation into value will be needed. In essence, an ongoing improvement loop will be critical to developing trust in the system and ensuring that BDA is delivering on the intended promise.

5 Conclusion

Similar to many other technological innovations, BDA's power does not eliminate the need for human insight or vision [3]. In fact, the managerial challenges underlying the unleash of this power could be even greater than the technical challenges of applying BDA systems and techniques in today's organizations and reaping the optimal benefits of the transition [23, 51, 52]. BDA offers an opportunity of converting the massive data that already exists in most enterprises into useful information and actionable knowledge. However, the opportunities offered by BDA necessitate the appropriate processes that must be used to interface with existing organizational systems or to develop new capabilities. As demonstrated in the BDAS-EPM, each BDA process can be aligned with existing business intelligence tools and analytics frameworks that are used to provide intelligent aids in the big data era for organizational processes.

In this chapter, we have conducted a selective review on the BDA systems evolution phases and techniques, frameworks, applications, and emerging trends to address big data challenges and opportunities. Along with the research, we developed an organized synthesis of BDA and presented an integrated BDAS-EPM to help organizations adopt or develop appropriate BDA solutions to derive desired impacts, advance the appropriateness, and increase the usefulness of big data in achieving their organizational goals and objectives. This paper also elicits a set of practical recommendations for data scientists and data architects along with executives and leaders in organizations who are pursuing strategic and operational advantages. Given the continuously growing prominence and complexity of the big data phenomenon, we encourage and anticipate future research efforts to establish updated conceptualizations and frameworks to cope with this complex yet critical subject.

References

1. Kesavan, S., Kushwaha, T.: Field experiment on the profit implications of merchants' discretionary power to override data-driven decision-making tools. Manag. Sci. **66**(11) (2020)
2. Mohamed, A., Najafabadi, M.K., Wah, Y.B., Zaman, E., Maskat, R.: The state of the art and taxonomy of big data analytics: view from new big data framework. Artif. Intell. Rev. **53**, 989–1037 (2020)
3. Chen, H., Chiang, R.H., Storey, V.C.: Business intelligence and analytics: from big data to big impact. MIS quart. **36**(4), 1165–1188 (2012)
4. Mashingaidze, K., Backhouse, J.: The relationships between definitions of big data, business intelligence and business analytics: a literature review. Int. J. Bus. Inf. Sys. **26**(4) (2017)
5. Ukhalkar, P.K., Phursule, R.N., Gadekar, D.P., Sable, N.P.: Business intelligence and analytics: challenges and opportunities. Int. J. Adv. Sci. and Tech. **29**(12s), 2669–2676 (2020)
6. Ranjan, J., Foropon, C.: Big data analytics in building the competitive intelligence of organizations. Int. J. Inf. Manag. **56** (2021). https://doi.org/10.1016/j.ijinfomgt.2020.102231
7. Davoudian, A., Liu, M.: Big data systems: a software engineering perspective. ACM Comput. Surv. (CSUR). **53**(5), 1–39 (2020)
8. Davenport, T., Bean, R.: The Quest to Achieve Data-Driven Leadership: a Progress Report on the State of Corporate Data Initiatives – Foreword. Special Report, New Advantage Partners (2022)
9. Davenport, T., Malone, K.: Deployment as a critical business data science discipline. Harvard Data Sci. Rev. (2021). https://doi.org/10.1162/99608f92.90814c32
10. Glass, R., Ramesh, V., Vessey, I.: An analysis of research in computing disciplines. Commun. ACM. **47**(6), 89–94 (2004)
11. Webster, J., Watson, R.: Analyzing the past to prepare for the future: writing a literature review. MIS Quart. **26**(2) (2002)
12. Brereton, P., Kitchenham, B.A., Budgen, D., Turner, M., Khalil, M.: Lessons from applying the systematic literature review process within the software engineering domain. J. Syst. Software. **80**(4), 571–583 (2007)
13. Ullah, F., Babar, M.A.: Architectural tactics for big data cybersecurity analytics systems: a review. J. Syst. Software. **151**, 81–118 (2019)
14. Grover, V., Chiang, R.H., Liang, T.P., Zhang, D.: Creating strategic business value from big data analytics: a research framework. J. manage. inf. syst. **35**(2), 388–423 (2018)
15. Wang, F., Raisinghani, M.S., Mora, M., & Forrest, J.YL: Effective decision support in the big data era: optimize organizational performance via BI&A. Int. J. Decis. Support Syst. Tech., 14(1), (2022)
16. Niu, Y., Ying, L., Yang, J., Bao, M., Sivaparthipan, C.B.: Organizational business intelligence and decision making using big data analytics. Inf. Process. Manag. **58**(6), 102725 (2021)
17. Holsapple, C., Lee-Post, A., Pakath, R.: A unified foundation for business analytics. Decis. Support. Syst. **64**, 130–141 (2014)
18. Phillips-Wren, G., Daly, M., Burstein, F.: Reconciling business intelligence, analytics and decision support systems: more data, deeper insight. Decis. Support. Syst. **146**, 113560 (2021)
19. Halevi, G., Moed, H.F.: The evolution of big data as a research and scientific topic: overview of the literature. Res. Trend. **1**(30), 2 (2012)
20. Defendi, H.G.T., da Silva Madeira, L., Borschiver, S.: Analysis of the COVID-19 vaccine development process: an exploratory study of accelerating factors and innovative environments. J. Pharm. Innovation. **17**(2), 555–571 (2022)
21. Rahman, M., Masum, M., Ullah, H., Wajed, S., Talukder, A.: A comprehensive review on COVID-19 vaccines: development, effectiveness, adverse effects, distribution and challenges. Virus Dis. **1-22** (2022)
22. Churchman, C.: Wicked problems. Manag. Sci. **14**(4), B-141–B-146 (1967)
23. Mikalef, P., Pappas, I.O., Krogstie, J., Giannakos, M.: Big data analytics capabilities: a systematic literature review and research agenda. Inf. Syst. e-Bus. Manag. **16**(3), 547–578 (2018)

24. SAS: Accessed from https://www.sas.com/en_us/insights/big-data/what-is-big-data.html#history (2022)
25. Davenport, T.H.: How strategists use "big data" to support internal business decisions, discovery and production. Strat. Leader. **42**(4), 45–50 (2014)
26. Delen, D., Ram, S.: Research challenges and opportunities in business analytics. J. Bus. Analytics. **1**(1), 2–12 (2018)
27. Magoulas, R., Swoyer, S.: AI Adoption in the Enterprise. O´ Reilly. Available from https://get.oreilly.com/rs/107-FMS-070/images/AI-Adoption-in-the-Enterprise-2020.pdf, Beijing (2020)
28. Phillips-Wren, G., Iyer, L., Kulkarni, U., Ariyachandra, T.: Business analytics in the context of big data: a roadmap for research. Comm. Assoc. Inf. Syst. **37**(23) (2015)
29. Martinez, I., Viles, E., Olaizola, I.G.: Data science methodologies: current challenges and future approaches. Big Data Res. **24**, 100183 (2021)
30. Jagatheesaperumal, S.K., Rahouti, M., Ahmad, K., Al-Fuqaha, A., Guizani, M.: The duo of artificial intelligence and big data for industry 4.0: applications, techniques, challenges, and future research directions. IEEE Internet Things J. (2021)
31. Maass, W., Parsons, J., Purao, S., Storey, V.C., Woo, C.: Data-driven meets theory-driven research in the era of big data: opportunities and challenges for information systems research. J. Assoc. Inf. Syst. **19**(12), 1253–1273 (2018)
32. Escobar, C.A., McGovern, M.E., Morales-Menendez, R.: Quality 4.0: a review of big data challenges in manufacturing. J. Intell. Manuf. **32**(8), 2319–2334 (2021)
33. Wang, J., Xu, C., Zhang, J., Zhong, R.: Big data analytics for intelligent manufacturing systems: a review. J. Manuf. Syst. **62**, 738–752 (2021)
34. Ahmed, I., Ahmad, M., Jeon, G., Piccialli, F.: A framework for pandemic prediction using big data analytics. Big Data Res. **25**, 100190 (2021)
35. Abbasi, A., Sarker, S., Chiang, R.H.: Big data research in information systems: toward an inclusive research agenda. J. Assoc. Inf. Syst. **17**(2), 1–32 (2016)
36. Kaffash, S., Nguyen, A.T., Zhu, J.: Big data algorithms and applications in intelligent transportation system: a review and bibliometric analysis. Int. J. Prod. Econ. **231**, 107868 (2021)
37. Jiang, D., Wang, Y., Lv, Z., Qi, S., Singh, S.: Big data analysis based network behavior insight of cellular networks for industry 4.0 applications. IEEE Trans. Industr. Inform. **16**(2), 1310–1320 (2019)
38. Dai, H.N., Wong, R.C.W., Wang, H., Zheng, Z., Vasilakos, A.V.: Big data analytics for large-scale wireless networks: challenges and opportunities. ACM Comput. Surv. (CSUR). **52**(5), 1–36 (2019)
39. Nargesian, F., Zhu, E., Miller, R.J., Pu, K.Q., Arocena, P.C.: Data lake management: challenges and opportunities. Proc. VLDB Endow. **12**(12), 1986–1989 (2019)
40. Choi, T.M., Wallace, S.W., Wang, Y.: Big data analytics in operations management. Prod. Oper. Manage. **27**(10), 1868–1883 (2018)
41. Osman, A.M.S.: A novel big data analytics framework for smart cities. Future Gene. Comput. Syst. **91**, 620–633 (2019)
42. Papadopoulos, T., Gunasekaran, A., Dubey, R., Altay, N., Childe, S.J., Fosso-Wamba, S.: The role of big data in explaining disaster resilience in supply chains for sustainability. J. Clean. Prod. **142**, 1108–1118 (2017)
43. López-Cuevas, A., Ramírez-Márquez, J., Sanchez-Ante, G., Barker, K.: A community perspective on resilience analytics: a visual analysis of community mood. Risk Anal. **37**(8), 1566–1579 (2017)
44. Williams, N., Ferdinand, N.P., Croft, R.: Project management maturity in the age of big data. Int. J. Managing Proj. Bus. **7**(2), 311–317 (2014)
45. Côrte-Real, N., Oliveira, T., Ruivo, P.: Assessing business value of big data analytics in European firms. J. Bus. Res. **70**, 379–390 (2017)
46. Yang, X., Ge, J.: Predicting student learning effectiveness in higher education based on big data analysis. Mobile Inf. Syst. (2022)

47. Ciampi, F., Demi, S., Magrini, A., Marzi, G., Papa, A.: Exploring the impact of big data analytics capabilities on business model innovation: the mediating role of entrepreneurial orientation. J. Bus. Res. **123**, 1–13 (2021)
48. Cetindamar, D., Shdifat, B., Erfani, E.: Understanding big data analytics capability and sustainable supply chains. Inf. Syst. Manage. **39**(1), 19–33 (2022)
49. Mangla, S.K., Raut, R., Narwane, V.S., Zhang, Z.J., Priyadarshinee, P.: Mediating effect of big data analytics on project performance of small and medium enterprises. J. Enterprise Inf. Manage. **34**(1), 168–198 (2021). https://doi.org/10.1108/JEIM-12-2019-0394
50. Galetsi, P., Katsaliaki, K., Kumar, S.: Big data analytics in health sector: theoretical framework, techniques and prospects. Int. J. Inf. Manag. **50**, 206–216 (2020)
51. Mikalef, P., Krogstie, J.: Examining the interplay between big data analytics and contextual factors in driving process innovation capabilities. Eur. J. Inf. Syst. **29**(3), 260–287 (2020)
52. McAfee, A., Brynjolfsson, E.: Big data: the management revolution. Harvard Bus. Rev. **90**(10), 61–67 (2012)

Big Data Adoption Factors and Development Methodologies: A Multiple Case Study Analysis

Ahmad B. Alnafoosi and Olayele Adelakun

1 Introduction

Data is one of the most valuable resources in the world for many organizations today [1]. Big data (BD) primarily provides the ability to access data's untapped potential by querying large sets of data. This paper defines BD as an information system characterized by high volume, velocity, and variety of data requiring analytics capabilities to query and transform it into value [2]. When properly implemented, BD benefits include product, workforce, outcome, pricing optimization, health, genomic analysis, user experience, and many other use cases and value propositions in many fields and industries [3]. Yet, many organizations with enormous amounts of data are not leveraging it correctly. The literature on BD shows that many organizations' failure rates in BD implementation are high [4, 5]. In this research, there are two main focal areas.

First, we strive to identify critical factors affecting the adoption of BD implementation in organizations [4]. Some factors affecting BD adoption include finding the appropriate use case to extract value, the challenge with security, the burden of regulation, BD's need for extensive infrastructure, and others. The semi-structured interview research method was used for data collection [6]. We interviewed a founder, four architects, three managers, and one engineer. All the interviewees have implemented BD or are part of a team adopting it in their organization. The interviewees are also from various organizations and industries. The technology-organization-environment (TOE) framework guided qualitative data analysis [7].

Second, we explored the organization's BD development methodologies which are overwhelmingly agile with mostly medium-size teams. The study also captured

A. B. Alnafoosi (✉) · O. Adelakun
Jarvis College of Computing and Digital Media, DePaul University, Chicago, IL, USA
e-mail: aalnafoo@depaul.edu

M. Mora et al. (eds.), *Development Methodologies for Big Data Analytics Systems*, Transactions on Computational Science and Computational Intelligence, https://doi.org/10.1007/978-3-031-40956-1_8

aspects of what worked well with agile, such as quick delivery, stakeholder participation, and incremental progress when technical knowledge is incomplete. It supplied a venue for continuous process improvement. Agile in BD also has drawbacks like the repeating conflict, feature interaction regression, divergence of BD specialization vs. generalization, and the scarcity of BD technical experience that can cause agile to run longer due to experimentation.

2 Background

There is so much data being generated. Over nine zettabytes of data were stored in 2016 [8]. It is estimated that only 3% of the data stored is analyzed [9]. There is substantial potential for analyzing a larger percentage of the data. Traditional analysis tools cannot manage these volumes, velocities, and varieties of data in a reasonable time. Increasing the adoption of BD can increase the amount of data analyzed.

BD technology offers the ability to analyze these large volumes, variety, and velocity even when the data exists on multiple data servers or locations. The economic benefits of using BD were estimated to be in the hundreds of billions of dollars in 2011 [10]. BD holds the promise of unlocking substantial value [11]. Per an *Economist* article, data is now the most valuable resource [1]. Yet, some industries discard 80–90% of their data [12]. This research intends to find the significant factors that enable or hinder BD adoption for organizations with data storage systems, thus contributing to unlocking the stored data value with BD.

There are several empirical studies on the factors affecting the adoption of BD [13–18]. BD is one of the leading technologies to realize and extract value from stored data [19]. Many organizations own their data storage systems. Yet few organizations adopt BD technologies to extract value from the stored data [5, 20, 21]. Also, there are no phenomenological studies on BD adoption from the individual adopter perspective that can provide meaning, experiences, insights, and interpretations for individuals within organizations adopting BD.

Once the decision to adopt BD is made, the development life cycle started to implement the BD solution. Development methodology can vary greatly depending on the criticality, the dynamism of the project, personnel expertise, culture, and the size of the development personnel involved [22]. Development methodologies can be categorized as plan-driven, lightweight, agile, or hybrids of these methodologies [23]. This research will explore how different organizations used different development methodologies, advantages, disadvantages, and lessons learned from that experience.

3 Literature Review

3.1 Big Data

"BD is the Information asset characterized by a high volume, velocity, and variety to require specific technology and analytical methods for its transformation into value" [2]. The main features of BD are data volume, data velocity, and data variety defined in 2001 [24, 25]. This definition is often called 3Vs [10]. Although these are the main features of BD, other researchers have identified features that distinguish BD from data storage and analytics technologies [26].

In 2011, the International Data Corporation (IDC) added Value as another V [10, 27]. Value is economically extracted from the other 3Vs, and this highlights the underlying reasons for BD. Bello-Orgaz et al. added data veracity to that list and extracted values from the stored data [28]. Conversely, variability (inconsistency of the BD) was added in 2012. Other researchers did not adhere to starting BD features with the letter V. For example, the complexity of connecting and linking the data is identified as BD features [29]. Others have added visualization making up 7Vs [15]. Although these are some of the distinguishing features of BD, they are also some of its challenges. A data storage system must manage the volume, velocity, and variety of data to extract value from data.

3.2 Previously Studied Big Data Adoption Factors

Several studies have examined the factors affecting the adoption of BD. These studies utilized multiple theories and frameworks. Some studies used a single theory or framework to explore these factors, like diffusion of innovation theory (DOI) [30], technology-organization-environment framework (TOE) [13, 17, 31], theory of planned behavior [32], technology acceptance model (TAM) [33], and maturity model [34]. Most studies have used combinations of theories and frameworks to study the BD adoption factors. Varieties surveyed in this literature review included DOI + TOE [35, 36], TOE+IT Fashion [5], DOI + TOE+ Institutional Theory [13], Resource-Based View + isomorphism [37], DOI + TOE+TAM [15, 20], and decision-making trial and evaluation lab (DEMATEL)-adaptive neuro-fuzzy inference systems (ANFIS) + TOE [18].

Multiple research methodologies have been applied, such as interviews [38], case studies [39, 40], and surveys [15, 17, 41]. These studies are also varied in geography. They are studied in China [41], Germany [36], India [17], Korea [42], Norway [15], Poland [43], Sweden [44], and the USA [45].

There is qualitative BD adoption research. However, none of the studies before took a phenomenological approach. The interpretive phenomenological analysis (IPA) approach allows access to the actual experiences of BD adopters with a detailed and personalized view of the experiences. It treats each adopter as a separate case study and then studies the commonality and differences between these experiences. This study explores new factors, replicates understudied factors, and gains novel insights into existing factors for BD adoption in organizations with data storage systems.

3.3 Development Methodologies

The software development methodology is the framework that structures, plans, and controls the software development process [46]. There are many ways to organize and execute these steps; thus, many methods have evolved over the years. These variations and adaptations in software development methods were trying to address the challenges of delivering quality, cost-effectiveness, and pace of delivery, resulting in multiple evolving software development methods. Each of these software methods has its strengths and weaknesses. The business need for delivering fast-paced results has driven IT in general and software development, in particular, to adopt agile, lightweight, and hybrid development methods to deliver in a shorter period with more ability to change the requirements. The pressure to deliver faster results is translated in multiple contexts (organization's size, age, industry, culture, IT skill sets, and others) into different development methodologies for each organization.

Development methodology varies greatly and evolves even within organizations depending on the criticality, the dynamism of the project, personnel expertise, culture, and the size of the development personnel involved [22]. There are multiple ways to classify software development methodologies [47]. Development methodologies can be categorized as plan-driven (e.g., waterfall, V-Model, incremental, etc.), lightweight (e.g., XP, feature-driven development, crystal clear, etc.), agile (e.g., Scrum, Kanban, lean, etc.), or a hybrid of these methodologies [23].

4 Methodology

The qualitative method of IPA is used in this research [48]. IPA explores how individuals in organizations adopting BD interpret their experience. IPA goes in-depth with data gathering, where each individual shares their experience of BD adoption. Then interviews are used to gather more insights from multiple participants who lived the same experience of adopting BD, albeit from diverse backgrounds. The researcher then conducts further analysis and examination.

The interpretive phenomenological analysis (IPA) is one of the most "participant-oriented" qualitative research approaches [49]. IPA captures participants' opinions, perceptions, and insights on the subject matter based on their real-world experiences. The researcher plays the role of interpreter of the participant's "sense-making" of the experience based on the collective subjective data [50]. Since this method's unit of analysis is an individual, it is also called the "idiographic approach" [51].

IPA captures the understanding, evaluation, and perception of each participant's essential aspects of the event and how they experience it [49, 51]. The researcher in IPA captures each participant's answers inquisitively, looks for common themes in the participants' responses, analyzes the responses in the research context, and places the responses in appropriate contexts or brackets. These units of meaning or contexts are compared across the participants. The researcher's placement is based on observing all the reactions, literature review, and familiarity with the subject matter. Based on that, the researcher will interpret the patterns of these responses of these similar lived experiences [49].

The diffusion of Innovation theory suggests that the adoption decision is not purely mechanical, technical, or even social [52]. In this research, the primary research question is concerned with finding factors influencing BD adoption in organizations with data storage systems as experienced by those who took part in that decision-making and implementation. This interpretive phenomenological analysis of BD adoption has not previously been conducted.

5 Procedure

IPA utilizes in-depth, semi-structured interviews with open-ended questions and then analyzes those interviews to gather rich first-person accounts and insights within and among multiple interviews [50]. First, the researcher captures the interview content. Each participant was interviewed using a semi-structured interview process. The interviews were conducted using a private online conferencing tool (Zoom) that recorded the conversation. The participants were first asked about their demographics. Then they were asked three pivotal, in-depth, open-ended, interactive, and follow-up questions. These questions investigated the technical, organizational, and environmental factors that affected BD adoption in your organization (TOE framework) [7]. The interviews as audio recordings were saved to the DePaul University private repository that is only accessible to the researcher. They were transcribed using a third-party transcription service (MS Office 365 cloud transcription). The participant of that interview verified each interview transcript.

6 Participants

IPA research is conducted on a small number of participants using purposeful sampling [53, 54]. The sampling focuses on selecting participants who lived similar experiences, which differs from other sampling methods [53]. The number of participants mentioned is between five and ten who experienced similar events (homogenous) [49]. This range of participants approximates a saturation point where no new insight is gained from adding new participants [51, 55]. IPA involves a detailed analysis of each case across cases; thus, having many cases is difficult.

For this research, the purposeful sampling focused on individuals within organizations involved in the decision and implementation of BD on their existing data storage systems. Selecting and recruiting participants for interviews is difficult, and one of the effective ways to improve participation in interview studies is word of mouth [55, 56]. This research used word-of-mouth recruitment, where the researcher's information is shared with other participants [49].

In this study, nine individuals who participated in adopting and developing BD within their organizations in the USA were interviewed. The participant's roles varied from architects (four), managers (three), engineers (one), and founders (one). The industries varied from IT (four), cloud (two), automotive (one), insurance (one), and media and entertainment (one). Seven participants worked in large organizations (>250 employees), 1 in a medium-sized organization (50 to 250 employees), and 1 in a small organization (<50 employees). Five of these organizations adopted and developed BD for internal use, while the other four adopted and developed it as a product to be sold. See Table 1.

7 Data Analysis

IPA data analysis consists of six steps. The first step is to read and reread each transcript multiple times. The goal is to immerse oneself in the data. One of the goals of this step is to make the participant's voice the focus of the analysis. Another goal is to slow down the analysis and allows for more absorption and reflection. It also provides for the ordering of ideas as the researcher moves forward [50].

Table 1 Participants' information

No.	Role in organization	Industry	Organization's size	Use case
1	Founder	IT	Medium	Developed BD as a product
2	Engineer	IT	Small	Developed BD as a product
3	Manager	Automotive	Large	Developed BD for internal use
4	Architect	Insurance	Large	Developed BD for internal use
5	Architect	IT	Large	Developed BD as a product
6	Manager	Cloud	Large	Developed BD for internal use
7	Architect	Media and entertainment	Large	Developed BD for internal use
8	Architect	IT	Large	Developed BD as a product
9	Manager	Cloud	Large	Developed BD for internal use

The second step is the initial noting of the transcript. The researcher keeps an open mind and explores and notes the text of the transcript, focusing on noting anything of interest. The goal is to produce a comprehensive note (not word by word but by relevance). This step describes the participant's explicit meaning, but the focus will change in the following steps. The notes at this step can be descriptive, linguistic, or conceptual. This free textual analysis can be one of the most time-consuming steps [50]. The researcher used NVivo 12 qualitative analysis software for this step, and the rest of the data analysis steps [57].

The third step is to develop emergent themes for each of the transcripts. The researcher uses the notes taken in the previous step to find emergent themes. If done correctly, the notes should closely relate to the main text, where the "Interoperative" part of IPA is applied. The researcher uses the parts of the interview scripts to develop a theme that can group these ideas. The researcher can be described as doing "double hermeneutics" because they are trying to make sense of the participant trying to make sense of the experience. This is done by producing concise statements of what is essential in direct line with the transcript statement [50].

The fourth step is to search for connections across emergent themes. Since not all ideas in an interview are chronological, there is a need to discover themes that can group these ideas. Not all emergent themes need to be incorporated. The goal is to connect the emergent themes in a way that points to the participants' most interesting and compelling accounts. Move the themes around and see if any clustering or connections can be drawn, even if it is not chronological. These themes can also be connected using abstraction, subsumption, polarization, contextualization, numeration, or function [50].

The fifth step is to repeat that process on the next participant's transcript [50]. The last step is to look for patterns across cases. This is where the researcher explores common themes across all participants. One can find which themes are more common, potent, or mentioned [50, 58].

8 Validity

Qualitative research has a distinct set of evaluation criteria than quantitative research. One qualitative research validity method is by Yardley [59]. IPA research methodology extends Yardley's qualitative research validity work [50, 59]. Researchers have used other qualitative validity approaches [58, 60]. Yardley's validity evaluation is based on four principles: sensitivity to context, commitment and rigor, transparency, coherence, and finally, impact and importance.

Per Yardley's work, sensitivity to context can be described as the researcher's "awareness of different perspectives and complex arguments that can be brought to bear on the subject provide the researcher with the scholastic tools to develop a more profound and far-reaching analysis" [59]. This research is part of a larger dissertation thesis that thoroughly examined the literature, current and previously used methodologies, participants, interview questions, and reporting for the BD adoption. That is also demonstrated in the purposeful sampling that pursued individuals who participated in BD adoption and captured their lived experiences individually [50].

Commitment and rigor are the other criteria for evaluating validity. Commitment can be described as the "degree of attentiveness to the participant during data collection and the care with which the analysis of each case is carried out" [50]. Close attention was given to all aspects of the interview and data analysis. The interview transcripts were given to the participants for review, and no serious issues were found. The data presented include the perspectives of all participants. Conversely, rigor refers to the "thoroughness of the study" [50]. This is manifested in selecting the sample, questions, interviews, and completeness of the analysis. The sample was chosen carefully to represent instances where BD was adopted in diverse sizes, industries, and job titles. The interview was semi-structured and interactive; the researcher asked probing questions and asked for clarification and characterization from the participants. The analysis was ideographic, where each interview script was read and noted and went beyond description to interpret and highlight the important aspects of individual interviews and the shared themes.

Transparency and coherence are described as clarity and cogency of the persuasiveness of the description and arguments presented [59]. Transparency is the clarity in which the steps of the IPA method, participant selection, interview, and data analysis are given. Coherence is a way to describe how the arguments are received and how the research adheres to the principle of the method.

Impact and importance are the final principles of Yardley's qualitative research validity. The research should leave the reader with the main themes and conclusions that enrich and influence. This study explores the inductive human element of BD adoption. The intent is for these insights to inform and assist IT practitioners and academics about the adoption factors. These adoption factors can be studied further on how they can be addressed to enable further adoption of BD.

9 Findings

9.1 Big Data Adoption Findings

The Challenge of BD Value: New Insights

Realizing perceived BD value remains a challenge to most interviewees, as 89% of the participant indicated, even when they successfully use BD. One participant stated: "The ability to link the data value to the business goals is, I think it is [sic], probably one of the biggest challenges." Several other participants repeated this challenge: "it's just that some people don't see the actual value of the BD." "A lot of smaller companies, where there is going to be a big value for them, [sic] it's difficult for them to see that." and "We have to convince our customers [sic] our value proposition."

This challenge of realizing the perceived value of BD has put many organizations in a dilemma about whether to adopt BD. BD is still relevant technology, with many organizations implementing it and successfully extracting value [61]. Conversely, many organizations need help discerning how to extract value from BD. As one participant described, "A lot of companies decided to look and see. Is this something we can do with our data [sic] because there's always value in data and processing it?" Another participant articulated: "Managers, directors, VPs think it's a trendy topic, and they want to explore it, but they don't know the benefit of it. They can't quantify it, and they have a hard time understanding why you need BD?" And another: "Until you actually run those jobs on the data, you don't know how much value you're going to get. So, it's a [sic] chicken and egg thing." And the question remains, "How do we link the benefits we get from managing big data sets?"

Some organizations attempt to be on the path of adoption or do limited adoption but fail to realize the value [62]. Some organizations are keen on storing data with the understanding that the data has value. The following participants stated, "They like to be able to collect more data because they feel like it might be valuable." "I think a lot of organizations have this idea that they would like to be able to extract some value from the data." Even with the data stored, is it in a form that its value can be extracted? "The question is your data in the form [sic] type Where you know [sic] processing would give you the most value from it."

Data is a necessary but not sufficient condition to build a viable BD (data storage + data + ability to analyze) [15]. Other organizations started to query the data but without well-defined use cases. Having the ability to query the data opens the door to extracting value. Conversely, a longer, iterative journey is needed to refine these queries. "They'll run queries. They'll be able to link datasets together with no problem, right? But what comes out of it is not, as it needs a lot of refinement [sic]." That journey of refinement to find value is unceasing.

Hirsch compared BD to oil in value and other aspects [11]. This comparison can be extended to various uses and processes to achieve them. There are over 6000 products that are made of petroleum [63]. Similarly, BD products/use cases can be as diverse as those of organizations [61]. To realize BD value, there is a need to identify the organization's use cases and design solutions that support these use cases.

The Challenge of Security (Old, New, and Unique)

Security as an essential BD adoption factor was mentioned by 89% of the participants, which is consistent with the literature [15, 64]. As one participant stated, "Concern number one was security." The participants did raise some novel points. One issue is that BD can access confidential information or violate the security policy. Structured data (e.g., databases or tables) can be secured by limiting access to the type of data to be analyzed. BD can access semi-structured or nonstructured, whereas secure access can be more obscured. That can be problematic, as one participant described, "I get a Social Security in a table, then you can easily classify

[sic]. Imagine if the data is buried in a Word document." Analyzed data can contain restricted access information that may not be correctly classified. One participant explained, "Some customer, external-facing systems will interact with that. That is where security comes in."

BD can interact with data-at-rest (data locality) [65] or move data and deal with data-in-transit [66]. This requires software tools to support security and ensure correct access, encryption, cryptography, and others. As a participant portrayed it, "Especially if it's dealing with any security algorithms or encryption or cryptography, they look for meeting certain standards."

Challenge of Managing Large Datasets

Managing datasets is a major adoption factor shown by 78% of the participants, as one of the pillars of BD traits is the volume (large datasets). This problem is only getting bigger. As one participant described, "The size of the data is getting bigger and bigger." That presents the technical and logistical challenge of adopting BD. As one participant described the ability to manage this challenge as the criteria for success, "Companies that are able to manage large datasets are the most successful."

As the datasets get larger, the impact on their data storage system and its ability to query the data is felt as latency. As another participant described it, "Datasets have grown to a size that is now so large and so [sic] IO operations." Adds another, "We've been focusing on managing the size of the datasets so that the queries run faster." Large datasets require specialized handling that is not only able to store current data but also grow with it.

One of BD's main features is the ability to process large volumes of data [25]. Some data needs to be transferred even with data locality (Xiaoqiang et al., 2017). This transfer of data can add significant utilization to the network bandwidth. One participant pointed out, "When you're processing terabytes or petabytes of data, then transferring the data from the storage system to the processing servers, [sic] becomes the bottleneck." Another participant reiterated that concern "As far as dealing with my customers, [sic] biggest factor for them on the technical side has always been bandwidth."

More Data Means More Privacy Concerns

Privacy with BD was a major concern for 78% of the participants. One participant described the privacy challenge as "the main barrier to just jumping [sic]." As the datasets are large and varied in BD, ensuring that queries adhere to privacy laws and regulations becomes a major challenge. "The data you called may be proprietary or may have different privacy regulations governing it." To protect privacy, additional measures and algorithms are needed. "Data that has privacy information adds overhead to the processing."

Data comes from diverse sources, formats, sizes, structures, and content which adds to the complexity of the privacy requirements. A document, picture, video, or audio file may contain confidential information, ensuring privacy is more complex than checking the title of the data. Other laws, like General Data Protection Regulation (GDPR), require additional protection and time constraints on stored and queried data.

Cost of Big Data

The cost was mentioned by 67% of the participants as one of the main challenges of adopting BD. They said developing and maintaining BD is costly; as one participant described, "The infrastructure cost was the biggest problem, and the data storage cost was the other biggest problem [sic]." For the participants who gained value from BD, the cost was justified, creating a "positive feedback loop" for those who made value from BD.

The Burden of Regulations

Regulation is a major factor affecting BD adoption, and 67% of the participants discussed it [13, 14, 36, 67]. One participant described BD as a liability in light of regulations "Just the existence of BD is almost in some ways a liability. Especially if you're talking about GDPR." One aspect that the participant repeated is the inconsistencies across regulations. One participant asserted, "The data may be proprietary or may have different privacy regulations governing it." Another participant confirmed the sentiment and said, "We cannot have any customer identifier information on a record for more than 45 days." That participant chose 45 days as the lower limit to simplify his requirements (instead of having different retention times for other data). Changing regulations in an evolving world is a major challenge for organization's BD.

IT Expertise in Big Data

IT expertise in BD was cited by 67% of the participants as one of the major adoption factors. As one participant mentioned, "Finding developers that are experienced Can be a barrier. Because it's not something that most people have experience [sic], and because it's such a new technology, it's probably not taught in many colleges." The lack of BD staff goes beyond developers and IT staff to data scientists. As one participant indicated, "Software companies don't have developers on staff, or maybe even necessarily data scientists on staff."

IT staff with traditional experience are having difficulties adopting BD technologies and their specialized requirements. As one participant said, "People are more familiar with SQL, databases, and relational databases in general. That's easier for them to comprehend. They have more experience in that. So, there's just a natural barrier to adoption just because not [sic] people are familiar with it." Pushing staff to adopt BD technologies may cause failures and friction. As one participant stated, "People have [sic] traditionally working on a traditional system. Now you have to push them to adapt to new challenges. Some people fail [sic]." That is also true with more recent graduates, as declared by another participant "The demographic that I work with is a lot younger people who just graduated college. Perhaps three to four years out of college is an example. They are less familiar with the other technologies out there."

Big Data Adoption Findings Summary

BD adoption factors explored in the semi-structured interviews were discussed above for insights and in-depth discussion. The tabulated summary is in Table 2.

9.2 *Big Data Development Methodology Findings*

Medium Development Team Size Is Common

The majority of these interviewees (78%) have medium-size teams (10–30 people) in their organizations that adopted BD. However, the Scrum standard team size is five to nine [68]. The scope of work for the BD project is larger than a small team can manage. A medium-size development team is a common choice. In one instance, the larger development team size was chosen to work on multiple parallel releases.

Table 2 Big data adoption factors findings summary

Adoption factor	Percentage of participants that discussed it (approx.)	Authors' summary
BD value	89%	Value even when success is a challenge Wait and see the stance on adoption Raw data does not mean value The ability to analyze data does not mean value either Find your BD use case and refine it, and you will find BD value
Security	89%	Big data means big security Data security at rest and in motion
Large datasets	78%	Large datasets require large data systems
Privacy	78%	Big data means big privacy concerns Privacy for multiple data formats
Cost	67%	Big data cost is a concern
Regulations	67%	Data is also a liability
IT expertise	67%	Big data expertise is scarce

Agile Development Method Is the Popular Option

All the interviewees used agile software development methodology, specifically Scrum agile development methodology. There were multiple variations in using Scrum regarding sprint duration, number of stories, points, and other aspects. As declared by a participant, "Sprint planning, Daily Scrum, Sprint Review worked well." The ability to deliver small increments of software at small time intervals (weeks) is paramount for all the interviewees. That is in addition to the ability to modify the software product based on these increments because of changing requirements. One participant described the agile process: "We had an ability to quickly pivot on ever changing requirements that were coming in."

Other interviewees emphasized the agile process's ability to deliver products that generate revenues because of its rapid cycle. Others mentioned that the ability of stakeholders to weigh in on the stories and their priorities helps in delivering desired product. As one participant cited, "Developers get a chance to weigh in on time estimates early." In two instances, there was insufficient technical expertise, so the agile method helped deliver parts of the product, grew developers' knowledge, and allowed experimentation. As one participant remarked, "many stakeholders without expertise in the space [sic] trying to adopt a solution that [sic] beyond a month or two." Another participant echoed the issue "Lack of expertise has the potential to create solutions that would have maintenance ramifications in the future." Others mentioned that agile helped with the continuous improvement of the product.

Agile Development Has Drawbacks Too

What could have worked better was a variation of the disadvantages of agile development. In two instances, there is a divergence in the main deliverance of the software development methodology between management and technical stakeholders. The business/management demands a quality product that delivers value quickly versus the development interest in a stable and reproducible process with technical debt paid. As one participant mentioned, "As a startup company, getting to revenues superseded the quality of the software." So, the push and pull between which stories to include is a constant struggle mentioned by multiple interviewees and hinders the ability to deliver consistent results. A notable drawback with agile was feature interaction. Agile methodology lacked in-depth feature interaction analysis. This caused regressions and needed further investigation. The good news in the capture instance is that developers captured the issue quickly (thanks to the agile method).

It is challenging to optimize the software methodology when the requirements diverge between optimization for specific use case versus general use BD. Expertise in BD is hard to come by as multiple interviewees expressed. Agile helped in certain ways of hedging the developers' knowledge gaps, which also prolonged the development process and delayed delivery. As BD's features become more complex and specialized, the challenge of having the right expertise with the optimal software development methodology grows more important.

Big Data Development Methodology Findings Summary

Software development methodology findings from the interviews using the IPA method are summarized in Table 3.

10 Study Limitations and Future Research Opportunities

The research is an exploratory qualitative study that provides novel phenomenological and qualitative insights into BD adoption factors for organizations with their own data storage systems. The purposeful sampling of BD adopters and developers with their storage systems has limited the sample. Although the researcher did not put or intend these as conditions, the sample is geographically limited to the US context. Also, there were no females found in this sample.

As BD matures, further studies on BD adoption and BD development methodologies are needed to answer finer research questions in these domains and for specific BD technologies (Hadoop, Spark, Hive, etc.). This research utilized qualitative research methodology. Other research methodologies in qualitative, quantitative, and mixed are needed to validate existing research and explore new aspects. This research was one of the first to use the IPA qualitative research method in the BD space. Other research is needed to explore other phenomena in BD space. The research was conducted as a generic big data system and was not specific to technology or address specific analytic issues. This might be explored in future studies.

11 Conclusion

BD promises to unlock one of the most valuable resources on the planet [1]. Data in many organizations being stored in various data storage systems are waiting to be analyzed, and not all of these data storage systems can evolve and adopt BD [20]. This research explored the factors affecting BD adoption in general in the literature review and identified gaps in unstudied and understudied factors that can be explored further. This study investigated the adoption of BD further by researching individuals in organizations that adopted BD with their storage systems. This investigation captured their individuals' perspectives using interpretive phenomenological analysis on the factors affecting BD adoption for these organizations. Among these factors that have specific nuance to BD are the challenge of perceiving and realizing BD value, security concerns, regulations, and network compatibility issues with BD. New insights and distinctive factors are captured on the unique challenges of adopting BD for organizations with their own data storage systems. Other factors based on the literature review revealed that they might be significant and must be investigated further.

Table 3 Big data development methodology findings summary

Development methodology finding	Percentage of participants that discussed it (approx.)	Summary
Size of development team	78%	Medium (10–30) Larger than the Scrum size of five to nine
Agile development method is popular	100%	All participants indicated they used a flavor of agile development methodology The ability of the agile method to deliver small increments in short (weeks) intervals was paramount Agile delivery allowed us to get quick wins and realize revenue Participation of stakeholders to determine stories and sprint is important Agile allows for incremental progress, which is important, especially when knowledge of the whole solution is not complete Agile helped in continuous process improvement
Agile development drawbacks	100%	Divergence of goals causes agile development conflict ongoing. Also, it produces inconsistent results Feature interaction regression is one of the drawbacks as not sufficient investigation was done Divergence of BD specialization versus general use BD Lack of technical knowledge about BD coupled with agile caused delivery to be delayed

This study also investigated BD development for the same participants using the same qualitative IPA methodology. All BD development used agile software development methodology in this sample. The majority of them (78%) used medium-size teams ranging from 10 to 30 developers. Agile development methodology provided the ability to deliver quick deliverables with the ability to pivot and change requirements. However, the interviewees indicated they used assorted flavors of agile development. That variation stems from sprint duration, story sizes, team sizes, and other factors to fit the organization's needs and culture. Multiple stakeholders had a say in the stories and the development requirements. One challenge in the BD domain is the technical expertise in BD. Agile development allows developers to start the project and make progress. However, the complete solution and the technical problems are not fully solved yet – this contrasts with a plan-driven approach, where a solution is architected beforehand. Agile also has drawbacks that manifest in BD software development. The divergence of goals and priorities caused constant and repeated struggles among stakeholders. Feature interaction regression was seen due to an incomplete investigation of the feature and how it affects previous ones. Technical expertise in BD is still a scarce item that can delay agile development due to incomplete knowledge and experimentation needed to bridge that.

The qualitative findings show the need to link BD adoption and its development to business goals, and in many instances, the value is found by the process of trials and refinements of the queries to big data. Hirsch compared big data to oil in terms of value and other aspects [11]. In their raw form, BD and oil both may have limited value. This comparison can also be extended to the various uses and processes to achieve them. There are over 6000 products made of petroleum [63]. Similarly, big data products/use cases can be as diverse as the organizations that use them [61]. To realize big data value, we need to identify the needs of the organizations and their customers and define these use cases with specific business and monetary objectives. Then, design a big data solution and development that supports these use cases.

References

1. Parkins, D.: Regulating the internet giants: the world's most valuable resource is no longer oil, but data. De Economist. https://www.economist.com/leaders/2017/05/06/the-worlds-most-valuable-resource-is-no-longer-oil-but-data. Accessed 24 Aug 2022
2. De Mauro, A., Greco, M., Grimaldi, M.: A formal definition of big data based on its essential features. Libr. Rev. **65**(3), 122–135 (2016). https://doi.org/10.1108/LR-06-2015-0061
3. Pramanik, P.K.D., Pal, S., Mukhopadhyay, M.: Healthcare big data: a comprehensive overview. In: IRMA, Research Anthology on Big Data Analytics, Architectures, and Applications, pp. 119–147. IGI Global (2022)
4. De Camargo Fiorini, P., Seles, B.M.R.P., Jabbour, C.J.C., Mariano, E.B., De Sousa Jabbour, A.B.L.: Management theory and big data literature: from a review to a research agenda. Int. J. Inf. Manag. **43**, 112–129 (2018)
5. Chen, H.M., Kazman, R., Matthes, F.: Demystifying big data adoption: beyond IT fashion and relative advantage. In: Proceedings of DIGIT workshop 2015, pp. 1–14. AIS eLibarary (2015)
6. Yin, R.K.: Case Study Research: Design and Methods, Applied Social Research Methods Series. Sage Publications, London (1994)
7. DePietro, R., Wiarda, E., Fleischer, M.: The context for change: organization, technology and environment. In: Tornatzky, L.G., Fleischer, M., Chakrabarti, A.K. (eds.) The processes of technological innovation, pp. 151–175. Lexington Books, Lexington (1990)
8. Westervelt, R.: Information-centric security: why data protection is the cornerstone of modern enterprise security programs. IDC (2023) https://docs.broadcom.com/doc/why-data-protection-is-the-cornerstone-of-modern-enterprise-security-programs-en. Accessed 7 May 2023
9. Reinsel, D., Gantz, J., Rydning, J.: Data Age 2025: The Evolution of Data to Life-Critical Don't Focus on Big Data. IDC (2023) https://www.seagate.com/files/www-content/our-story/trends/files/idc-seagate-dataage-whitepaper.pdf. Accessed 7 May 2023
10. Chen, M., Mao, S., Liu, Y.: Big data: A survey. Mob. Netw. Appl. **19**(2), 171–209 (2014). https://doi.org/10.1007/s11036-013-0489-0
11. Hirsch, D.: The glass house effect: big data, the new oil, and the power of analogy. Maine Law Rev. **66**(2), 374–395 (2014)
12. Hilbert, M.: Big data for development: a review of promises and challenges. Dev. Policy Rev. **34**(1), 135–174 (2016). https://doi.org/10.1111/dpr.12142
13. Agrawal, K.: Investigating the determinants of big data analytics (BDA) adoption in asian emerging economies. In: Proceedings of AMCIS 2015. pp 1–18 (2015)
14. Mahesh, D.D., Vijayapala, S., Dasanayaka, S.W.S.B.: Factors Affecting the Intention to Adopt Big Data Technology: A Study Based on Financial Services Industry of Sri Lanka. Paper pre-

sented at the 2018 Moratuwa Engineering Research Conference (MERCon), Moratuwa, Sri Lanka (2018)
15. Nguyen, T., Petersen, T.E.: Technology Adoption in Norway: Organizational Assimilation of Big Data. Thesis, Norwegian School of Economics (2017)
16. Salleh, K.A., Janczewski, L.J.: An Implementation of Sec-TOE Framework: Identifying Security Determinants of Big Data Solutions Adoption. In: Proceedings of International Conference On Information Resource Management (2016)
17. Verma, S., Bhattacharyya, S.S.: Perceived strategic value-based adoption of big data analytics in emerging economy: a qualitative approach for Indian firms. J. Enterp. Inf. Manag. **30**(3), 354–382 (2017). https://doi.org/10.1108/JEIM-10-2015-0099
18. Yadegaridehkordi, E., Hourmand, M., Nilashi, M., Shuib, L., Ahani, A., Ibrahim, O.: Influence of big data adoption on manufacturing companies' performance: an integrated DEMATEL-ANFIS approach. Technol. Forecast. Soc. Change. **137**, 199–210 (2018). https://doi.org/10.1016/j.techfore.2018.07.043
19. Nair, G.: Development of a Real-Time Business Intelligence (BI) Framework Based on Hex-Elementization of Data Points for Accurate Business Decision-Making. Thesis, Westren Sydny University (2019)
20. Ajimoko, O.J.: Exploring the Cloud-Based Big Data Analytics Adoption Criteria for Small Business Enterprises. Dissertaion, Colorado Technical University (2017)
21. Dubey, R., Gunasekaran, A., Childe, S.J., Wamba, S.F., Papadopoulos, T.: The impact of big data on world-class sustainable manufacturing. Int. J. Adv. Manuf. Technol. **84**(1–4), 631–645 (2016). https://doi.org/10.1007/s00170-015-7674-1
22. Boehm, B., Turner, R.: Using risk to balance agile and plan-driven methods. Computer. **36**(6), 57–66 (2003). https://doi.org/10.1109/MC.2003.1204376
23. Boehm, B., Turner, R.: People factors in software management: lessons from comparing agile and plan-driven methods. Crosstalk- J. Def. Software Eng. Dec. **105** (2003)
24. Laney, D.: 3D Data management: Controlling data volume, velocity, and variety. Gartner. https://studylib.net/doc/8647594/3d-data-management%2D%2Dcontrolling-data-volume%2D%2Dvelocity%2D%2Dan... Accessed 7 May 2023
25. Russom, P.: Big data analytics, TDWI best practices report. TDWI. Fourth quarter 2011, 1–35 (2011). https://vivomente.com/wp-content/uploads/2016/04/big-data-analytics-white-paper.pdf Accessed 7 May 2023
26. Chen, H., Chiang, R.H.L., Storey, V.C.: Business intelligence and analytics: from big data to big impact. MIS Q. **36**(4), 1165–1188 (2012). https://doi.org/10.2307/41703503
27. Gantz, J., Reinsel, D.: Extracting value from chaos. IDC, http://www.kushima.org/wp-content/uploads/2013/05/DigitalUniverse2011.pdf. Accessed 7 May 2023
28. Bello-Orgaz, G., Jung, J.J., Camacho, D.: Social big data: recent achievements and new challenges. Inform. Fusion. **28**, 45–59 (2016). https://doi.org/10.1016/j.inffus.2015.08.005
29. Hood-Clark, S.F.: Influences on the Use and Behavioral Intention to Use Big Data. Dissertaion, Capella University (2016)
30. Micheni, E.M.: Diffusion of big data and analytics in developing countries. Int. J. Eng. Sci. **4**(8), 44–50 (2015)
31. Nam, D.W., Kang, D., Kim, S.H.: Process of big data analysis adoption: Defining big data as a new IS innovation and examining factors affecting the process. In: Proceedings of the Annual Hawaii International Conference on System Sciences, 2015 March, pp. 4792–4801. IEEE (2015)
32. Esteves, J., Curto, J.: A risk and benefits behavioral model to assess intentions to adopt big data. J. Intell. Stud. in Bus. **3**(3), 37–46 (2013)
33. Lombardo, G.: Predicting the Adoption of Big Data Security Analytics for Detecting Insider Threats. Dissertation, Capella University (2018)
34. Olszak, C. M., Mach-Król, M.: Conceptual Framework for Assessing organization's Readiness to Big Data Adoption (2018). https://www.preprints.org/manuscript/201808.0335/v1. Accessed 7 May 2023

35. Agrawal, K.P.: The assimilation of big data analytics (BDA) by Indian firms: a technology diffusion perspective. In: Proceedings for 3rd Biennial Conference of the Indian Academy of Management (IAM), 2013. IIMA Institutional Repository Home, India (2013)
36. Bremser, C.: Starting Points for Big Data Adoption. In: Proceeding of ECIS. AIS eLibrary (2018)
37. Kwon, O., Lee, N., Shin, B.: Data quality management, data usage experience and acquisition intention of big data analytics. Int. J. Inf. Manag. **34**(3), 387–394 (2014). https://doi.org/10.1016/j.ijinfomgt.2014.02.002
38. Skaale, D.K., Rygh, E.: Big Data Technology Adoption through Digitalization in Yara International ASA. Thesis, University of Stavanger, Norway (2018)
39. Dremel, C., Herterich, M.M., Wulf, J., Brenner, W., Herterich, M.M., Waizmann, J.C.: How AUDI AG established big data analytics in its digital transformation. MIS Q. Exec. **16**(2), 81–100 (2017)
40. Gong, Y., Janssen, M.: Enterprise Architectures for Supporting the Adoption of Big Data, pp. 505–510. ACM International Conference Proceeding Series, F128275 (2017)
41. Wang, L., Yang, M., Pathan, Z.H., Salam, S., Shahzad, K.: Analysis of influencing factors of big data adoption in Chinese enterprises using DANP technique. Sustainability. **10**(11), 1–16 (2018). https://doi.org/10.3390/su10113956
42. Kim, M.-K., Park, J.-H.: Identifying and prioritizing critical factors for promoting the implementation and usage of big data in healthcare. Inf. Dev. **33**(3), 257–269 (2017). https://doi.org/10.1177/0266666916652671
43. Mach-Król, M.: Big Data Analytics in Polish Companies–Selected Research Results, vol. 85. ICT Management for Global Competitiveness and Economic Growth in Emerging Economies (ICTM) (2017)
44. Zanabria, V., Mlokozi, D.: Big Data Analytics for Achieving Smart City Resilience Key Factors for Adoption. Thesis, Lund University (2018)
45. Ghosh, B.: Exploratory study of organizational adoption of cloud based big data analytics. J. Inf. Syst. Appl. Res. **11**(3), 4–14 (2018)
46. Li, J.P., Nneji, G.U., Ukwuoma, C.C., Dike, I.D., Nneii, R.I.: Design of an improved cost efective electronic locking system. In: 2018 IEEE 9th Annual Information Technology, Electronics and Mobile Communication Conference (IEMCON), pp. 493–499. IEEE (2018)
47. Blum, B.I.: A taxonomy of software development methods. Commun. ACM. **37**(11), 82–94 (1994). https://doi.org/10.1145/188280.188377
48. Smith, J.A., Jarman, M., Osborn, M.: Doing interpretative phenomenological analysis. Qualitative health psychology: theories and methods. Sage Publications Ltd (1999)
49. Alase, A.: The interpretative phenomenological analysis (IPA): a guide to a good qualitative research approach. Int. J. Educ. Lit. Stud. **5**(2), 9–19 (2017). https://doi.org/10.7575/aiac.ijels.v.5n.2p.9
50. Smith, J.A., Flowers, P., Larkin, M.: Interpretative phenomenological analysis: theory, method and research. Qual. Res. Psychol. **6**(4), 346–347 (2009). https://doi.org/10.1080/14780880903340091
51. Brocki, J.M., Wearden, A.J.: A critical evaluation of the use of interpretative phenomenological analysis (IPA) in health psychology. Psychol. Health. **21**(1), 87–108 (2006). https://doi.org/10.1080/14768320500230185
52. Rogers, E.M.: Diffusion of Innovations. Free Press, New York (2003)
53. Smith, J.A., Shinebourne, P.: In: Cooper, H., Camic, P.M., Long, D.L., Panter, A.T., Rindskopf, D., Sher, K.J. (eds.) Interpretative phenomenological analysis, pp. 73–82. APA Handbook of Research Methods in Psychology, 2. Research Designs: Quantitative, Qualitative, Neuropsychological, and Biological (2012). https://doi.org/10.1037/13620-005
54. Gauci, M.G.: WASP (write a scientific paper): interpretative phenomenological analysis: its attraction and relevance to the medical field. Early Hum. Dev. **133**, 52–56 (2019). https://doi.org/10.1016/j.earlhumdev.2019.03.012

55. Parthasarathy, R.: Empirical Assessment of the Role of Technology-Related Factors and Organization-Related Factors in Electronic Medical Records Implementation Success. Dissertaion, DePaul University (2017)
56. Van Hoye, G., Van Hooft, E.A., Lievens, F.: Networking as a job search behaviour: a social network perspective. J. Occup. Organ. Psychol. **82**(3), 661–682 (2009). https://doi.org/10.1348/096317908X360675
57. Brown, C., Smith, P., Arduengo, N., Taylor, M.: Trusting telework in the federal government. Qual. Rep. **21**(1), 87–101 (2016)
58. Rivituso, J.: Cyberbullying victimization among college students: an interpretive phenomenological analysis. J. Inf. Syst. Educ. **25**(1), 71–76 (2014)
59. Yardley, L.: Dilemmas in qualitative health research. Psychol. Health. **15**(2), 215–228 (2000). https://doi.org/10.1080/08870440008400302
60. Creswell, J.W., Miller, D.L.: Determining validity in qualitative inquiry. Theory Pract. **39**(3), 124–130 (2000). https://doi.org/10.1207/s15430421tip3903_2
61. Eggers, J., Hein, A.: Turning big data into value: a literature review on business value realization from process mining. In: ECIS 2020 Proceedings. AIS eLibrary (2020)
62. Barham, H.: Achieving competitive advantage through big data: a literature review. In: Proceedings of 2017 Portland International Conference on Management of Engineering and Technology (PICMET), pp. 1–7. IEEE (2017)
63. Abutu, O.P.: Consequences of the January 2012 oil subsidy removal in Nigeria. J. Bus. Retail Manage. Res. **8**(2), 24–29 (2014)
64. Motau, M., Kalema, B.M.: Big data analytics readiness: A south African public sector perspective. In: Proceedings of 2016 IEEE International Conference on Emerging Technologies and Innovative Business Practices for the Transformation of Societies (EmergiTech), pp. 265–271. IEEE (2016)
65. Xiaoqiang, M., Xiaoyi, F., Jiangchuan, L., Hongbo, J., Kai, P.: vLocality: revisiting data locality for map reduce in virtualized clouds. IEEE Netw. **31**(1), 28–35 (2017). https://doi.org/10.1109/MNET.2016.1500133NM
66. Liu, Y., Katramatos, D.: A software defined network design for analyzing streaming data in transit. In: Proceedings of the ACM Workshop on Systems and Network Telemetry and Analytics, 2019, pp. 25–28. ACM (2019)
67. Sun, S., Cegielski, C.G., Jia, L., Hall, D.J.: Understanding the factors affecting the organizational adoption of big data. J. Comput. Inf. Syst. **58**(3), 93–203 (2018). https://doi.org/10.1080/08874417.2016.1222891
68. Alqudah, M., Razali, R.: A review of scaling agile methods in large software development. Int. J. Adv. Sci., Eng. Inf. Technol. **6**(6), 828–837 (2016). https://doi.org/10.18517/ijaseit.6.6.1374

Detection of Breast Cancer in Mammography Using Pretrained Convolutional Neural Networks with Fine-Tuning

Cesar Muñoz-Chavez, Hermilo Sánchez-Cruz, Humberto Sossa-Azuela, and Julio Ponce-Gallegos

1 Introduction

Latin American women have higher rates of breast cancer incidence and mortality compared to women in developed countries. The incidence rate of breast cancer is 80% for women over 44 years old, and the mortality rate is 86% [1]. Young women are also affected by breast cancer, with rates as high as 15% in less developed countries such as Mexico [2]. Low- and middle-income countries have a mortality to incidence ratio that is considerably higher, between 60% and 75%, compared to high-income countries [3]. Early detection of breast cancer is crucial for better survival rates, and digital screening mammography is commonly used by radiologists for analysis, diagnosis, and categorization of breast cancer [4, 5].

Although mammography is currently one of the most reliable methods for detecting breast cancer, medical experts face challenges in interpreting mammogram images, which could lead to an incorrect diagnosis. Current methods of image classification were investigated, and those that use artificial intelligence stood out the most [6, 7]. Artificial intelligence (AI) has advanced quickly in recent years in a variety of sectors, including image processing. Since images are one of the most crucial sources of information for activities involving human intelligence, AI has been widely used in image processing [8, 9]. In this research, AI techniques were found to be beneficial, particularly in the form of computer-aided diagnosis (CAD)

C. Muñoz-Chávez · H. Sánchez-Cruz (✉) · J. Ponce-Gallegos
Centro de Ciencias Básicas, Universidad Autónoma de Aguascalientes,
Aguascalientes, Mexico
e-mail: hermilo.sanchez@edu.uaa.mx

H. Sossa-Azuela
Centro de Investigación en Computación, Instituto Politécnico Nacional,
Mexico City, Mexico

M. Mora et al. (eds.), *Development Methodologies for Big Data Analytics Systems*, Transactions on Computational Science and Computational Intelligence, https://doi.org/10.1007/978-3-031-40956-1_9

as an alternative to radiologists' time-consuming and inaccurate double-reading procedure. Some of these CADs utilize machine learning algorithms that take up the information from complex patterns [10–12]. However, traditional classifiers based on handcrafted features are considered to be complex and time-consuming, especially for tasks involving feature extraction and selection [11, 12].

Recent studies have shown that deep learning algorithms are efficient on tasks including image segmentation, detection, and classification in a range of computer vision and image processing disciplines [13–15]. Deep learning is a type of machine learning that is inspired by the structure and function of the brain. Deep learning has lately received a lot of attention for classification problems, particularly in the field of medical image analysis [16]. Therefore, for this research, current methods of image classification were analyzed, and based on the state-of-the-art, deep convolutional neural networks (DCNNs) were found to perform well on the image classification task [16–18]. These models require a large number of images, which is why we searched for well-known state-of-the-art datasets. Two datasets were used in this study: CBIS-DDSM (Curated Breast Imaging Subset of DDSM) [19] and Mini-MIAS (Mammographic Image Analysis Society) [20]. Transfer learning with fine-tuning was used to accomplish this research [21, 22], where a model built for one task is used for another and the model's output is modified to meet the new task [23, 24].

This chapter presents the contributions of a research study aimed at exploring the effectiveness of deep learning algorithms for classifying mammogram images. In particular, the study compared the performance of two deep learning models, ResNet-50 and EfficientB7. To enhance the quality of the images, the researchers applied various image processing techniques, including CLAHE, unsharp masking, and a median filter. Additionally, they used a data augmentation algorithm from a library called Albumentation to increase the number of training images and improve the robustness of the convolutional neural network (CNN). The performance of each model was evaluated using five metrics, including accuracy, precision, recall, F1-score, and confusion matrix.

The rest of this chapter is organized as follows. A review of pertinent literature is presented in Sect. 2. The methodology used and the differences between each dataset are presented in Sect. 3, along with full disclosure of all the steps used to accomplish the goal of classification tasks. Experimental results comparing every model and metric are described in Sect. 4. Finally, Sect. 5 provides some conclusions.

2 Previous Works

Deep convolutional neural network (DCNN) models have become increasingly popular in recent years due to their exceptional performance in various computer vision tasks, including image classification, segmentation, and detection. Many different models have been proposed and utilized in research studies. However, researchers have obtained varied results, and it is important to analyze these results to identify areas where DCNN models can be improved.

The research by Rampun et al. [25] focuses on developing an automated method for breast pectoral muscle segmentation in mediolateral oblique mammograms. To achieve this, they employed a convolutional neural network (CNN) inspired by the holistically nested edge detection (HED) network, which is capable of learning complex hierarchical features and resolving spatial ambiguity in estimating the pectoral muscle boundary. The CNN is also designed to detect "contour-like" objects in mammograms. The study utilized several datasets, including MIAS, INBreast, BCDR, and CBIS-DDSM, to evaluate the performance of the proposed method. An ensemble approach was employed by Altameem et al. [26], in which the Gompertz function was used to build fuzzy rankings of the base classification techniques and the decision scores of the base models were adaptively combined to construct final predictions. Using Inception V4, ResNet 16, VGG 11, and DenseNet 121, as well as other deep CNN models, a deep learning approach using convolutional neural networks (CNNs) was used to classify breast cancer histopathological images from the BreaKHis dataset. The approach introduced by Wei et al. [27] enables the use of high-resolution histopathological images as input to existing convolutional neural networks (CNNs) without requiring complex and computationally expensive modifications to the network architecture. This is achieved through the extraction of image patches from the high-resolution image, which are then used to train the CNN. The final classification is obtained by combining the predictions from these patches. This method allows existing CNNs to be used for histopathological image analysis without the need for extensive modifications or the development of a new architecture. Additionally, a network was trained and validated by Auccahuasi et al. [28] using a database of images containing microcalcifications classified as benign and malignant from mammographic images of MIAS. One recent study introduced "double-shot transfer learning," which is a revolutionary method built on the idea of transfer learning. This strategy, presented by Alkhaleefah et al. [29], significantly improved categorization accuracy. The following research, proposed by Charan et al. [30], uses deep learning and neural networks for the classification of normal and abnormal breast detection in mammogram images using the Mammograms-MIAS dataset, which contains 322 mammograms with 189 images of normal breasts and 133 images of abnormal breasts. The study used a convolutional neural network (CNN) and obtained promising experimental results that suggest the efficacy of deep learning for breast cancer detection in mammogram images. Saber et al. [31] used pretrained convolutional neural networks (CNNs) to detect and classify breast tumors in the INbreast dataset using mammography images. The proposed model preprocesses the images to improve image quality and reduce computation time and then transfers the learned parameters from the CNNs to improve the classification. Another recent study proposes, by Qasim et al. [32], the use of a convolutional neural network (CNN) to detect breast cancer in mammography images by classifying them as noncancerous or cancerous abnormalities using the DDSM dataset. A set of mammogram images is preprocessed using histogram equalization, and the resulting images are used as a training source for the CNN. The proposed system, called

BCDCNN, is compared to the MCCANN system, and the results show that BCDCNN has higher classification accuracy and a higher resolution compared to other existing systems.

The use of pretrained models is good for discovering new models for specific tasks; a new method proposed by Montaha et al. [33] utilizing the fine-tuned VGG-16 model called BreasNet-18 is introduced. This methodology used the CBIS-DDSM dataset, and preprocessing was applied to these images. Artifact removal was the name of the initial stage of the preprocessing. It made use of techniques like binary masking, morphological opening, and detecting the largest contour. "Remove line" is the name of the second stage. The following techniques were used on certain images that included a bright, straight line attached to the breast contour: in-range operation, Gabor filter, morphological operations, and invert mask. Some algorithms, like gamma correction, CLAHE, and Green Fire Blue, enhance images in the third stage. After the preprocessing, the data augmentation technique is used to acquire additional images and solve issues with over- and under-fitting. Finally, the BreastNet-18 model was used to evaluate the classification problem of four classes. Additionally, Allugunti et al. [34] show a computer-aided diagnostic (CAD) method is recommended to classify three classes for the methodology using certain traditional methods (cancer, no cancer, and noncancerous). Convolutional neural networks (CNNs), support vector machines (SVM), and random forests (RF) were the three classifiers they employed. They used a dataset with a total of 1000 images from the Kaggle website. Following the test of each classifier, it was discovered that CNN performed better, achieving an accuracy of 99.6%.

The study by Shen et al. [35] introduces a novel deep learning algorithm for detecting breast cancer on mammograms. The algorithm utilizes an "end-to-end" training approach, which decreases the dependence on lesion annotations and enables the use of image-level labels. The model is trained on two separate datasets, the Digital Database for Screening Mammography (CBIS-DDSM) and the INbreast database, and the results indicate excellent performance with high accuracy on both heterogeneous mammography platforms. These findings have the potential to improve clinical tools and decrease the incidence of false-positive and false-negative screening results, which can lead to more accurate diagnoses and improved patient outcomes. The research conducted by Khamparia et al. [36] presented an approach for classifying mammograms using deep learning models. The authors experimented with various models, such as a pretrained VGG model, a residual network, and a mobile network, to determine their effectiveness. They found that their fine-tuned VGG16 model, with data augmentation and pretrained ImageNet weights, outperformed the other models in terms of accuracy. They utilized the DDSM dataset for their experiments, and their approach yielded an accuracy of 88.30% and an AUC value of 93.30%. Hameed et al. [37] proposed a deep learning methodology for the accurate detection of cancerous and noncancerous tissue using pretrained convolutional neural networks (CNNs). They used both the VGG16 and VGG19 models as is and with modifications to enhance performance. The study collected 544 whole slide images (WSIs) from 80 patients with breast cancer from the pathology division of Colsanitas Colombia University in Bogota, Colombia. The images were normalized, and the data was divided into 80% for training and 20% for testing. They used a data augmentation strategy during training and achieved an accuracy and F1-score of 95.29%. This study has significant implications for the development of tools to aid in the accurate diagnosis of breast cancer, potentially improving patient outcomes. Additionally, the three stages of the Multi-View Feature Fusion (MVFF) methodology proposed by Nasir Khan et al. [38] are as follows: In the first step, mammography is binary-classified as abnormal or normal, and in the second, mass and calcification are classified. The final step is to classify the condition as malignant or benign. They utilized the CBIS-DDSM dataset for this investigation. The AUC values for mass and categorization were 93.20% and 0.84% for malignant and benign tumors, respectively. A new computer-aided detection (CAD) system is proposed for classifying benign and malignant mass tumors in breast mammography images using deep learning and segmentation techniques. The CAD system uses two segmentation approaches, one involving manual determination of the region of interest (ROI) and the other using threshold- and region-based techniques. AlexNet, a deep convolutional neural network (DCNN) used for feature extraction and fine-tuning to classify two classes, Ragab et al. present the performance in their paper [39]. VGGNet models that have been adjusted may perform better if classifications of masses and calcifications from mammography are performed using transfer learning. Xi et al. presented research about it [40]. The research by Hepsa et al. [41] showed that breast biopsies based on mammography and ultrasound results have a high rate of being diagnosed as benign (40–60%), which can lead to negative impacts such as unnecessary operations, fear, pain, and cost. To address this, they apply deep learning using convolutional neural networks (CNNs) to classify abnormalities in mammogram images as benign or malignant using two databases: Mini-MIAS and BCDR. While Mini-MIAS has valuable information such as the location and radius of the abnormality, BCDR does not. Initially, accuracy, precision, recall, and F1-score values range from 60 to 72%. To improve results, the authors implement preprocessing methods including cropping, augmentation, and balancing image data. They create a mask to find regions of interest in BCDR images and observe an increase in classification accuracy from 65% to around 85%.

3 Material and Methods

3.1 CNN Architectures

In this research, four popular CNN architectures were evaluated for their suitability for the classification of mammograms at different stages. These architectures have been widely cited in recent studies and were customized for this specific task using the suggested methodology. The evaluation of different CNN architectures is important in identifying the best model for a particular task, as different architectures have different strengths and weaknesses. By comparing the performance of these architectures, we can determine which one is most suitable for the task at hand. The customization of the architectures for mammogram classification involved adjusting various hyperparameters, such as the learning rate and batch size, to optimize performance. This process is crucial for achieving the best possible results and improving the accuracy of mammogram classification, which can have important implications for breast cancer diagnosis and treatment.

VGG19

Simonyan et al. [42] contribute a thorough evaluation of networks of increasing depth using an architecture with 3 × 3 convolution filters. They indicate the model's ability to significantly improve 16–19 weight layers deep in both models.

Very deep convolutional neural network layers, totaling 19 layers, are used in the VGG19 architecture. It consists of multiple fully connected layers that are followed by a string of convolutional and max pooling layers. The network has been trained on a sizable dataset of photos in order to learn to recognize a wide range of objects and scenarios. The network is intended to be used for image classification tasks. The use of tiny convolutional filters with 3 × 3 pixels is one of the distinguishing characteristics of the VGG19 design. Fine-grained characteristics can then be learned by the network from the input photos, which is helpful for tasks like object recognition.

VGG19 has achieved good results on a variety of image classification tasks, which suggests that it may be a reliable and robust model for this type of problem.

ResNet-50 and ResNet152

He et al. [43] introduced ResNet, a CNN architecture that uses skip connections to enable residual function learning. The ResNet models include ResNet-34 A, ResNet-34 B, ResNet-34 C, ResNet-50, ResNet-101, and ResNet-152. An ensemble of these models achieved a 3.57% error rate on the ImageNet test set and won first place in the 2015 ILSVRC classification challenge. In this study, ResNet-50 and ResNet-152 were used for testing.

The ResNet architecture allows for the learning of residual functions through skip connections, which skip over layers that are not essential for the current task. This improves learning efficiency and speed. ResNet is known for its ability to train very deep networks without the vanishing gradient problem, thanks to the use of skip connections. ResNet-50 has an efficient and simple architecture, making it faster to train and easier to implement compared to other models. This feature is especially useful for tasks that involve processing a large amount of data, such as mammogram classification.

EfficientNetB7

Tan et al. [44] introduced a family of models called EfficientNets. EfficientNet is an efficient model that can achieve state-of-the-art accuracy on ImageNet and is commonly used for image classification transfer learning tasks. This architecture was formed by leveraging a multi-objective neural architecture search that optimizes accuracy as well as floating point operations per second (FLOPS).

The main idea behind EfficientNet is to scale up CNNs in a more efficient manner. Conventional CNNs increase the depth and width of the network to improve accuracy, but this also increases the number of parameters and computation required. EfficientNet, instead, proposes to scale the network up in a more balanced way by also increasing the resolution of the input image. This allows the network to improve the accuracy while keeping the computational cost constant.

In order to scale up CNNs in a more organized way, this model suggests a novel model scaling technique that makes use of a straightforward but incredibly powerful compound coefficient. Our method uniformly scales each dimension with a fixed set of scaling coefficients, in contrast to existing approaches that arbitrarily scale network dimensions like width, depth, and resolution.

EfficientNetB7 has been designed to be highly efficient in terms of both accuracy and resource usage. It has achieved state-of-the-art performance on a number of image classification and object detection benchmarks and has been widely used in a variety of applications.

3.2 Datasets

The Curated Breast Imaging Subset of DDSM, also known as CBIS-DDSM [19], is a subset of images that have been selected and curated by radiologists with specialized training from the original DDSM dataset. These images are stored in the standard DICOM format, which is commonly used for storing medical images such as CT and MRI scans.

Another curated mammographic dataset that is widely available is the Mammographic Imaging Analysis Society (MIAS) dataset [20]. Both of these datasets are often used for training and testing machine learning algorithms for tasks such as image classification, object recognition, and segmentation. In addition, the Mini-MIAS dataset is often used as a benchmark for image compression techniques.

Both CBIS-DDSM and Mini-MIAS datasets are available for free and provide numerous images that can be used for medical image analysis research. The CBIS-DDSM dataset contains over 2620 scanned film mammography studies, while the Mini-MIAS dataset contains over 322 images. The location of the tumor has already been indicated for both datasets, making them particularly useful for breast cancer detection and diagnosis research.

3.3 Experiment Environment

The Python programming language was utilized in this study and was run on a workstation equipped with an NVIDIA Geforce RTX 3070ti 8GB graphics card, Ryzen R9 5900x processor, 32 of DDR4 3200Mhz RAM, and an XPG Spectrix 512GB SSD. TensorFlow and Keras were employed in this research, both of which are open-source libraries designed for deep learning applications [45].

TensorFlow was launched by Google to aid in the development of deep learning models, while Keras is a neural network library that was written in Python. To achieve faster and more accurate results, we utilized the GPU, which necessitated the installation of the Deep Neural Network library (cuDNN) [46] and Compute Unified Device Architecture (CUDA).

4 Methodology

This section describes the data preprocessing steps, data selection process, data augmentation strategy, and CNN architectures used for the classification tasks in this research.

Deep convolutional neural network (DCNN) models require a significant amount of data for training to achieve good performance [47]. To ensure accurate diagnosis and better outcomes, it is crucial to use trustworthy datasets. Deep learning has shown that more data can improve results. We searched through many sources to find a dataset that could provide the information we needed, recognizing that the lack of data could create issues. Ultimately, we identified two of the best datasets for mammography images used in state-of-the-art research: Mini-MIAS and CBIS-DDSM. We utilized images from both datasets in this investigation. Mini-MIAS contains 322 images, 133 of which show abnormalities (63 benign and 51 malignant), and the remaining 208 do not. In contrast, CBIS-DDSM comprises 2620 scanned film mammography studies. Once we obtained these datasets, we divided them into four different classification problems: normal and abnormal for Mini-MIAS; normal and abnormal for CBIS-DDSM; masses and calcifications for CBIS-DDSM; and finally, masses, calcifications, and normal for CBIS-DDSM. We tackled each of these classification problems in four stages (Fig. 1).

To achieve good performance, the deep convolutional neural network (DCNN) models required a significant amount of data, which is often not readily available. Therefore, transfer learning and fine-tuning techniques were applied to improve the performance of the models, even with limited data availability.

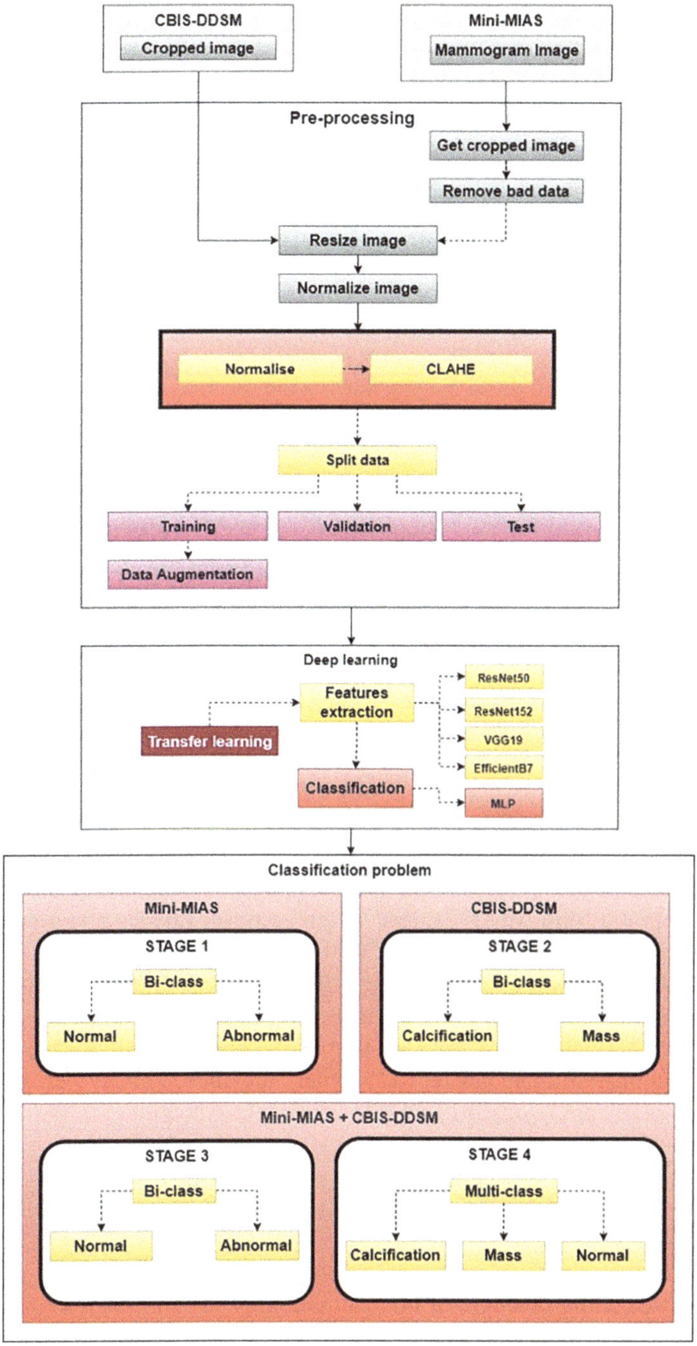

Fig. 1 Proposed methodology for this research

4.1 Stage 1

In the first stage, it involved a comparison of images without any abnormalities to those with tumors, both benign and malignant. Preprocessing of the final images with anomalies received special consideration at this stage because an algorithm was necessary to crop the images using the coordinates provided on the website page. Since it was only necessary to take random crops from images without any anomalies, there were no issues with the normal images.

4.2 Stage 2

In stage 2, a binary classification is applied to distinguish between the two types of mammography anomalies: masses and calcifications. The dataset is divided into masses (benign, benign without callback, and malignant) and calcifications (benign, benign without callback, and malignant), making CBIS-DDSM suitable for this scenario. To solve this classification problem, benign masses are combined with malignant masses and benign calcifications with malignant calcifications. Images labeled as "benign without callback" were not used at this stage.

4.3 Stage 3

In stage 3, something different was attempted. It was not possible to compare healthy mammograms without anomalies to abnormal mammograms in CBIS-DDSM since there is no category for healthy images. To address this, images were obtained from the Mini-MIAS dataset, which includes various abnormalities such as architectural distortion, calcification, well-defined or confined masses, spiculated masses, ill-defined masses, and asymmetry. The Mini-MIAS dataset also enables us to determine the severity of the abnormality (benign or malignant). However, to perform the binary classification, normal images were required. A random cut was made on each mammogram labeled as "normal," and these images were used alongside abnormal images from CBIS-DDSM for the classification.

4.4 Stage 4

In the final stage, the potential of this methodology for a multiclass problem was tested using labeled images of masses, calcifications, and normal images. As CBIS-DDSM does not include healthy or normal images, images from the Mini-MIAS dataset were used instead. As a result, the output changed from a binary problem to a trinary problem due to the difference in data.

This methodology has been utilized, as depicted in Fig. 1. The primary difference is that CBIS-DDSM does not require an algorithm to crop images, unlike Mini-MIAS, which needed it. Cropped images can be obtained directly from the website's archives.

4.5 Preprocessing

As previously stated, we examined four stages that utilized Mini-MIAS and CBIS-DDSM image databases. The first database, Mini-MIAS, contained images in PGM format, which we obtained from the website. However, we stored the images in PNG format because it preserves the quality of edited images. While downloading the images, we encountered an issue: The digital mammograms had a resolution of 1024 × 1024, which could hinder processing. Luckily, the necessary information to identify anomalies was available on the Mini-MIAS website, and we developed an algorithm to crop the images based on their coordinates.

To identify regions in images with anomalies, we used the coordinates to obtain each cropped image. However, several images that contained anomalies lost sharpness and quality when expanded, resulting in their exclusion from the evaluation. Conversely, healthy images were easier to obtain and were randomly selected for patches with a resolution of 112 × 112. Using an intercubic interpolation algorithm, we resized each image (normal and abnormal) to the standard 224 × 244 size after obtaining the ROI.

The CBIS-DDSM dataset, which contains cropped images of mammograms with masses and calcifications, was downloaded. The images had different sizes, so we created an algorithm to automatically resize and extract them from the folders to a standard size of 244 × 244 using an intercubic interpolation algorithm. For each lesion (masses and calcifications), we combined benign and malignant images to test whether the architectures used in this study could distinguish between them. The dataset contained 1555 images of masses and 1331 images of calcifications.

After obtaining the cropped images, we normalized each one to ensure that the pixel values ranged from 0 to 255, as the original mammography image had a resolution of 16 bits and pixel values ranging from 0 to 65,535. We used the CLAHE technique to enhance image contrast, limiting contrast amplification to reduce noise amplification [47–49]. A clip limit of 0.01 was used for CLAHE. Finally, we compared the results of two different approaches, namely, normalize (NO) and CLAHE.

Data augmentation was necessary after preprocessing to increase the amount of data available [50–54]. We used the Albumentation library [55] to develop an algorithm that employed horizontal and vertical flips, with the option of adding rotations. Researchers can increase the number of rotational samples required by adjusting the sample variable (Fig. 2).

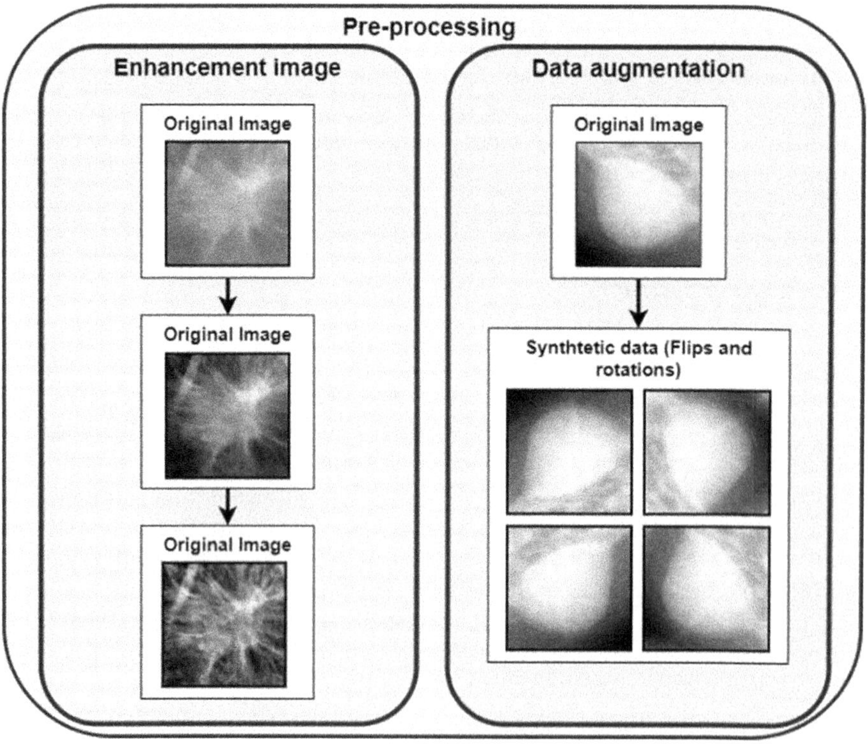

Fig. 2 To preprocess the images, several steps were taken. Firstly, the images were cropped and then normalized by adjusting the pixel values to fit within a certain range. After that, the contrast of the images was enhanced using the CLAHE technique. To increase the number of images available for the training phase, data augmentation was used to create synthetic data

In the conducted research, transfer learning models were utilized, and ImageNet weights were used to achieve better results. By employing pretrained models with learned features, the models were fine-tuned on the mammography image classification task. The ImageNet dataset, which is a large-scale dataset used for pretraining deep neural networks for image classification tasks, was chosen due to its millions of images and thousands of object categories, making it a valuable resource for transfer learning [56, 57]. The models were initialized with weights pretrained on ImageNet, which allowed them to leverage the learned features and adapt to the mammography image classification task more efficiently [21–24].

5 Results and Evaluations

The preprocessed dataset was divided into three parts: 80% for training data, 10% for validation data, and 10% for test data. The ResNet-50 and EfficientNetB7 neural networks were used for classification with the following parameters settings: Adam optimizer, with a learning rate of 0.0001 [58], batch size set to 32, and binary and categorical cross-entropy used as loss functions [59]. The training process was set to run for 20 epochs.

For the classification neural network, we employ these layers in order to improve the outcome (Fig. 3).

5.1 Metrics

It is important to understand the metrics, including the confusion matrix which shows the number of true positive, true negative, false-positive, and false negative predictions made by the model. True positive (TP) refers to the number of instances that were correctly classified as positive, while true negative (TN) refers to the number of instances that were correctly classified as negative. False positive (FP) refers

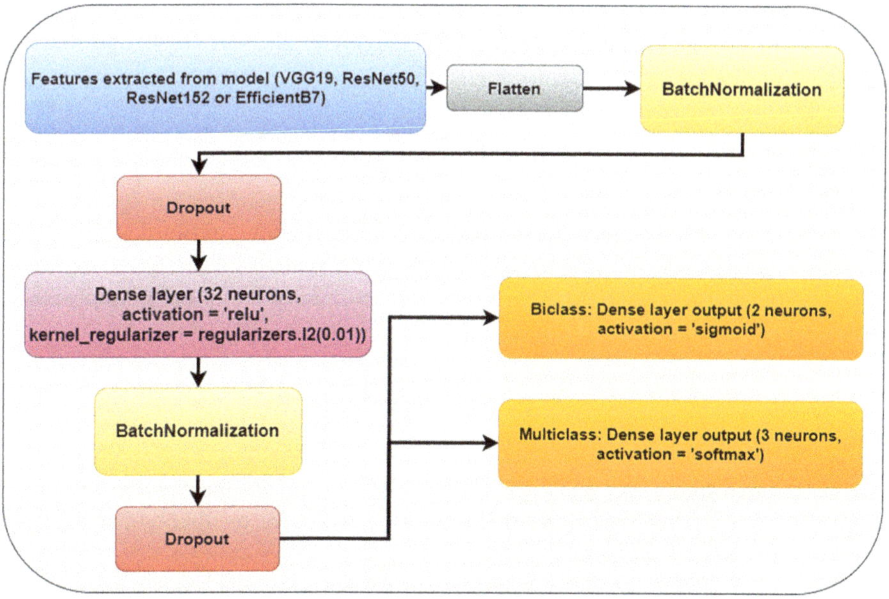

Fig. 3 Using a neural network to classify the features extracted from the models (VGG19, ResNet-50, ResNet152, or EfficientB7). Multiple layers were used, such as the dropout layer [60], batch normalization layer [61], and dense layers. The dense layer that classified the features extracted was assigned a 0.01 for the regularizer l2 value [62]

to the number of instances that were incorrectly classified as positive, and false negative (FN) refers to the number of instances that were incorrectly classified as negative.

Evaluating the performance of a machine learning model requires the use of various metrics, with accuracy being one of the most commonly used. This metric measures the percentage of correct predictions made by the model, making it simple and intuitive. However, accuracy can be misleading in cases where the classes in the dataset are imbalanced, meaning that one class is significantly more prevalent than others (Eq. 1).

$$\text{Accuracy} = \frac{\text{TP} + \text{TN}}{\text{TP} + \text{TN} + \text{FP} + \text{FN}}, \tag{1}$$

Precision is another crucial metric that measures the proportion of true positives, or correct predictions of the positive class, among all positive predictions made by the model. This metric is particularly relevant in tasks where minimizing false positives is essential, such as medical diagnosis or spam detection (Eq. 2).

$$\text{Precision} = \frac{\text{TP}}{\text{TP} + \text{FP}}, \tag{2}$$

Conversely, recall measures the proportion of true positives among all actual positive examples in the dataset. This metric is especially valuable in tasks where detecting as many positive examples as possible is critical, such as fraud detection or cancer screening (Eq. 3).

$$\text{Recall} = \frac{\text{TP}}{\text{FN} + \text{FP}}, \tag{3}$$

To achieve a balance between precision and recall in a model, a useful metric is the F1-score. This metric is the harmonic mean of precision and recall, providing a more comprehensive evaluation of a model's performance by taking both precision and recall into account (Eq. 4).

$$\text{F1} - \text{score} = 2 * \frac{\text{Precision} * \text{Recall}}{\text{Precision} + \text{Recall}}, \tag{4}$$

The AUC (area under the curve) metric is used to evaluate the performance of a binary classifier. It is a measure of the classifier's ability to distinguish between positive and negative classes.

AUC is calculated by plotting the true positive rate (TPR) against the false-positive rate (FPR) at various classification thresholds. The true positive rate is the proportion of positive cases that are correctly identified as positive, while the false-positive rate is the proportion of negative cases that are incorrectly classified as positive.

It is essential to select the right metric for the task at hand because various metrics may highlight various elements of model performance. Model selection, hyperparameter adjustment, and model evaluation are a few of the uses for deep learning metrics. Deep learning metrics in this research is used to evaluate the performance of a deep learning model on a particular task or problem. They provide a way to quantitatively measure how well the model is able to solve the problem and can be used to compare the performance of different models or to track the progress of a model as it is being trained.

Overall, the purpose of using deep learning metrics is to help you understand how well your model is performing and to identify areas where it may be underperforming. This can help you to fine-tune your model and improve its performance.

5.2 *Tables*

For the training set, data augmentation was used on the training set to introduce variations to the model by applying changes to the datasets, which can increase the robustness of machine learning and reduce training costs.

To increase the amount of data available for training, various data augmentation techniques were applied to the Mini-MIAS and CBIS-DDSM datasets. The specific techniques used for each stage are summarized in Table 1 for stage 1 images in the Mini-MIAS dataset, Table 2 for stage 2 images made synthetically using both datasets, Table 3 for stage 3 images with lesions labeled as "masses and calcifications" using only the CBIS-DDSM dataset, and Table 4 for stage 4 images with normal images labeled for lesions using both datasets for the multiclass problem. Before the DCNN could use the dataset, the number of images in each class needed to be balanced to ensure that the model could learn from all classes equally. This was especially important for Tables 2 and 4, which required greater increases in normal images because there were fewer of them compared to the other classes.

Table 5 shows the outcomes of using the normalized dataset, including performance metrics such as accuracy, precision, recall, F1-score, and AUC. Table 6 shows the results of using the dataset with the CLAHE preprocessing technique and

Table 1 Using the technique of data augmentation to the training dataset of stage 1

Without data augmentation			
Class	Training set	Validation set	Test set
Normal	165	22	22
Tumor	82	11	11
With data augmentation			
Class	Training set	Validation set	Test set
Normal	4648	720	22
Tumor	4648	560	11

Table 2 Using the technique of data augmentation to the training dataset of stage 2

Without data augmentation			
Class	Training set	Validation set	Test set
Normal	209	20	22
Abnormal	2308	288	290
With data augmentation			
Class	Training set	Validation set	Test set
Normal	18,464	20	22
Abnormal	18,464	288	290

Table 3 Using the technique of data augmentation to the training dataset of stage 3

Without data augmentation			
Class	Training set	Validation set	Test set
Masses	1244	155	156
Calcifications	1064	133	134
With data augmentation			
Class	Training set	Validation set	Test set
Masses	24,880	155	156
Calcifications	24,880	133	134

different pretrained models, comparing the performance of each stage. The tables include overall test results as well as precision, recall, F1-score, and AUC results for each class. The ResNet-50 and EfficientNetB7 models achieved excellent results, with up to 99% accuracy attained when fine-tuning for stage 3. The VGG19 model also produced good results, especially when images were normalized with a range of 0–255, as shown in Tables 5 and 6.

5.3 *Comparison with Previous Works*

In this section, we compare our model with a few recent studies that were previously mentioned. The comparison is shown in Table 6, which displays the best results of our proposed models, namely, VGG-19, ResNet-50, ResNet-152, and EfficientNet-B7, compared to those of earlier studies with a common focus. It is noteworthy that EfficientNet-B7 performed the best in a binary classification between tumor images and healthy images. Additionally, Table 7 presents further information.

Table 4 Using the technique of data augmentation to the training dataset of stage 4

Without data augmentation			
Class	Training set	Validation set	Test set
Normal	209	22	22
Masses	1244	155	156
Calcifications	1064	133	134
With data augmentation			
Class	Training set	Validation set	Test set
Normal	19,904	20	22
Masses	19,904	155	156
Calcifications	19,904	133	134

6 Conclusions and Future Work

In conclusion, four pretrained deep convolutional neural networks (DCNNs) were compared for their effectiveness in classifying mammogram images using fine-tuning. The study utilized images from two datasets, Mini-MIAS and CBIS-DDSM, and ROIs were obtained by cropping images to help identify objects of interest with more accuracy. Transfer learning and fine-tuning were employed to improve the models' efficiency compared to state-of-the-art.

The four phases of the DCNNs were evaluated to determine their performance in an unrelated task, and ImageNet weights were incorporated to optimize the models. The third stage using improved ResNet-50 and EfficientNetB7 models generated remarkable results compared to state-of-the-art models. EfficientNetB7 is considered a better choice due to its high accuracy and efficiency, outperforming other models on various tasks.

However, the equipment used in this study had limitations, and the authors hope that future studies will use more specialized equipment to obtain quicker results and better comparisons. The authors also plan to compare the pathology of the images among normal, benign, and malignant and explore the possibility of applying the same networks to DCNN with a smaller input. Additionally, they plan to create a multimodal convolutional neural network and apply this classification task using different information from each DCNN to obtain various outcomes and integrate them for a final result that is more accurate and varied.

Acknowledgments César Eduardo Muñoz Chavez wants to thank CONACYT for the support of this research. Hermilo Sánchez-Cruz was partially supported by Universidad Autónoma de Aguascalientes, under grant PII22-5. Humberto Sossa thanks CONACYT and IPN under grants FORDECYT-PRONACES 6005 and SIP 20220226 for the financial support.

Table 5 The values obtained as a result of the training are shown in this table. Each stage's final result is observed along with each DCNN only with normalized images

		DCNN performance when using normalized images				
DCNN	Class	Accuracy	Precision	Recall	F1-score	AUC
Mini-MIAS						
	Normal		0.9545	0.9545	0.9545	
VGG19	Tumor	0.93939	0.9091	0.9091	0.9091	0.9318
	Overall		0.9091	0.9091	0.9091	
	Normal		0.9545	0.9545	0.9545	
ResNet-50	Tumor	0.9393	0.9091	0.9091	0.9091	0.9318
	Overall		0.9091	0.9091	0.9091	
	Normal		0.9130	0.9545	0.9333	
ResNet152	Tumor	0.9091	0.9	0.8181	0.8571	0.8863
	Overall		0.9	0.8181	0.8571	
	Normal		0.9565	1	0.9777	
EfficientB7	Tumor	0.9696	1	0.9091	0.9523	0.9545
	Overall		1	0.9091	0.9523	
CBIS-DDSM						
	Calcification		0.8875	0.8656	0.8787	
VGG19	Masses	0.8896	0.8875	0.9102	0.8987	0.8879
	Overall		0.8897	0.8896	0.8896	
	Calcification		0.9147	0.8805	0.8939	
ResNet-50	Masses	0.9068	0.9006	0.9294	0.9148	0.9050
	Overall		0.9006	0.9294	0.9148	
	Calcification		0.9076	0.8805	0.8939	
ResNet152	Masses	0.9034	0.9	0.9230	0.9113	0.9018
	Overall		0.9	0.9230	0.9113	
	Calcification		0.9166	0.9029	0.9097	
EfficientB7	Masses	0.9172	0.9177	0.9294	0.9235	0.9162
	Overall		0.9177	0.9294	0.9235	
	Normal		0.9829	0.9829	0.9965	
VGG19	Abnormal	0.9807	0.9444	0.7727	0.85	0.8846
	Overall		0.9802	0.9807	0.9798	
	Normal		0.9965	0.9965	0.9965	
ResNet-50	Abnormal	0.9935	0.9545	0.9545	0.9545	0.9755
	Overall		0.9545	0.9545	0.9545	
	Normal		0.9931	1	0.9965	
ResNet152	Abnormal	0.9935	0.9989	0.9090	0.9523	0.9545
	Overall		0.9090	0.9523	0.9931	
	Normal		0.9931	1	0.9965	
EfficientB7	Abnormal	0.9935	0.9989	0.9090	0.9523	
	Overall		0.9989	0.9090	0.9523	
	Normal		0.8636	0.8636	0.9266	0.9266

(continued)

Table 5 (continued)

		DCNN performance when using normalized images				
VGG19	Calcification	0.9253	0.8731	0.8731	0.8731	0.8888
	Masses		0.8974	0.8974	0.8974	0.8974
	Overall		0.8846	0.8846	0.884	0.9042
	Normal		0.9987	0.9545	0.9767	0.9772
ResNet-50	Calcification	0.9006	0.9055	0.8582	0.8812	0.8953
	Masses		0.8841	0.9294	0.9062	0.9038
	Overall		0.9014	0.9006	0.9004	0.9255
	Normal		0.9988	0.8636	0.9268	0.9318
ResNet152	Calcification	0.9006	0.8776	0.9104	0.8937	0.9074
	Masses		0.9090	0.8974	0.9032	0.9038
	Overall		0.9020	0.9006	0.9008	0.9143
	Normal		1	0.7722	0.8717	0.8863
EfficientB7	Calcification	0.9102	0.875	0.9402	0.9064	0.9195
	Masses		0.9337	0.9038	0.9185	0.9198
	Overall		0.9132	0.9102	0.91007	0.9086

Table 6 The values obtained as a result of the training are shown in this table. Each stage's final result is observed along with each DCNN using the CLAHE technique

		DCNN Performance using CLAHE				
DCNN	Class	Accuracy	Precision	Recall	F1-score	AUC
Mini-MIAS						
	Normal		0.9545	1	0.9767	
VGG19	Tumor	0.9696	1	0.9166	0.9565	0.9583
	Overall		0.9166	0.9565	0.9545	
	Normal		0.8333	0.8695	0.9091	
ResNet-50	Tumor	0.9091	0.9091	0.9523	0.9302	0.8928
	Overall		0.9091	0.8333	0.8695	
	Normal		0.8333	0.9091	0.9130	
ResNet152	Tumor	0.9393	0.913	1	0.9545	0.9166
	Overall		1	0.8333	0.9091	
	Normal		0.9166	0.9166	0.9523	
EfficientB7	Tumor	0.9393	0.9523	0.9523	0.9523	0.9345
	Overall		0.9166	0.9166	0.9166	
CBIS DDSM						
	Calcification		0.8617	0.7910	0.8249	
VGG19	Masses	0.8448	0.8323	0.8910	0.8606	0.8410
	Overall		0.8459	0.8448	0.8441	
	Calcification		0.8759	0.8955	0.8856	
ResNet-50	Masses	0.8931	0.908	0.8910	0.8996	0.8932

(continued)

Table 6 (continued)

		DCNN Performance using CLAHE				
	Overall		0.90844	0.8910	0.8996	
	Calcification		0.9076	0.8805	0.8939	
ResNet152	Masses	0.9034	0.9	0.9230	0.9113	0.9018
	Overall		0.9	0.9230	0.9113	
	Calcification		0.8978	0.9179	0.9077	
EfficientB7	Masses	0.9137	0.9281	0.9102	0.9190	0.9140
	Overall		0.9281	0.9102	0.9190	
	Normal		0.9830	1	v	
VGG19	Abnormal	0.9839	1	0.7727	0.8717	0.8863
	Overall		0.7727	0.8717	0.9830	
	Normal		0.9989	0.9989	1	
ResNet-50	Abnormal	1	0.9989	0.9989	1	1
	Overall		0.9989	1	0.9989	
	Normal		0.9931	1	0.9965	
ResNet152	Abnormal	0.9935	1	0.9090	0.9523	0.9545
	Overall		0.9090	0.9523	0.9931	
	Normal		1	0.9989	1	
EfficientB7	Abnormal	1	0.9989	0.9989	1	1
	Overall		1	0.9989	1	
	Normal		1	0.9090	0.9523	0.95454
VGG19	Calcification	0.8557	0.856	0.7985	0.8262	0.8486
	Masses		0.8383	0.8974	0.8668	0.8621
	Overall		0.8573	0.8557	0.8554	0.8884
	Normal		1	0.9090	0.9523	0.9545
ResNet-50	Calcification	0.9166	0.9029	0.9029	0.9029	0.9149
	Masses		0.9177	0.9294	0.9235	0.9230
	Overall		0.9171	0.9166	0.9167	0.9308
	Normal		0.9090	0.9090	0.9090	0.9510
ResNet152	Calcification	0.8974	0.8931	0.8731	0.8830	0.8972
	Masses		0.8993	0.9166	0.9079	0.9070
	Overall		0.8973	0.8974	0.8973	0.9184
	Normal		1	0.9090	0.9523	0.9545
EfficientB7	Calcification	0.8974	0.9126	0.8582	0.8846	0.8982
	Masses		0.8734	0.9294	0.9006	0.8974
	Overall		0.8992	0.8974	0.8973	0.9167

Table 7 Comparison of this methodology to others found in the literature

Author	Database	Model	Accuracy	AUC
Auccahuasi et al.	MIAS	Custom CNN	94%	None
Alkhaleefah et al.	CBIS-DDSM, MIAS, BCDR,	Multiple fine-tuned models	96.49%	0.994%
Saber et al.	INBreast	VGG16 and VGG19	97.1%	0.988%
Qasim et al.	Mini-MIAS	Custom CNN	99.4%	
Shen et al.	CBIS-DDSM, INBreast	ResNet-50 and VGG16		0.91%
Khamparia et al.	DDSM	Modified VGG Alexnet, VGG16, VGG19, MVGG, MobileNet, and ResNet-50	94.3%	
Nasir Khan et al.	CBIS-DDSM, MIAS	VGGNet, GoogleNet, ResNet	92.29%	0.93%
Ragab et al.	DDSM, CBIS-DDSM	Custom AlexNet with SVM	87.2%	0.94%
Our proposed model	CBIS-DDSM, Mini-MIAS	VGG19, ResNet-50, ResNet151, and EfficientNetB7	99.98%	0.99%

References

1. Villarreal-Garza, C., Aguila, C., Magallanes-Hoyos, M.C., Mohar, A., Bargalló, E., Meneses, A., Cazap, E., Gomez, H., López-Carrillo, L., Chávarri-Guerra, Y., Murillo, R., Barrios, C.: Breast cancer in young women in Latin America: an unmet, growing burden. Oncologia. **18**(12), 1298–1306 (2013). https://doi.org/10.1634/theoncologist.2013-0321
2. Villarreal-Garza, C., Mesa-Chavez, F., Plata de la Mora, A., Miaja-Avila, M., Garcia-Garcia, M., Fonseca, A., de la Rosa-Pacheco, S., Cruz-Ramos, M., García Garza, M.R., Mohar, A., Bargallo-Rocha, E.: Prospective study of fertility preservation in young women with breast cancer in Mexico. J. Natl. Compr. Cancer Netw, 1–8 (2021). https://doi.org/10.6004/jnccn.2020.7692
3. Chávarri-Guerra, Y., Villarreal-Garza, C., Liedke, P.E., Knaul, F., Mohar, A., Finkelstein, D.M., Goss, P.E.: Breast cancer in Mexico: a growing challenge to health and the health system. Lancet. Oncol. **13**(8) (2012). https://doi.org/10.1016/S1470-2045(12)70246-2
4. Houssein, E.H., Emam, M.M., Ali, A.A., Suganthan, P.N.: Deep and machine learning techniques for medical imaging-based breast cancer: a comprehensive review. Expert Syst. Appl. **167** (2021). https://doi.org/10.1016/j.eswa.2020.114161
5. Pisano, E.D., Yaffe, M.J.: Digital mammography. Radiol. **234**(2), 353–362 (2005). https://doi.org/10.1148/radiol.2342030897
6. Wang, J., Zhu, H., Wang, S.H., Zhang, Y.D.: A review of deep learning on medical image analysis. Mobile. Netw. Appl. **26**, 351–380 (2021). https://doi.org/10.1007/s11036-020-01672-7
7. Dong, S., Wang, P., Abbas, K.: A survey on deep learning and its applications. Comput. Sci. Rev. **40** (2021). https://doi.org/10.1016/j.cosrev.2021.100379
8. Zhang, X., Dahu, W.: Application of artificial intelligence algorithms in image processing. J. Vis. Commun. Image Represent. **61**, 42–49 (2019). https://doi.org/10.1016/j.jvcir.2019.03.004
9. Hassabis, D., Kumaran, D., Summerfield, C., Botvinick, M.: Neuroscience-inspired artificial intelligence. Neuron. **95**(2), 245–258 (2017). https://doi.org/10.1016/j.neuron.2017.06.011

10. Bagchi, S., Huong, A.: Signal processing techniques and computer-aided detection systems for diagnosis of breast cancer – a review paper. Ind. J. Sci. Technol. **10**(3) (2017). https://doi.org/10.17485/ijst/2017/v10i3/110640
11. Batchu, S., Liu, F., Amireh, A., Waller, J., Umair, M.: A review of applications of machine learning in mammography and future challenges. Oncologia. **99**(8), 483–490 (2021). https://doi.org/10.1159/000515698
12. Mohanty, A.K., Senapati, M.R., Beberta, S., Lenka, S.K.: Texture-based features for classification of mammograms using decision tree. Neu. Comput. Appl. **23**(3–4), 1011–1017 (2013). https://doi.org/10.1007/s00521-012-1025-z
13. Guo, Y., Liu, Y., Oerlemans, A., Lao, S., Wu, S., Lew, M.S.: Deep learning for visual understanding: a review. Neurocomputing. **187**, 27–48 (2016). https://doi.org/10.1016/j.neucom.2015.09.116
14. Janiesch, C., Zschech, P., Heinrich, K.: Machine learning and deep learning. Electr. Mark. **31**, 685–695 (2021). https://doi.org/10.1007/s12525-021-00475-2
15. LeCun, Y., Bengio, Y., Hinton, G.: Deep learning. Nature. **521**(7553), 436–444 (2015). https://doi.org/10.1038/nature14539
16. Fourcade, A., Khonsari, R.H.: Deep learning in medical image analysis: a third eye for doctors. J. Stomatol. Oral. Maxillofac. Surg. **120**(4), 279–288 (2019). https://doi.org/10.1016/j.jormas.2019.06.002
17. Mamoshina, P., Vieira, A., Putin, E., Zhavoronkov, A.: Applications of deep learning in biomedicine. Mol. Pharm. **13**(5), 1445–1454 (2016). https://doi.org/10.1021/acs.molpharmaceut.5b00982
18. Esteva, A., Robicquet, A., Ramsundar, B., Kuleshov, V., DePristo, M., Chou, K., et al.: A guide to deep learning in healthcare. Nat. Med. **25**(1), 24–29 (2019). https://doi.org/10.1038/s41591-018-0316-z
19. Lee, R.S., Gimenez, F., Hoogi, A., Miyake, K.K., Gorovoy, M., Rubin, D.L.: A curated mammography data set for use in computer-aided detection and diagnosis research. Sci. Data. **4** (2017). https://doi.org/10.1038/sdata.2017.177
20. Suckling, J.P.: The mammographic image analysis society digital mammogram database. Digital. Mammo., 375–386 (1994)
21. Tan, C., Sun, F., Kong, T., Zhang, W., Yang, C., Liu, C.: A survey on deep transfer learning. ICANN. **11141**, 270–279 (2018). https://doi.org/10.48550/arXiv.1808.01974
22. Zhuang, F., Qi, Z., Duan, K., Xi, D., Zhu, Y., Zhu, H., Member, S., Xiong, H., He, Q.: A comprehensive survey on transfer learning. Proc. IEEE. **109**, 43–76 (2019). https://doi.org/10.48550/arXiv.1911.02685
23. Falconi, L.G., Perez, M., Aguilar, W.G., Conci, A.: Transfer learning and fine tuning in breast mammogram abnormalities classification on CBIS-DDSM database. Adv. Sci. Technol. Eng. Syst. **5**(2), 154–165 (2020). https://doi.org/10.25046/aj050220
24. Pan, S.J., Yang, Q.: A survey on transfer learning. IEEE Trans. Knowl. Data Eng. **22**(10), 1345–1359 (2010). https://doi.org/10.1109/TKDE.2009.191
25. Rampun, A., et al.: Breast pectoral muscle segmentation in mammograms using a modified holistically-nested edge detection network. Med. Image Anal. **57**, 1–17 (2019). https://doi.org/10.1016/j.media.2019.06.007
26. Altameem, A., Mahanty, C., Poonia, R.C., Saudagar, A.K.J., Kumar, R.: Breast cancer detection in mammography images using deep convolutional neural networks and fuzzy ensemble modeling techniques. Diagnostics. **12**(8), 1812 (2022). https://doi.org/10.3390/diagnostics12081812
27. Wei, B., Han, Z., He, X., Yin, Y.: Deep Learning Model Based Breast Cancer Histopathological Image Classification, pp. 348–353. Int. Conf. Cloud. Comput. Big. Data. Anal (2017). https://doi.org/10.1109/ICCCBDA.2017.7951937
28. Auccahuasi, W., Delrieux, C., Sernaqué, F., Flores, E., Moggiano, N.: Detection of Microcalcifications in Digital Mammography Images, Using Deep Learning Techniques,

Based on Peruvian Casuistry, pp. 1–4. E-Health. Bioeng. Conf (2019). https://doi.org/10.1109/EHB47216.2019.8969906

29. Alkhaleefah, M., Shang-Chih, M.A., Chang, Y.L., Huang, B., Chittem, P.K., Achhannagari, V.P.: Double-shot transfer learning for breast cancer classification from x-ray images. Appl. Sci. **10**(11), 3999 (2020). https://doi.org/10.3390/app10113999
30. Charan, S., Khan, M.J., Khurshid, K.: Breast Cancer Detection in Mammograms Using Convolutional Neural Network, pp. 1–5. iCoMET (2018). https://doi.org/10.1109/ICOMET.2018.8346384
31. Saber, A., Sakr, M., Abo-Seida, O.M., Keshk, A.: Tumor detection and classification in breast mammography-based on fine-tuned convolutional neural networks. Int. J. Comput. Inf. **9**(1), 74–84 (2022). https://doi.org/10.21608/IJCI.2021.103605.1063
32. Qasim, K.R., Ouda, A.J.: An accurate breast cancer detection system based on deep learning CNN. MLU. **20**(1), 984–990 (2020). https://doi.org/10.37506/mlu.v20i1.499
33. Montaha, S., Azam, S., Rafid, A.K.M.R.H., Ghosh, P., Hasan, M.Z., Jonkman, M., De Boer, F.: BreastNet18: a high accuracy fine-tuned VGG16 model evaluated using ablation study for diagnosing breast cancer from enhanced mammography images. Biology. **10**(12), 1347 (2021). https://doi.org/10.3390/biology10121347
34. Allugunti, V.R.: Breast cancer detection based on thermographic images using machine learning and deep learning algorithms. In.t J. Eng. Comput. Sci. **4**(1), 56–49 (2022). https://doi.org/10.33545/26633582.2022.v4.i1a.68
35. Shen, L., Margolies, L.R., Rothstein, J.H., Fluder, E., McBride, R., Sieh, W.: Deep learning to improve breast cancer detection on screening mammography. Sci. Rep. **9**(1), 12495 (2019). https://doi.org/10.1038/s41598-019-48995-4
36. Khamparia, A., Bharati, S., Podder, P., Gupta, D., Khanna, A., Phung, T.K., Thanh, D.N.H.: Diagnosis of breast cancer based on modern mammography using hybrid transfer learning. Multidimens. Syst. Signal. Process. **32**(2), 747–765 (2021). https://doi.org/10.1007/s11045-020-00756-7
37. Hameed, Z., Zahia, S., Garcia-Zapirain, B., Aguirre, J.J., Vanegas, A.M.: Breast cancer histopathology image classification using an ensemble of deep learning models. Sensors. **20**(16), 4373 (2020). https://doi.org/10.3390/s20164373
38. Khan, H.N., Shahid, A.R., Raza, B., Dar, A.H., Alquhayz, H.: Multi-view feature fusion based four views model for mammogram classification using convolutional neural network. IEEE Access. **7**, 165724–165733 (2019). https://doi.org/10.1109/ACCESS.2019.2953318
39. Ragab, D.A., Sharkas, M., Marshall, S., Ren, J.: Breast cancer detection using deep convolutional neural networks and support vector machines. PeerJ. **7**, 6201 (2019). https://doi.org/10.7717/peerj.6201
40. Xi, P., Shu, C., Goubran, R.: Abnormality Detection in Mammography using Deep Convolutional Neural Networks, pp. 1–6. MeMeA (2018). https://doi.org/10.1109/MeMeA.2018.8438639
41. Hepsa, P.U., Özel, S.A., Yazıcı, A.: Using Deep Learning for Mammography Classification, pp. 418–423. UBMK (2017). https://doi.org/10.1109/UBMK.2017.8093429
42. Simonyan, K., Zisserman, A.: Very Deep Convolutional Networks For-Large-Scale Image Recognition, pp. 1–14. ICLR (2015). https://doi.org/10.48550/arXiv.1409.1556
43. He, K., Zhang, X., Ren, S., Sun, J.: Deep Residual Learning for Image Recognition, pp. 770–778. CVPR (2016). https://doi.org/10.48550/arXiv.1512.03385
44. Tan, M., Le, Q.V.: EfficientNet: Rethinking Model Scaling for Convolutional Neural Networks, pp. 6105–6114. ICML (2019). https://doi.org/10.48550/arXiv.1905.11946
45. Géron, A.: Hands-on Machine Learning with Scikit-Learn, Keras, and TensorFlow Concepts, Tools, and Techniques to Build Intelligent Systems, vol. 2. Shroff Publishers (2019)
46. Chetlur, S., Woolley, C., Vandermersch, P., Cohen, J., Tran, J., Catanzaro, B., Shelhamer, E.: cuDNN: Efficient Primitives for Deep Learning. ArXiv (2014). https://doi.org/10.48550/arXiv.1410.0759

47. Szegedy, C., Liu, W., Jia, Y., Sermanet, P., Reed, S., Anguelov, D., Erhan, D., Vanhoucke, V., Rabinovich, A.: Going Deeper with Convolutions, pp. 1–9. CVPR (2015). https://doi.org/10.1109/CVPR.2015.7298594
48. Zuiderveld, K.: Contrast Limited Adaptive Histogram Equalization, pp. 474–485. Graphics Gems (1994). https://doi.org/10.1016/B978-0-12-336156-1.50061-6
49. Sharma, J., Rai, J.K., Tewari, R.P.: Identification of Pre-processing Technique for Enhancement of Mammogram Images, pp. 115–119. MedCom (2014). https://doi.org/10.1109/MedCom.2014.7005987
50. Iswardani, A., Hidayat, W.: Mammographic image enhancement using digital image processing technique. IJCSIS. **16**(5), 222–226 (2018). https://doi.org/10.48550/arXiv.1806.11496
51. Shorten, C., Khoshgoftaar, T.M.: A survey on image data augmentation for deep learning. J. Big. Data. **6**, 60 (2019). https://doi.org/10.1186/s40537-019-0197-0
52. Oza, P., Sharma, P., Patel, S., Adedoyin, F., Bruno, A.: Image augmentation techniques for mammogram analysis. J. Imaging. **8**(5), 141 (2022). https://doi.org/10.3390/jimaging8050141
53. Oyelade, O.N., Ezugwu, A.E.: A deep learning model using data augmentation for detection of architectural distortion in whole and patches of images. BSPC. **65**, 102366 (2021). https://doi.org/10.1016/j.bspc.2020.102366
54. Wang, J., Perez, L.: The Effectiveness of Data Augmentation in Image Classification Using Deep Learning (2017). https://doi.org/10.48550/arXiv.1712.04621
55. Buslaev, A., Iglovikov, V.I., Khvedchenya, E., Parinov, A., Druzhinin, M., Kalinin, A.A.: Albumentations: fast and flexible image augmentations. Information. **11**(2), 125 (2020). https://doi.org/10.3390/info11020125
56. Russakovsky, O., Deng, J., Su, H., et al.: ImageNet Large Scale Visual Recognition Challenge. Int. J. Comput. Vis. **115**, 211–252 (2015). https://doi.org/10.3390/info11020125
57. Deng, J., Dong, W., Socher, R., Li, L.-J., Li, K., Fei-Fei, L.: ImageNet: A Large-Scale Hierarchical Image Database, pp. 248–255. CVPR (2009). https://doi.org/10.1109/CVPR.2009.5206848
58. Kingma, D.P., Lei Ba, J.: ADAM: a method for stochastic optimization. ICLR. (2015). https://doi.org/10.48550/arXiv.1412.6980
59. Badr, E.A., Joun, C., Nasr, G.E.: Cross Entropy Error Function in Neural Networks: Forecasting Gasoline Demand, pp. 381–384. In FLAIRS-02 Proceedings (2002)
60. Srivastava, N., Hinton, G., Krizhevsky, A., Salakhutdinov, R.: Dropout: a simple way to prevent neural networks from overfitting. JMLR. **15**(1), 1929–1958 (2014)
61. Ioffe, S., Szegedy, C.: Batch normalization: accelerating deep network training by reducing internal covariate shift. ICML. **37**, 448–456 (2015). https://doi.org/10.48550/arXiv.1502.03167
62. Ng, A.Y.: Feature Selection, L1 vs. L2 Regularization, and Rotational Invariance. ICML (2004). https://doi.org/10.1145/1015330.1015435

Challenges and Opportunities of Intercompany Big Data Analytics in Supply Chains

J. Kallisch, Jorge Marx-Gómez, and C. Wunck

1 Introduction

Big data analytics (BDA) and analysis techniques have already found their way into companies and their production lines in the past. In addition to reducing inventories, scrap quantities, and storage costs, these techniques and methods enable manufacturing companies to better plan shifts and machine assignments and improve the forecasting accuracy of expected sales quantities to customers. Examples of the use of big data analytics can therefore be observed today in almost every area, from marketing to sales to accounting. Thus, the number of companies using BDA is also increasing, and the number of available applications and providers of BDA is increasing, while the costs of their integration and deployment are decreasing. This development enables small and medium-sized companies to benefit from BDA without employing their own data analysts.

In the meantime, using these methods and tools between companies is not happening. Numerous studies and surveys show that data exchange between companies is rare. As a rule, only ERP-level information is exchanged between companies in supply chains. There is no exchange of data from production systems or even cross-company analysis of these. The reasons for the lack of such distributed analyses between the, already for a long time, cooperating companies are manifold. This chapter explains these reasons in the following Sect. 2 and the challenges of integrating BDA. Section 4 will display the possible benefits of networking the cooperating companies' data stores.

J. Kallisch (✉) · C. Wunck
University of Applied Science Emden/Leer, Emden, Germany
e-mail: Jonas.kallisch@hs-emden-leer.de

J. Marx-Gómez
Department of Informatics, University of Oldenburg, Oldenburg, Germany

M. Mora et al. (eds.), *Development Methodologies for Big Data Analytics Systems*, Transactions on Computational Science and Computational Intelligence, https://doi.org/10.1007/978-3-031-40956-1_10

After displaying these scenarios and opportunities, we will present examples of concepts and architectures that can close the data gaps between the connected companies.

2 State of the Art of Supply Chain Data Exchange

Supply chains (SCs) have become a regular part of global economic architecture [1]. An SC can be defined as a system of organizations supplying a product or service to a consumer over different production stages. In the case of industrial SCs, most of these organizations have to deal with the management of the SC, the logistics of the distribution of goods within the SC, and all kinds of manufacturing issues. Organizational activities have challenged the problem of managing these SCs, but digitalization has brought further technology options to automatize the business layer communication between the different members [2]. The construct of the relation between companies in SCs is displayed in Fig. 1. As the figure shows, the basic relation between these companies is a supplier/vendor and customer relation, where information and goods are exchanged. Inside these SCs, processes like purchasing or tracking have been automatized in the past to gain value from data exchange, like faster reaction times, establishing just-in-time or even just-in-sequence production.

Besides these examples that apply to analytics and BDA in SCs, there is an invisible gap between the companies, as the intercompany exchange is divided into two layers, the information flow and the flow of goods, as presented in Fig. 1. The SC management is often managed at the business layer with the support of an enterprise resource planning (ERP) system. The decisions made influence the flow of goods by creating orders at the operations planning level. Therefore, the usage of automated business connections in SCs has changed how companies work together – as it should be obvious, the current state of intercompany data exchange and analysis.

This architecture's issue is that the SC management is separated from the data generated by the shop floor infrastructure [4]. There are quite a few issues in this state of the art of intercompany data exchange in SCs.

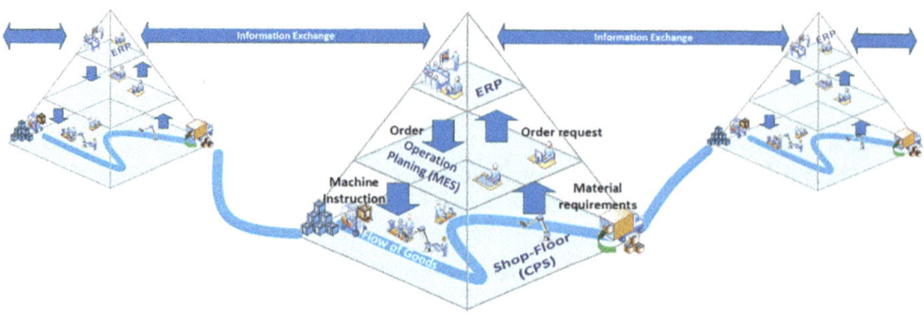

Fig. 1 Layers of supply chain communication [3]

One of the problems arising from the lack of data exchange is the need for multiple entries of exactly identical data [5]. An example is the quality inspection carried out in most companies at goods receipt, production, and goods issue. The determined data of the different companies and stages show a high degree of correspondence; often, even the exact identical parameters are checked. One reason for this is the lack of communication between the suppliers and buyers of the products on the shop floor level. Using data exchange could lead to considerable cost and effort savings and improve the quality processes in the entire group of cooperating companies using data analysis procedures. These assumptions are supported by studies, such as [6], which have shown that data-driven cross-departmental quality analysis procedures within a company deliver better results than control at individual workstations. Similar results are shown in the survey of companies [7, 8].

Since an SC can be defined as a series of interconnected shop floors, it can be assumed that the advantages of networking workstations within companies also arise when networking workstations of different organizations [3].

In addition to this possibility of increasing supply chain efficiency, the use of BDA across companies might be able to make it possible to identify causes of errors that cannot be traced back to causes in a single manufacturing infrastructure [9]. The complexity of a supply chain leads to nonobvious correlations between production conditions of individual workstations, and the analysis of individual segments of a production infrastructure does not lead to an optimization of the overall production but to an optimization for local maximization [9]. These problems and the optimization potentials are not identifiable for the individual companies in the analysis of their own process data.

Why do companies not regularly exchange data with their suppliers and customers? The reasons for this are manifold and concern both technical factors and concerns as well as trust issues.

3 Challenges for Big Data Integration in Supply Chain

Issues and challenges in using BDA inside an SC can be divided into two main categories. Technical challenges are the first of these categories. We will now describe some of the major issues that companies have to face when they try to use BDA on intercompany data [10].

The first technical challenge to the exchange of data between companies is the fact that the companies interacting with each other do not have common data structures. Thus, the databases cannot be directly linked to each other [11]. A complex and company-specific ETL process would be necessary to aggregate the data and therefore perform a central analysis. Although the BDA offers numerous tools that simultaneously enable the analysis of different data structures, the underlying understanding of the data would still have to be provided beforehand. For example, the distinction between different companies' data may lie in the structure and the way it is collected. For example, data may come from direct machine data collection, i.e., the relatively precise sensor information or manual data collection. This can vary from company to company, even for measuring the exact same parameter. These differences in how data is generated are particular challenges, as most analysis algorithms do not support different weightings or reliability of data points.

Another technical obstacle is the use of labeled data. The obstacle is that between the different, usually evolved, data architectures, labels are not necessarily used for identical objects and states [10]. Most algorithms will easily overcome this challenge if the data points are sufficiently correlated. The challenge is whether the different partners' labels are captured from the same point of view and whether there is a similar underlying structure. If the labels of the partners differ too much, it might be necessary to choose one of them. The label of the end user would be particularly suitable for this purpose.

The challenge of unclear labels is complicated because the connections between the different supply chain participants in most supply chains are not always obvious. Therefore, before analyzing the supply chain data, the data that is actually relevant to the process must first be identified and combined.

As the last barrier mentioned, the author concludes that the major reason for the lack of intercompany data exchange, especially in SMEs, lies in the fact that many companies fear the risk of data theft [12]. These companies might risk losing control of their intellectual property. In some cases, this could thread the existence of the affected company because competitors could use the information to improve their own production. Therefore, the fear of many companies of sharing data from their manufacturing systems is justified and understandable. This problem is solvable with trust and legal actions that protect the SC intellectual property. In the interview series, SME executives revealed that some of their methods and intellectual property are the only reason they can compete with other low-price competitors. It should be mentioned that the fear of data theft is not only on third parties that steal the data from outside but also on some SC members, who might try to get advantages by giving access to the data to their suppliers' competitors. This correlates with the results of an interview series on data sharing in SC taken by Södergren and Wallén [1]. Other studies also found that the trust between companies in an SC is an enabler for sharing information with each other, and a lack of this trust affects the collaboration between the partners [13, 14].

The author's hypothesis is that the current deficiencies in data exchange in SCs are the result of a lack of trust between the cooperating companies. This hypothesis could be proven with examples of technical architectures, which support an intercompany data exchange. On the other hand, it could be falsified, if there would be existing examples of companies exchanging data from their production system. If the hypothesis could be proven ride, it could be stated that the application of cross-company analyses and BDA is particularly difficult, as this issue could not be overcome by technical solutions alone. However, there are ample reasons to address the problem, as the benefits of networking are significant. We will describe these benefits in more detail in the next section of this chapter.

4 Benefits of Intercompany Big Data Analytics

To show the potential of an analysis of cross-company data structures, the authors conducted a structured literature review. The results of that review are some existing data exchange relationships and their successes. This chapter will show existing examples of BDA in different processes and tasks within SCs, to demonstrate the impact of connected data analysis between companies. Later we will explain why these examples are not falsifying the hypothesis of the chapter.

4.1 Management and Planning

The use of BDA in manufacturing companies does not start in the manufacturing processes but already in the planning of the manufacturing processes and the design of the production lines. Today, BDA can provide precise suggestions for the layout of manufacturing systems from simulation data and digital twins of the manufacturing systems. Kumar, Sigh, and Lamba showed in their paper how BDA could evaluate multidimensionally optimized layouts and accordingly improve the quality of decisions when planning new manufacturing systems [15]. In the future, BDA could also optimize existing layouts and evaluate the planning of shifts, staffing of workstations, and the selection of new manufacturing equipment. This can have a positive influence on the quality of decisions as well as shorten planning times by reducing manual planning processes. In this way, companies can react more agilely to changing business environments by using BDA.

Furthermore, some publications show that planning in risk consideration and assessment can be improved using BDA [16]. Examples can be found here, for example, in the area of lending, the evaluation of site security in relation to environmental risks or also in the evaluation of delivery times. Therefore, a company's risk management can benefit from using BDA, especially in an increasingly complex SC. Risk management of a manufacturing company also has to consider likely risks, like heavy traffic along the supply routes or temporary power outages.

Identifying these rapidly occurring problems and creating solutions are very personnel-intensive without the use of BDA and require the ability to bring a lot of information together [17]. For this reason, the use of BDA in this area has become quite common. Further risk evaluations, for example, concerning natural disasters or the COVID-19 pandemic, are also conceivable in principle but more realistic in assessing possible consequences. Personal evaluations of company risks, for example, supply chain dependencies and risks, are the core of the solution here. Possible results of such analyses would be, for example, to show the dependencies on certain goods, to identify the risk of possible changes in the law, or to make geographical concentration risks transparent. This results in opportunities to reduce these dependencies and risks.

Another planning perspective of BDA usage is the possibility of monitoring the manufacturing system and the whole SC state [18]. This option allows the SC members' management to integrate a kind of early warning system to detect manufacturing problems like condition monitoring, quality prediction, or fault detection. Studies, like the design of an analysis system by Syafrudin, Alfian, Firiyani, and Rhee, have found that the methods of BDA can improve management decisions in the manufacturing system [19]. They suggest that especially the growing number of IoT devices improve the value of BDA for the manufacturing sector, as they are able to deliver even more data from the companies shop floor. Other studies show similar results, and many commercial solutions are starting to integrate real-time manufacturing monitoring in their management information systems.

4.2 Logistics

It can be described that the data stores in the logistics sector are significantly more interconnected than the rest of the company's processes. Studies have shown that the exchange and analysis of logistics data can improve the organization's performance [20–22]. Examples of the usage of BDA in logistics tasks can be categorized into the categories of distribution and supply.

With regard to distribution, the view of data-driven planning of transports, in particular, has been a source of optimization in the past. Possibilities such as just-in-time delivery of goods or even just-in-sequence delivery, i.e., the delivery of goods matched to their use in the recipient's production, have only become possible through the exchange of data between suppliers, logistics service providers, and the recipients of the goods [23]. In addition, further optimizations are constantly being achieved in this area through exchanging and using data with BDA procedures. The monitoring and control of thousands of goods movements, the recognition of sources of disruption, and the advance planning of machine occupancy are just a few examples of the enormous added value that BDA has provided manufacturing companies in the past through the analysis of logistics processes. A simple example is the planned time window for unloading at a ramp. If a truck misses its allocated time slot, very long waiting times result. Nowadays, another vehicle already on site is brought forward for unloading [24].

In the future, however, this application can be made much more flexible and dynamic by collecting real-time data. For example, if the estimated arrival times of all trucks are known, an algorithm can quickly and efficiently determine the appropriate time windows and the optimal sequence quickly and efficiently. Furthermore, the collection of data also enables improved use of existing resources, such as in terms of trucks or personnel.

In addition to these opportunities to improve in- and outbound logistics processes, BDA's analytical methods have enabled companies to improve the quality of their forecasts for demand, inventory levels, and necessary purchasing decisions. The improvements in these areas of supplying tasks are mainly due to company-generated and public data sources [25]. With BDA, many companies have been enabled to reduce storage capacity and focus more on constructing value-adding supply networks than on control of completing individual transactions. While there are some examples of supply chains sharing sales data and forecasts in the distribution sector, these are based on the limited data exchange described above at the ERP level.

4.3 Production

Since, as described earlier, an SC can be described as a set of interconnected manufacturing assets, it can be assumed that the analytical algorithms used in the shop floor infrastructure of the individual manufacturing companies can also be used throughout the system. Examples of this include detecting and predicting errors within the manufacturing process [26]. This can then be used to derive adjustments, such as different machine assignments. Another example of the benefits of BDA in manufacturing is the ability to deconstruct the manufacturing infrastructure and identify path dependencies, not on a model approach but on the individual path of each component. Studies in manufacturing companies show the enormous potential of this deep analysis of the entire manufacturing system [27].

It should be noted here that the advantage of analyzing the manufacturing infrastructure becomes stronger with the complexity of the system since, at the same time, the number of available data sources allows for a more accurate picture and, at the same time, rescheduling with conventional methods of manufacturing planning becomes no longer feasible.

Finally, one of the main advantages of cross-company data analysis is that it strengthens the resilience of an SC. This can already be seen through the networking of logistics data, as the earlier availability of information accelerates decision-making processes, and the quality of decisions is positively influenced by more and more detailed information. Several studies have described these effects and can be transferred, albeit with limitations, to the analysis of manufacturing infrastructures.

4.4 Discussion

This part of the chapter found that many companies share data in different areas with their customers or suppliers to apply BDA and improve the value of their cooperation. From the authors' point of view, the identified data exchange relationships and areas in which cross-company data analyses are carried out have in common that they only concern support processes. Data from the core processes which create value are only shared to a small extent or not at all. Therefore, the hypothesis that the lack of trust results in a lack of intercompany data analysis is not falsified. There are technical and organizational reasons for this state.

First, it should be mentioned that the step of networking the company management level, or its decision support systems, and the logistics level is easier than that of the shop floor systems. The reason for this lies, on the one hand, technically, in the structure of the data in the systems to be networked and, on the other hand, in the economic importance of the processes to be networked. From a technical point of view, the data structures of the ERP and logistics systems are more similar than those of the manufacturing infrastructure, which has usually grown over time and is heterogeneously structured. On the other hand, the ERP level collects the data in a uniform structure, which can then be easily exchanged via a single interface. In addition, the data linked at the ERP level only contains information from the support processes, which usually does not allow any conclusions to be drawn about the actual core processes and intellectual property. Thus, the exchange of this data does not endanger the companies' business models and competitive position.

Nevertheless, sharing sensitive data that map the core processes also results in some opportunities for the companies. In addition to the examples mentioned for optimizations from individual manufacturing and from the linking of data from logistics, the linking of manufacturing data also offers the possibility for suppliers to create market exit barriers and thus strengthen their position within the supply chain. The barrier here is, on the one hand, that data exchange is not restricted in one direction and, on the other hand, that mutual investment is required in creating the structures for exchange. These investments ensure a mutual interest in maintaining the value-added partnership. The added values resulting from the investments can also further increase the bond, as possibly less expensive competitors cannot offer them.

5 Technical Concepts to Connect Data Stores Securely

After the chapter has evaluated which challenges and opportunities come with using BDA inside of SC, the authors will describe technical solutions to connect the different companies' data to apply BDA to the overall stages generated data. All presented concepts result from a literature review of existing solutions for connecting company data stores.

In addition to the concepts described in this section, the in-house development of a data integration platform for exchanging data in a supply chain is also conceivable. However, there are two arguments against this approach. Firstly, this would have to be implemented for each new supplier integration, which can make it difficult for new companies to join the supply chain. Secondly, the challenges mentioned, in particular the lack of trust on the part of the companies, speak against the feasibility of an in-house solution. In addition, the challenge of creating a cross-company data analysis platform is difficult for most small and medium-sized enterprises.

5.1 Data Spaces

Franklin, Halevy, and Maier have defined data spaces as the next step in the evolution of data integration architectures [28]. The evolutionary step is that data spaces combine storing data with services to merge data from different sources to extract information. The key to this approach is integrating data from different domains and mapping their different data elements. Therefore, data spaces can represent an SC as a number of related data sources which can be connected. This data connection allows the whole system to improve its value. Similar to an SC, the data spaces' value depends on the level of compatibility – mapping and matching – between the different suppliers [29].

One implementation of these data spaces is the International Data Spaces (IDS) [30]. The IDS is a system of data providers interacting on a platform. Each participant can be a provider or user of data and has the right to negotiate about the rights of the data. The platform allows the connection of the data on a contract base. The model of the IDS is displayed in Fig. 2.

The model consists of data providers and consumers holding their data in their physical store locations. However, these physical storages do not interact directly with each other. The broker controls the interaction, which offers the participants two main services. Firstly, the broker lists and categorizes the data delivered by the data providers. This service enables search companies to find the data they need. The companies can search based on labeled data or based on the types of data they already have – e.g., a certain type of application or asset. If the system includes a data provider with the needed data, the broker connects the two – or more – companies. As the data can be traded anonymously, it is possible to buy data from trusted sources without knowing the company supplying the data.

Eventually, a service provider intermediating between the entities is part of the model. This trusted service provider could be a nongovernmental organization, a lawyer, or an IT service provider, which is not involved in the SC. This organization or company could also deliver the service of analyzing the data provided by the SC members. Another opportunity of the IDS concept is that individual providers can offer services to companies within the system. One possible service could therefore

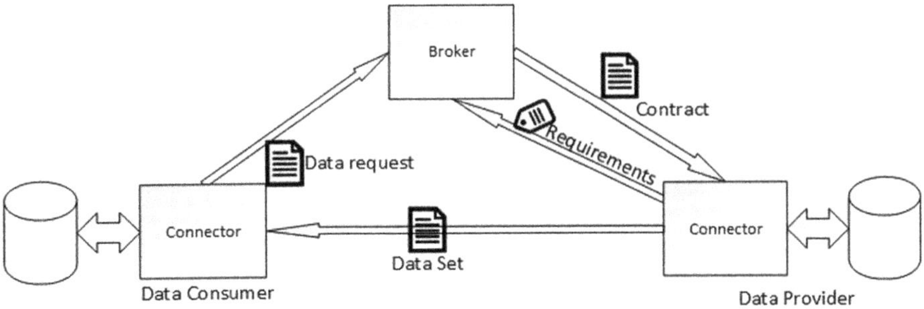

Fig. 2 Role model of data spaces [3]

be a secure and anonymous SC BDA. With a view to the definite needs of small and medium-sized enterprises in the field of data analysis, such a provider could bring them significant improvement. Considering its properties, the IDS is a possible option for integrating BDA into SCs. In particular, standardized provision and data transfer are valuable functions for SMEs and could close the gap described.

5.2 GAIA-X

The European GAIA-X project can also be described as a solution for connecting different kinds of data stores. These data stores might also be IDS infrastructure elements. It contributes an architecture concept that allows sharing of data in a public catalog, where everybody can see the available data but can only access it when granted [31]. Similar to IDS brokers, GAIA-X enables companies to give permission for sharing data on an individual level.

The GAIA-X foundation guarantees data sovereignty as a service. This means that participants have the capability to fully self-determinate their data exchange and sharing. The secure exchange is realized by a function called data contract transaction. This service initiates a handshake between the data provider and the requesting party. The service validates the contract, and if the content is valid and both parties have confirmed the transaction, the data contract service distributes the data contract to both companies. After that, the requesting company can access the requested data and may analyze it. Data distribution is observed by a function called data exchange logging, which enables companies to restrict their data usage to a certain extent or for a specific purpose.

The GAIA-X model allows data sharing in a secure and customizable way but still needs to exchange the data to analyze them inside the SC. A very interesting part of the solution is how the catalog combines data identification and services by self-description.

Summed up, the GAIA-X foundation provides a reliable, effective, and secure solution for sharing data. The author believes that GAIA-X does not fulfill the requirement of protecting data ownership but can work as a platform for analyzing intercompany data with BDA. Especially standardized interfaces for displaying and providing data of the individual companies, provided by the GAIA-X Federated Catalogue, make it possible to develop BDA algorithms within an SC without having to organize the security of the data exchange by themselves. This is possible because the Federated Catalogue requires the companies to describe the delivered data and deliver them in a specific way. This standardization allows the fast integration of new data sources without creating an extensive ETL process. In addition to this benefit, the use of the GAIA-X Federated Catalogue provides the opportunity to participate in future data-driven business models, such as manufacturing as a service.

5.3 *Federated Learning*

Another identified possibility to use BDA in an SC is federated learning (FL). FL is a concept to analyze datasets distributed over different devices connected via a central station [32]. It can be divided into horizontal and vertical FL [33]. The difference between these two types of FL is the selection of elements they share. As Fig. 2 shows, horizontal FL shares features, e.g., temperature measuring or other kinds of data points, but not the samples – a concrete case of measurement – while vertical FL shares samples but not features.

The more common case of FL is horizontal FL, as it is used in mobile devices to improve their ability to analyze their user data without transferring them. A horizontal FL starts with an initial algorithm created on a data sample. In the second step, the model is decomposed into sub-models, matching the data elements of the different storages. The different data stores, e.g., smartphones, then train the model on their own data. In the fourth and last step, the results of the individual models can be transferred to the central application to improve the model. The result is that the different data stores can improve the analytical models of their data without sharing with each other. The limitation of the horizontal FL is that it requires similar data structures on different devices.

In SCs that do not share common data architectures between different companies, vertical FL can be used. As shown in Fig. 3, the data stores in a vertical FL model are not sharing the same features – data structures – but the different data stores share the same samples. For example, two companies in the same city might not collect the same data but collect the data from the same customers. If these companies share a common interest, they could combine their data and use it to improve the quality of their prediction algorithms.

Looking at the application of BDA within an SC, it can be stated that FL supports distributed analysis methods well and is also already seen by some scientists as part of BDA methods. Although the majority of studies on the use of FL in manufacturing companies are limited to the question of how IoT sensors can be integrated into

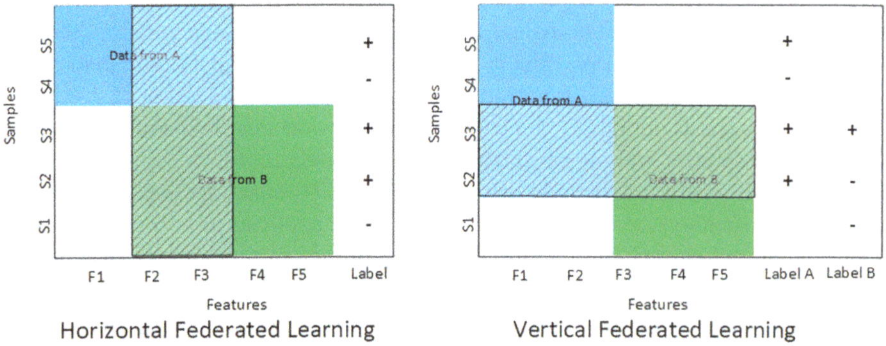

Fig. 3 Horizontal and vertical federated learning [34]

production control, the possibilities for use in SCs are conceivable. In contrast to data spaces and GAIA-X, FL also offers the advantage that companies can benefit from the cross-enterprise analysis without actually having to aggregate data. This could be a key argument for the use of FL in cross-enterprise BDAs and address the reservations of enterprises.

It should be mentioned that the usage of FL comes with a downside that results from the uncertain data sources and their individual viability. The problem is called "unbalanced clients." This means that some participants of the SC can contribute more to the whole system than others. The issue with this is that a federated learning architecture cannot balance this different feature relevance without exchanging datasets [31]. It should be mentioned that Zhang et al. have found that selecting an adaptive number of local training rounds for each party can lead to better models, but this also increases the danger of data leakage.

In summary, FL can provide a basis for using BDA in a cross-company context. Problems of FL have been described, but the added value of ensuring the protection of data sovereignty can be crucial for many companies. Therefore, FL may be able to bridge the gap. It remains unclear how significantly the results of using BDA in FL differ from those on an aggregated data platform.

6 Discussion and Further Work

In this chapter, we asked why there is a lack in data sharing and cross-company analytics between companies that are actually already collaborating in supply chains. To do this, we described the technical and organizational challenges identified in the literature review and interview series. Looking at already established cross-company analytics, the "Benefits of Intercompany Big Data Analytics" section highlighted the advantages of these. These are clearly evident but are generally only applied in support processes. From the authors' point of view, the reason for this is the fear of losing data sovereignty. Our hypothesis is that the main reason for this state of the art is that the companies have a lack of trust to their suppliers and customers. Therefore, they try to avoid sharing data from their core processes. As the networked areas of the companies do not affect any core value creation, the risk to companies is manageable. Since the connected areas had shown that cross-company big data analytics offers significant potential, the last section of this chapter evaluates various options for achieving secure networking of data resources. The concepts shown offer various advantages, such as standardized access, secure interfaces, or, with regard to federated learning, the certainty that data is not exchanged directly. In summary, there is a lot of potential in the application of big data analytics in supply chains. The hurdles are also high, but the existing applications in support processes show that networking is possible. In the future, the concepts and processes described should be evaluated and tested in studies.

References

1. Södergren, F., Cartling Wallén, M.: Creating Value Through Information Sharing: Exploring the Transition Towards a Digital Supply Chain. https://www.diva-portal.org/smash/record.jsf?pid=diva2:1654203 (2022) Accessed 2 Apr 2023
2. Marmolejo-Saucedo, J.A., Hurtado-Hernandez, M., Suarez-Valdes, R.: Digital twins in supply chain management: a brief literature review. In: Vasant, P., Zelinka, I., Weber, G.-W. (eds.) Intelligent Computing and Optimization, pp. 653–661. Springer, Basel (2020)
3. Kallisch, J., Wunck, C.: Options for connecting decentralized data infrastructure to improve Supply-Chain decision making without giving up individual data property, pp. 19–21. Paper Presented at the 53rd Annual Conference of the Decision Sciences Institute, Houston (2022)
4. Radanliev, P., de Roure, D.C., Nurse, J.R.C., Montalvo, R.M., Burnap, P.: Cyber risk at the edge: current and future trends on cyber risk analytics and artificial intelligence in the industrial internet of things and industry 4.0 supply chains. Cybersecurity Award, Springer Open, London (2019)
5. Alcácer, V., Cruz-Machado, V.: Scanning the industry 4.0: a literature review on Technologies for Manufacturing Systems. Eng. Sci. Technol. Int. J. **22**(3), 899–919 (2019)
6. Baihaqi, S., Sohal, A.S.: The impact of information sharing in supply chains on organisational performance: an empirical study, vol. 24, pp. 743–758. PPC (2013). https://doi.org/10.1080/09537287.2012.666865
7. Sohel, A., Schroeder, R.G.: The impact of electronic data interchange on delivery performance, pp. 16–30. In: Production and Operations Management (2001)
8. Stentoft, J., Jensen, K. W., Philipsen, K., Haug, K.: Drivers and Barriers for Industry 4.0 Readiness and Practice: A SME Perspective with Empirical Evidence. Paper presented at 52nd Annual Hawaii International Conference on System Sciences, Maui, 8–11 January 2019, https://scholarspace.manoa.hawaii.edu/handle/10125/59952
9. Kozlenkova, I.V., Hult, G.T.M., Lund, D.J., Mena, J.A., Kekec, P.: The role of marketing channels in supply chain management. J. Retail. **91**(4), 586–609 (2015)
10. Arunachalam, D., Kumar, N., Kawalek, J.P.: Understanding big data analytics capabilities in supply chain management: unravelling the issues, challenges and implications for practice. Transp. Res. E: Logist. Transp. Rev. **114**, 416–436 (2017)
11. Han, D., Kwon, I.G., Bae, M., Sung, H.: Supply chain integration in developing countries for foreign retailers in Korea: Walmart experience. Comput. Ind. Eng. **43**, 111–121 (2002)
12. Colicchia, C., Creazza, A., Noè, C., Strozzi, F.: Information sharing in supply chains: a review of risks and opportunities using the systematic literature network analysis. Supply Chain Manag. **24**(1), 5–21 (2019)
13. Panahifar, F., Byrne, P.J., Salam, M.A., Heavey, C.: Supply chain collaboration and firm's performance. J. Enterp. Inf. Manag. **31**(3), 358–379 (2018)
14. Chen, Z., Huang, L.: Digital twins for information-sharing in remanufacturing supply chain: a review. Energy J. **220** (2019)
15. Kumar, R.K., Singh, S.P., Lamba, K.: Sustainable robust layout using Big Data approach: A key towards industry 4.0. J. Clean. Prod. **204** (2018)
16. Mikavica, B., Kostic-Ljubisav, A., Radonjic, V.: Big Data: Challenges and Opportunities in Logistics Systems, pp. 21–23. Paper Presented at 2nd Logistics International Conference, Belgrade (May 2015)
17. Mageto, J.: Big data analytics in sustainable supply chain management: a focus on manufacturing supply chain. Sustain. Supply Chain Innov. Oper. Manag. **13**, 22 (2021)
18. Moktadir, M.A., Ali, S.M., Paul, S.K., Shukla, N.: Barriers to big data analytics in manufacturing supply chains: a case study from Bangladesh. CAIE. **128**, 1063–1075 (2019)
19. Syafrudin, M., Alfian, G., Latif Fitriyani, N., Rhee, J.: Performance analysis of IoT-based sensor, big data processing, and machine learning model for real-time monitoring system in automotive manufacturing. J. Sens. **18**, 9 (2018)

20. Fawcett, S.E., Waller, M.A.: Considering supply chain management's professional identity: the beautiful discipline (or, "we Don't cure cancer, but we do make a big difference"). J. Bus. Logist. **34**(3), 183–188 (2013)
21. Groves, W., Collins, J., Gini, M., Ketter, W.: Agent-assisted supply chain management: analysis and lessons learned. Decis. Support. Syst. **57**, 274–284 (2014)
22. Ivan Varela, R., Tjahjono, B.: Big Data Analytics in Supply Chain Management: Trends and Related Research. Paper presented at 6th International Conference on Operations and Supply Chain Management, Bali, 10–13 December 2014
23. Trebilcock, B.: The big picture on big data. Supply Chain Manag. **53–57** (2013)
24. Govindan, K., Cheng, T.C.E., Mishra, N., Shukla, N.: Big data analytics and application for logistics and supply chain management. Transp. Res. E: Logist. Transp. Rev. **114**, 343–349 (2018)
25. McKinsey Global Institute: Big Data: The Next Frontier for Innovation, Competition, and Productivity. Washington (2011)
26. Ji, W., Wang, L.: Big data analytics based fault prediction for shop floor scheduling. J. Manuf. Syst. **43**, 187–194 (2017)
27. Cochran, D.S., Kinard, D., Bi, Z.: Manufacturing system design meets big data analytics for continuous improvement. Procedia CIRP. **50**, 647–652 (2016)
28. Franklin, M., Halevy, A., Maier, D.: From databases to dataspaces. SIGMOD Rec. **34**(4), 27–33 (2005). https://doi.org/10.1145/1107499.1107502
29. Sarma, A.D., Dong, X., Halevy, A.Y.: Data modeling in dataspace support platforms. In: Borgida, A.T., Chaudhri, V.K., Giorgini, P., Yu, E.S. (eds.) Lecture Notes in Computer Science, pp. 122–138. Springer, Berlin (2009)
30. Fraunhofer-Gesellschaft: International Data Spaces. https://www.fraunhofer.de/en/research/lighthouse-projects-fraunhofer-initiatives/international-data-spaces.html. Accessed 4 Mar 2023
31. GAIA-X Foundation: GAIA-X: Technical Architecture. https://www.data-infrastructure.eu/GAIAX/Redaktion/EN/Publications/gaia-x-technical-architecture.html. Accessed 4 Mar 2023
32. Zhang, J., Guo, S., Qu, Z., Zeng, D., Wang, H., Liu, Q., Zomaya, A.Y.: Adaptive vertical federated learning on unbalanced features. IEEE Trans. Parallel Distrib. Syst. **33**(12), 4006–4018 (2022)
33. Yang, Q., Liu, Y., Chen, T., Tong, Y.: Federated machine learning: concept and applications. ACM Trans. Intell. Syst. Technol. **12**(2), 1–19 (2019)
34. Kallisch, J. Wunck, C.: Development of a Prototype for a Process Support and Analysis Platform for Small and Medium-sized Enterprises. Paper presented at 35th International Conference on Computer Applications in Industry and Engineering 89, 17–19 (October 2022)

From Big Data to Big Insights: A Synthesis of Real-World Applications of Big Data Analytics

Mahesh S. Raisinghani, Efosa C. Idemudia, and Fen Wang

1 Introduction

The notion of big data conceptualized by Doug Laney is a dataset with large volume, high speed, and high diversity that requires a new style of processing to facilitate decision-making and exploring knowledge and optimization of techniques [5]. Big data analytics (BDA) can be conceptualized as the analysis of detailed, dynamic, low-cost, massive, and varied datasets to deliver sophisticated solutions [6, 7]. These data can be classified into web-based data, sensor-based data, demographic data, transactional data, and machine-generated data [8]. The primacy of BDA has often been attributed to its ability to convert data-scarce decisions into data-rich decisions and to provide competent simulations for problems in various fields [9].

M. S. Raisinghani (✉)
Department of Business & Economics, Texas Woman's University, Denton, TX, USA
e-mail: mraisinghani@twu.edu

E. C. Idemudia
Department of Information Systems and Supply Chain Management, Howard University, School of Business, Washington, DC, USA
e-mail: Efosa.Idemudia@howard.edu

F. Wang
Department of Information Technology & Administrative Management, Central Washington University, Ellensburg, WA, USA

M. Mora et al. (eds.), *Development Methodologies for Big Data Analytics Systems*, Transactions on Computational Science and Computational Intelligence, https://doi.org/10.1007/978-3-031-40956-1_11

1.1 Characteristics of Big Data

The concept of BDA overarches several data-intensive approaches to the analysis and synthesis of large-scale data [10, 11]. Such large-scale data derived from information exchange among different systems is often termed "big data" [12, 13]. Although it is referred to as "big" data, its importance is associated with its ability to capture small details about the subject being studied [14, 15]. Big data was initially characterized by volume, velocity, and variety known as 3Vs [16], with veracity and value [17, 18] added as additional characteristics at a later stage. Kitchin [9] summarized the characteristics of big data with seven "Vs" as follows:

(a) Volume (size). A large amount of data is a primary characteristic of big data.
(b) Variety (complexity). Big data includes structured, semi-structured, and unstructured
 data in different formats, such as text, image, audio, video, and sensor data, among others.
(c) Velocity (speed/rate of generation of data). Big data handles high rates of data inflow and processes the data in real time.
(d) Veracity (quality). Big data accumulates detailed data that is exhaustive in scope and trustworthy.
(e) Value (knowledge). Big data offers in-depth information about a topic of discussion and the worth of the data being extracted.
(f) Variability (flexibility). Big data provides support for the constantly changing nature of data by offering structured, semi-structured, and unstructured data and extensionality (the addition of new data fields) and scalability (expansion in size).
(g) Valence (connectedness). Big data connects common fields to conjoin different datasets.

In addition to being data-driven, BDA is highly applied and can leverage challenges and opportunities presented by big data and domain-specific analytics needed in many high-impact application areas [5, 19]. Several domains such as marketing and logistics are using big data to make better decisions and gain competitive advantage. One of the real-world examples cited in this chapter includes the telecommunication industry that is using social network/big data analysis to listen to and understand the needs and wants of customers worldwide [20]. The social network/big data analysis enables communication industry and companies to offer communication plans, prices, functions, services, and features that are tailored to personalize an individual customer. To date, the marketing pressures and competitions are forcing telecommunication companies to be more creative and innovative [20]. Thus, using the social network analysis, telecommunication companies can manage churn, exceed customers' expectations, improve cross-sell/profits and technology transfer, manage online/viral campaigns, better identify customers, and gain competitive insights. Other benefits that business analytics offer to the telecommunication industry are retaining customers, decreasing costs, fine-tuning pricing models, improving customer satisfaction, acquiring new customers, and understanding the role of social media in customer loyalty [20].

Another prominent example is the health industry that is using analytics to enhance information reporting and visualization through data exploration and discovery [20]. Data analytics can be used to understand the relationship in data and identify trends and patterns relating to cancer vaccines [2, 3]. In addition, several companies are using data analytics to explore which drugs are the most effective and why? Leading hospitals in the world are using business intelligence and data analytics to show current performance measured against standards [20].

BDA also provide the opportunity to explore large volume of data visually to help users identify the root causes of medical problems and then provide the basis for best available solutions. For example, Tableau's business intelligence application gives users the great opportunity to create daily, monthly, or yearly dashboards that help users to improve day-to-day decision-making drastically and significantly. In addition, visualizing data about patient wait-times helped leading hospital globally to improve the overall waiting time and to improve the availability of drugs, medications, doctors, nurses, and beds. Thus, it helps hospital globally to save money from supply chain and more efficiency in processes. In the medical disciplines where data analytics driven research has been applied successfully, novel and creative research directions have been identified that help advance the clinical and biological studies. Researchers worldwide are using data mining techniques to identify novel patterns and paving the road toward a disease-free society and community. Other technologies such as cloud, IOT, and artificial intelligence can serve as platforms and tools for BDA and are useful to underline the extensibility of BDA.

The rest of the chapter is structured as follows: Next we take a closer look at the application of big data analytics in the healthcare, retail, and telecommunication industries. These three industries were chosen based on the authors' experience and/or research interest to explore the impact of BDA in these three industries. This is followed by a discussion of the implications for theory and practice. The chapter ends with some future research directions and the conclusion section.

2 Application of Big Data Analytics in the Healthcare Industry

In the healthcare industry, decisions made in medical facilities are highly data-driven, and medical facilities are moving toward data-based healthcare, together with its benefits. As a current real-world example, the National Institutes of Health is funding a massive research collaboration project between USF, Cornell, and ten other institutions to collect voice data and develop an artificial intelligence (AI) system that could diagnose people based on their speech. The team will start by collecting the voices of people with conditions in five areas: neurological disorders, voice disorders, mood disorders, respiratory disorders, and pediatric disorders like autism and speech delays. These will be stored in large open source databases. The ultimate goal is an app that could help bridge access to rural or underserved communities, by helping general practitioners refer patients to specialists. In the long term, iPhones or Alexa could detect changes in your voice, such as a cough, and advise you to seek medical attention. To get there, researchers have to start by amassing data, since the AI can only get as good as the database it's learning from. There are a few roadblocks such as the Health Insurance Portability and Accountability Act of 1996 (HIPAA), the law that regulates medical privacy, isn't really clear on whether researchers can share voices. The goal is to create large-scale healthcare databases for precision medicine [21]. Big data plays an important role in the current digital era due to the significant advancement of healthcare technologies [22]. Healthcare data generally incorporate electronic medical records (EMRs) such as patient's medical history, physician notes, clinical reports, biometric data, and other medical data related to health. All these data together result in healthcare big data [23]. Applications of BDA in healthcare are gradually increasing with the growing volume of big data in this context [6, 10]. Among the possible sources of big data in healthcare are heterogeneous and multispectral observations, such as patient demographics [24], treatment history [25], and diagnostic reports [26].

Mehta and Pandit [27] suggest that such data may be structured (e.g., genotype, phenotype, or genomic data) or unstructured (e.g., clinical notes, prescriptions, or medical imaging). Implementing data in healthcare often requires the generation and collection of real-time data [28] of high quality [22]. Decision-makers in healthcare organizations are able to take meaningful action based on valuable insights derived from big data [22, 29].

Healthcare organizations deploy technologies to cope with the changing nature of big data [30–32]. Moreover, big data in healthcare can be employed to connect different fields to comprehensively study a disease [31, 32]. In sum, all of the characteristics of big data mentioned above are observable in the context of healthcare. Figure 1 illustrates the big data analytics (BDA) applications in the healthcare industry.

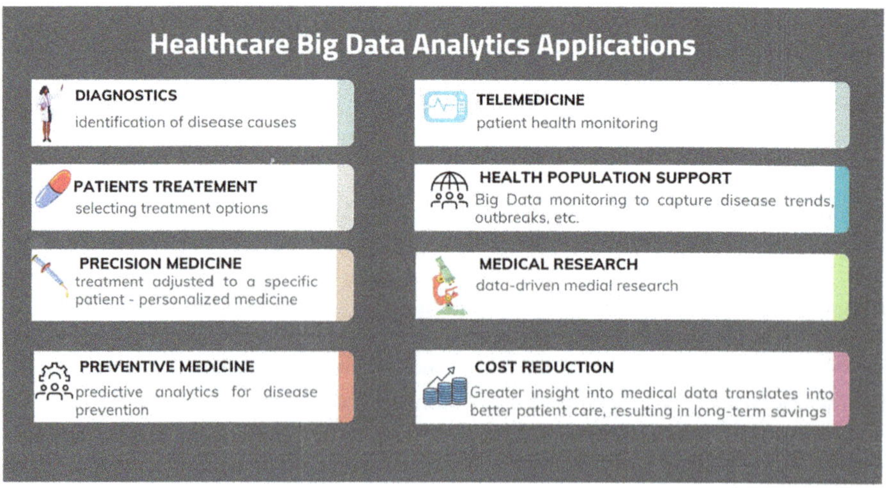

Fig. 1 Big data analytics applications in healthcare [6]

BDA mainly perform three types of analytics, viz., descriptive analytics, predictive analytics, and prescriptive analytics. From a perspective of the healthcare industry, the descriptive analytics analyzes the data gathered in order to interpret, understand, summarize, and visualize significant health-related information and facilitates to explore insights and allows healthcare practitioners to understand what is happening in a given situation [33–35]. In the context of healthcare data, the descriptive analytics provide current and historical data to identify trends and relationships. On the other hand, predictive analytics assist healthcare stakeholders to identify the healthcare services and respond appropriately according to the requirements of patients. It also enables clinicians to be capable of making patient-related decisions on the basis of system predictions [33–35]. The statistical tools for predictive analytics include Hadoop/MapReduce. Prescriptive analytics is a hybrid of descriptive and predictive analytics [36] that incorporates clinical and genomic data to provide appropriate diagnoses and treatments to be provided by healthcare providers in the future [33–35, 37, 38].

3 Application of Big Data Analytics in the Retail Industry

Retail is an advanced and progressive industry sector where the revolution of big data first emerged and has been leveraging analytics strategies and techniques for decades. A typical example would be the quantification of the impact of user-generated content on consumers' purchase expenditures or the development of decision support systems for social media brands and competitive analysis of markets. More recently, businesses have been using vast amounts of data for segmenting customers, identifying emerging marketing trends, improving managerial decision-making, driving more sales, and developing new revenue-making strategies [34, 35]. These data are pulled from the web, such as online searches, posts, and messages, as well as from local stores, such as customer movement in the store and fitting rooms. Unlike traditional sales transaction records collected from various legacy systems of the 1980s or 1990s, the big data that businesses today collect from various sources are less structured and often contain rich customer opinion and behavioral information [19].

Given the sheer volume of data that retailers can collect on their customers, online and/or offline, retailing is by definition a big data industry [39]. Along the process, colossal opportunities and challenges emerged from the ongoing, and rapidly accelerating, big data revolution. Various analytical techniques have been developed for social media customer opinion analysis and customized recommender systems, such as text analysis and sentiment analysis, association rule mining, database segmentation and clustering, anomaly detection, and graph mining, to ensure increased sales and profit margins, improved inventory management, speedier and more personalized promotions, and differentiated and value-added services [40, 41]. Figure 2 summarizes popular big data applications implemented, large volume

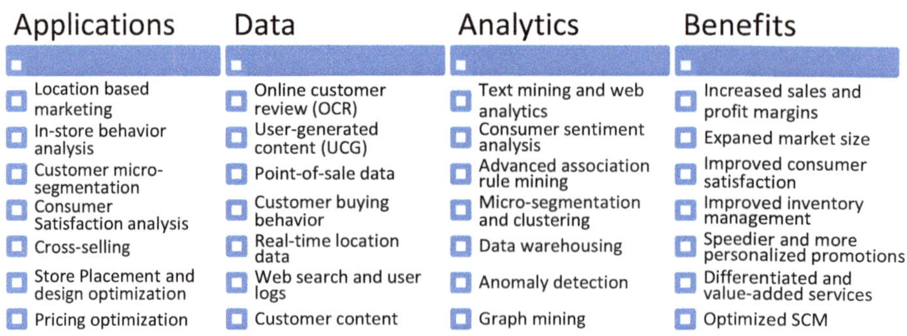

Fig. 2 Summary of big data applications in retail

of data gathered, analytics tools and techniques utilized, and desired benefits achieved within the retail sector.

One pertinent example is the retail giant Walmart, the largest retailer in the world with more than two million employees and annual sales of around $450 billion, which has been a data-driven company since the 1990s. The company uses innovative big data analytics tools and techniques that allowed the retailer to peer into its massive databases of previous transactions to anticipate market demand, identify customer buying patterns, predict future buying trends, and drive business performance [31, 42, 43]. Major Internet firms, such as Amazon, eBay, Google, and Facebook, also continue to lead the development of web analytics, cloud computing, and fog computing. This enhances and complements the cloud by bringing the data processing closer to a cluster of IoT devices. This in turn results in faster analytics and insights; and social media platforms that offer substantial opportunities for researchers and practitioners to "listen" to the voice of the market from a vast number of business constituents [5, 31]. Nevertheless, there is still tremendous potential across the industry for businesses to expand and improve their use of bid data for insights, particularly given the increasing ease with which they can collect information on their consumers, suppliers, and inventories [31, 39]. The recent coronavirus pandemic (COVID-19) has also impacted global economies significantly and unsettled several value networks [44]. Analyzing the implications and effects of such disruptions on businesses is an important application aspect of data analytics, particularly in the area of retail supply chain management (SCM) [45, 46].

One area of great interest of research on COVID-19 is evaluating consumer satisfaction in pandemic times. For instance, by capturing and evaluating consumer data and sentiments available on the open web, Brandtner et al. [45] investigate the impact of COVID-19 on the customer end of retail supply chains (SC) in physical grocery shopping in Austria. Big data from the open web in the form of online customer review (OCR) data or user-generated content (UCG) are examined using text-mining-based analysis. Compared to earlier studies, the study conducted by Brandtner et al. [45] is both novel in terms of the data collection and analysis method applied and in terms of the larger volume of consumer reviews (over 533,000) and additional textual comments (over 153,000) analyzed. Their findings indicate a general and significant decline in consumer satisfaction and a high interconnection of the strictness of measures and decreases in consumer satisfaction. In addition, the authors identify the major service quality factors and product types that affected consumers the most during the pandemic. Such findings can be very helpful to guide demand planning in the course of retail SCM, product management, and store planning and store management. Undoubtedly, the usage of big data offers huge potential for analyzing consumer satisfaction especially in the current times.

4 Application of Big Data Analytics in the Telecommunication Industry

Telecommunication industries are implementing big data analytics daily to deal with vast volume of data [20]. Big data analytics give telecommunication industry competitive edge and improve firm performance [47]. Figure 3 illustrates, generally, the sequential steps that are involved in big data analytics. Big data analytics can be used to forecast outcomes and results and enhance telecommunication industry's processes, strategies, and competitive advantages [20, 47]. Big data analytics, appropriate organizational resources, technological support, and training enhance competitive advantages and provide invaluable insight [47]. Telecommunication industry has access to massive amount of data and large volume of subscribers. Users of smartphones are using their smartphones for daily multiple tasks relating to phone calls, testing, GPS navigation, online courses, and social media.

Telecommunication industries are implementing big data analytics to reduce costs; increase sales/profits, value creation, product innovation, decision-making processes, process development, and overall performance; manage daily risks; and increase supply chain visibility [24]. Al-Alwan et al. [48] investigate the impact of big data on the quality of decision-making. Big data has a positive impact on

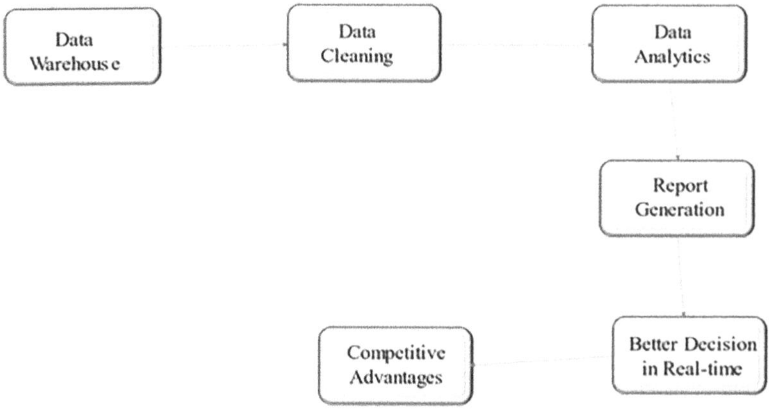

Fig. 3 Steps for big data analytics

decision quality [48]. Volume, varity, and velocity have a positive and significant influence on telecommunication industry decision quality [48, 49]. To date, telecommunication industries are implementing machine learning and data mining techniques for sales prediction analysis. Wisesa et al. [50] discuss that some of the big data analysis projects that the telecommunications industry is implementing for sales prediction are exploration analysis, outlier detection, forecasting/trends, prediction, generalized linear model, decision tree, gradient boosted trees, random forest, and transformation. Ahmad and Mustafa [49] discuss how telecommunication firms in Jordan are using artificial intelligence. Artificial intelligence (AI), big data analytics, and business intelligence positively influence both transforming capability and digital transformation.

Telecommunication companies such as mobile service providers are using and implementing data warehousing and analytics for competitive advantages relating to exceeding customer services and plan pricing [20]. Some of the challenges that telecommunication companies are facing are retaining customers, exceeding customer expectations, decreasing cost, improving customer satisfaction, acquiring new customers, fine-tuning pricing models, and understanding the roles of social media relating to customer loyalty, increased sales, and profits. Figure 4 shows the data warehouse model and how telecommunication companies are using data warehouse to adapt and predict risks. In addition, the data warehouse model has the datasets for (1) technical risks, (2) natural calamities, (3) financial risks, (4) economic/political risks, (5) customer/competitor risks, (6) organizational risks, (7) statutory clearance risks, and (8) other risk. The arrows in Fig. 4 are bidirectional because the data warehouse has to adapt to any changes in all the constructs in Fig. 4. Telecommunication companies are implementing highly targeted data analytics and business intelligence techniques for competitive advantages [20].

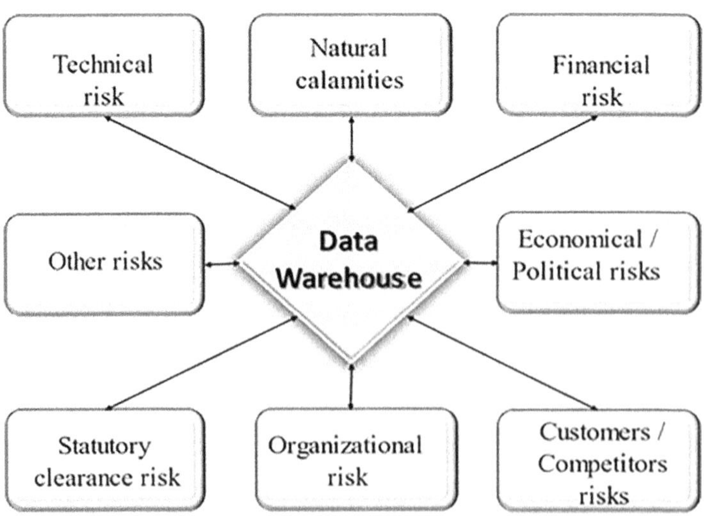

Fig. 4 Data warehouse model

Telecommunication leaders are using dashboards and analytics to provide business-to-consumer services, business-to-business services, and consumer-to-consumer services. The current volatile global economic environment and the global pandemic pose significant challenges to telecommunication companies [2, 3].

To exceed customers' expectations and to stay ahead of the competition, telecommunication companies recognize that top executives must be equipped with the tools and skills for managing daily business decisions by eliminating primitive, traditional, manual, and time-consuming processes [20]. Big data analytics and decision support systems are helping telecommunication companies to provide real-time results and outputs [2, 3]. Telecommunication companies are developing enterprise data warehouse using tetra data to hold big and massive data. These data warehouses are updated in real time. In addition, these data warehouses are used to create and develop top executive dashboard that provides real-time daily knowledge and insights. Some of the software used in creating and developing enterprise data warehouse are Cognos 8, Oracle, SAS, SAS Enterprise Miner, SPSS, R, and so forth. The executive dashboard helps decision-makers to make and take real-time decisions that are useful and user friendly. The dashboard and analytics provide useful daily and intraday snapshots of key performance indicators relating to exceeding customers' expectations and assessing internal operation performance [20]. The dashboard is helping telecommunication's customers to clearly understand and comprehend the data through visual display. The visual display allows for more focused review of the daily data with less cognitive processing effort and time. Telecommunication companies worldwide are using enterprise data warehouse to gain insights and understanding through analytics relating to analyzing customer profiles, behaviors, and sales interactions to exceed customers' expectations [20]. The dashboards and analytics help to enhance cluster analysis and segmentation for value-added services and products. Saudi Telecom is implementing business intelligence and analytics to serve over 160 million customers in the Middle East, Africa, and South Asia [20]. Telecommunication companies are using information visualization for network usage/statistics, billing, service analytics, customer calls, payment, and monitoring and tracking customers' behaviors [20]. Younus et al. [51] argue that artificial intelligence, big data analytics, and business intelligence have a positive and significant effect on digital transformation of UAE telecommunication firms. Big data predictive analytics provide competitive strategies that have a positive and significant effect on strategic alliance performance of telecommunication sector in Pakistan [52]. Big data predictive analytics and competitive strategies have a positive and significant effect on strategic alliance performance of telecommunication companies [52].

Diaz-Aviles et al. [53] discuss strategies telecommunication operators are implementing to improve customer service, reduce churn rate, improve customer experience, and stimulate revenue growth. To date, people all over the world are depending on their mobile phone more and more relating to GPS navigation, WhatsApp, voice and text over data, social media communication, online education, etc. [20]. Telcos are using data analytics, machine learning, and data mining to extract customer insights, to measure customer satisfaction, and to address poor customer experience

[2, 3]. Some of the strategies to measure customer satisfaction are (1) calls and complaints to the telco's care center, (2) providing low ratings, (3) surveys-based input, and (4) churning [20]. Telecommunication industry is using data mining techniques in Hadoop for cloud and big data cluster analysis [21]. The main steps of data analytics are data collection, merge data, clean data, data analysis, and report generation [54]. Data analytics involve machine learning algorithms [54]. Data mining techniques can be used to measure quantitative performance and qualitative indicator [48]. The two main data mining jobs in the telecommunication industry are clustering and forecasting [54]. Examples of some big data analytics are (1) statistical approach, (2) classification approach, (3) clustering approach, and (4) nearest neighbor approach [48]. Data quality, system quality, and service quality have a positive and significant effect on perceived benefits of telecommunication companies [55]. By synthesizing prior research, we develop the big data analytics model as illustrated in Fig. 5 that shows that big data analytics can be used to predict (1) technical risks, (2) natural calamities, (3) financial risks, (4) economic/political risks, (5) customer/competitor risks, (6) organizational risks, (7) statutory clearance risks, and (8) other risk.

Keshavarz et al. [47] discuss how big data analytics pillars can be used to improve targeting, profitability, and firm performance. Telecommunication industries are using big data analytics for (1) management capacity, (2) technology capacity, (3) talent capacity, (4) innovation capacity, and (5) knowledge capacity [24]. Telecommunication industries are using big data analytics for competitive advantages through connectivity, compatibility, maintainability, and modularity. Keshavarz et al. [47] present that in telecommunication industry, big data analytics management capacity includes planning, investment, coordination, control, monitoring, review, and evaluation. Big data analytics talent capability includes technical

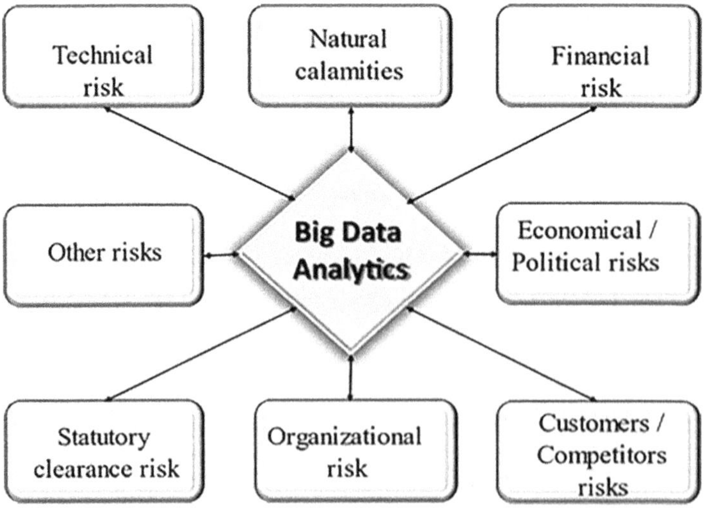

Fig. 5 Big data analytics model

knowledge, technology management, business knowledge, and relational knowledge [47].

Data mining techniques give telecommunication industry the great opportunity to extract useful information from raw data for better decision-making and competitive advantages. Data mining techniques help to improve efficiency, profits, and sales and to reduce costs in the telecommunication industry. Younus et al. [51] discuss that some of the benefits of big data analytics to UAE telecommunication firms are (1) reducing cost, (2) making faster and better decisions, and (3) developing and marketing new products and services. The features of artificial intelligence to UAE firms are (1) eliminating dull and boring tasks, (2) data ingestion, (3) facial recognition and chatbots, (4) imitating human cognition, (5) preventing natural disasters, and (6) futuristics [54]. Some of the benefits of business intelligence are (1) quick real-time analysis, (2) business metrics report, (3) fast review, (4) selling opportunity identification, (5) decision-making process, and (6) easy information sharing [51].

5 Implications for Research and Practice

There are many implications of our study for research and practice. First, research should investigate how big data positively influence healthcare, retail, and telecommunication industries. To date, companies all over the world are using big data analytics to (1) gain new customers, (2) retain customer, and (3) exceed customers' expectations. Big data analytics can be used to track, monitor, and determine which drug and medications are the most useful and effective in treating specific diseases and illnesses. All industries all over the world are using data analytics to make better decisions. One strategy telecommunication, retail, and healthcare companies are implementing to positively influence gaining new customers, retaining customers, and exceeding customers' expectation is to react fast with success relating to customers' service request. Business intelligence and analytics are used to trace the step-by-step processes to understand the points of failure, success, acceleration, workflow measure, performance indicators, and improving/exceeding customers' expectations. In addition, companies are using visualization tools to see and view trends and to correct and address issues and problems before these issues and problems become crisis.

Second, healthcare, retail, and telecommunication industries are using big data analytics to reduce cost. Worldwide big data analytics are reducing cost through automation and by improving real-time decision-making. Our book chapter opens the door for research to investigate how big data analytics directly or indirectly influence the different types of industries that exist. Telecommunication, healthcare, and retail companies are implementing cost reduction strategies to have competitive advantages. Some of the strategies for cost reduction are automation relating to marketing and communication. The goal of data warehousing and analytics relating to telecommunication is to improve productivity, value-added services, quality, and real-time response.

Third, healthcare, telecommunication, and retail companies are using big data analytics for customer acquisition and social networking. Our book chapter opens the doors for research to investigate how big data analytics directly and indirectly influence customer acquisition and social networking. Worldwide, telecommunication, healthcare, and retail companies are implementing business intelligence, data warehousing, and analytics to improve subscriber recharge, expand new customer acquisition, and improve profits and sales. Telecommunication, retail, and healthcare companies are using social network to improve customer service, sales, and marketing. In addition, telecommunication companies are using social media to better understand and influence customers and online visitors' behaviors.

5.1 Future Research Directions

With the prominent value proposition, big data has also created new challenges for businesses and decision-makers across many different industries and jobs. Businesses are collecting vast amounts of data more frequently, yet they are still not grasping the potential of all the data. Businesses in this global environment needs adequate methods and tools to fully address the opportunities brought by big data and achieve its greatest potential. Accordingly, BDA has become their weapon of choice, and future research of BDA applications in the various fields has virtually endless possibilities with an aggressive push toward Web 2.0 and Internet of Things (IoT) or the industrial Internet. Using the healthcare field, for example, BDA could also be used for studies related to the spread of pandemics, the efficacy of the treatment, enabling international comparative analyses, and so forth. Research on the application of BDA in healthcare that encompasses five perspectives, i.e., health awareness among the general public, interactions among stakeholders in the healthcare ecosystem, hospital management practices, treatment of specific medical conditions, and technology in healthcare service delivery, captures the interplay among the process of health data accumulation, derivation of the insights from the data, and application of these insights to healthcare [12, 13]. However, since data can originate from a variety of sources, such as diagnostic reports, hospital registers, and patient history, there is a research gap in data accumulation processes in healthcare.

Future research directions in the BDA domain can include (1) building on a particular finding in existing research; (2) addressing a flaw in the current research; examining (or testing) a theory (framework or model) either (3) for the first time or (4) in a new context, location, and/or culture; (5) reevaluating; and (6) expanding a theory (framework or model). Future research can assess the efficacy of big data analytics in the healthcare domain and explore the potential benefits that technologies such as augmented reality, artificial intelligence, machine learning, and quantum computing offer to healthcare delivery [12, 13]. The authors recommend that scholars study the application of BDA in product and service industries such as automobiles/electric vehicles, banking and financial institutions, social media, and the supply chain management in technology (e.g., semiconductors, sensors, networking hardware/software, tablets, PCs, and so forth) industry. In addition, the efficacy of these new technologies can be assessed and evaluated, as it can guide and inform innovation, best and next practices, and future research directions.

6 Conclusion

In this big data age, an ever-increasing number of companies are attempting to leverage on these data to exploit new opportunities and gain an in-depth understanding of hidden values. Diverse industries such as healthcare, financial, education, sports, retail, and manufacturing, among others, are using big data to make better decisions and gain competitive advantage. From an application perspective, global executives and leaders of organizations are facing these new challenges of effective decision-making, and they desire practical BDA solutions that can help them convert the big data into strategic big insights and impacts.

In this book chapter, the data and analytics characteristics, potential impacts, and illustrative case studies within each prominent domain are discussed. By carefully analyzing the real-world applications through a selective review and synthesis, this chapter provides insights and understanding on how different industries and organizations are and can be using BDA to facilitate more effective and efficient organizational decision-making. Practical suggestions and theoretical implications, including strategic recommendations and ethical considerations, are also discussed to help the BDA researchers and practitioners adopt or develop the appropriate BDA techniques and solutions to derive the intended impact in the new era. Continued research efforts to investigate and establish optimal BDA solutions are anticipated in the future.

References

1. Du, J., Dong, L.U., Wang, T., Yuan, C., Fu, R., Zhang, L., et al.: Psychological symptoms among frontline healthcare workers during COVID-19 outbreak in Wuhan. Gen Hosp Psychiat. **67**, 144 (2020)
2. Idemudia, E.C., Iyamu, T., Ndayizigamiye, P., Shaanika, I.N. (eds.): Using information technology advancements to adapt to global pandemics. IGI Global (2022)
3. Idemudia, E.C.: Handbook of Research on IT Applications for Strategic Competitive Advantage and Decision Making. IGI Global, Hershey (2020)
4. Loebbecke, C., Galliers, R.D.: CAIS Rebuttal for "Five Ethical Issues in the Big Data Analytics Age" by Richardson et al. (2019). CAIS. **49**(1), 22 (2021)
5. Chen, C.P., Zhang, C.Y.: Data-intensive applications, challenges, techniques and technologies: a survey on big data. Inform Sci. **275**, 314–347 (2014)
6. Kamble, S., Gunasekaran, A., Goswami, M., Manda, J.: A systematic perspective on the applications of big data analytics in healthcare management. IJHM. **12**(3), 226–240 (2019)

7. Kaur, P., Sharma, M., Mittal, M.: Big data and machine learning based secure healthcare framework. Procedia Comput Sci. **132**, 1049–1041 (2018)
8. Lee, I.: Big data: dimensions, evolution, impacts, and challenges. Bus Horizons. **60**(3), 293–303 (2017)
9. Kitchin, R.: Big data, new epistemologies and paradigm shifts. Big Data Soc. **1**(1), 12 (2014)
10. Galetsi, P., Katsaliaki, K., Kumar, S.: Big data analytics in health sector: theoretical framework, techniques and prospects. Int. J. Inf. Manag. **50**, 206–216 (2020)
11. Mergel, I., Rethemeyer, R.K., Isett, K.: Big data in public affairs. Public Admin Rev. **76**(6), 928–937 (2016)
12. Khanra, S., Dhir, A., Mäntymäki, M.: Big data analytics and enterprises: a bibliometric synthesis of the literature. Enterp Inform Syst. **14**(6), 737–768 (2020a)
13. Khanra, S., Dhir, A., Islam, N., Mäntymäki, M.: Big data analytics and enterprises: a bibliometric synthesis of the literature. Enterp Inform Syst. **14**(7), 3878–3912 (2020b)
14. George, G., Haas, M.R., Pentland, A.: Big data and management. Acad. Manag. J. **57**(2), 321–326 (2014)
15. McAfee, A., Brynjolfsson, E., Davenport, T.H., Patil, D.J., Barton, D.: Big data: the management revolution. Harvard Bus Rev. **90**(10), 60–68 (2012)
16. Laney, D.: 3D data management: controlling data volume, velocity, and variety. META Group Res. Note. **6** (2001)
17. Nazir, S., Nawaz, M., Adnan, A., Shahzad, S., Asadi, S.: Big data features, applications, and analytics in cardiology-a systematic literature review. IEEE Access. **7**, 143742 (2019)
18. Sarkar, B.K.: Big data for secure healthcare system: a conceptual design. Complex Intell. Syst. **3**(2), 133–115 (2017)
19. Chen, H., Chiang, R.H.L., Storey, V.C.: Business intelligence and analytics: from big data to big impact. MIS Quart. **36**(4), 1165–1188 (2012)
20. Sharda, R., Delen, D., Turban, E.: Business Intelligence: a Managerial Perspective on Analytics. Prentice Hall Press (2013)
21. Acosta, C.M. & Weiner, L.: Artificial intelligence could soon diagnose illness based on the sound of your voice. Retrieved from https://www.npr.org/2022/10/10/1127181418/ai-app-voice-diagnose-disease (2022)
22. Wang, X., Wang, Y., Gao, C., Lin, K., Li, Y.: Automatic diagnosis with efficient medical case searching based on evolving graphs. IEEE Access. **6**, 53307–53318 (n.d.)
23. Raja, R., Mukherjee, I., Sarkar, B.K.: A Systematic Review of Healthcare Big Data, Scientific Programming (2020). https://doi.org/10.1155/2020/5471849
24. Malik, M.M., Abdallah, S., Ala'raj, M.: Data mining and predictive analytics applications for the delivery of healthcare services: a systematic literature review. Ann. Oper. Res. **270**, 287–312 (2018)
25. Ozminkowski, R.J., Wells, T.S., Hawkins, K., Bhattarai, G.R., Martel, C.W., Yeh, C.S.: Big data, little data, and care coordination for medicare beneficiaries with medigap coverage. Big data. **3**(2), 114–125 (2015)
26. Amirian, T., Haghighi, M., Mostaghimi, P.: Pore scale visualization of low salinity water flooding as an enhanced oil recovery method. Energ & Fuel. **31**(12), 13133–13143 (2017)
27. Mehta, N., Pandit, A.: Concurrence of big data analytics and healthcare: a systematic review. Int. J. Med. Inform. **114**, 57–65 (2018)
28. Tang, Z., Kang, B., Li, C., Chen, T., Zhang, Z.: GEPIA2: an enhanced web server for large-scale expression profiling and interactive analysis. Nucleic Acids Res. **47**(W1), W556–W560 (2019)
29. Prasser, F., Spengler, H., Bild, R., Eicher, J., Kuhn, K.A.: Privacy-enhancing ETL-processes for biomedical data. Int. J. Med. Inform. **126**, 72–81 (2019)
30. Harerimana, G., Jang, B., Kim, J.W., Park, H.K.: Health big data analytics: a technology survey. IEEE Access. **6**, 65661–65678 (2018)
31. Zhang, H., Zang, Z., Zhu, H., Irfan Uddin, M., Asim Amin, M.: Big data-assisted social media analytics for business model for business decision making system competitive analysis. Inform. Process. Manag. **59**(1), 102762 (2022)
32. Zhang, R., Simon, G., Yu, F.: Advancing Alzheimer's research: a review of big data promises. Int. J. Med. Inform. **106**, 48–56 (2017)

33. Phillips-Wren, G., Iyer, L.S., Kulkarni, U., Ariyachandra, T.: Business analytics in the context of big data: a roadmap for research. CAIS. **37**(1), 23 (2015)
34. Watson, H.J.: Tutorial: big data analytics: concepts, technologies, and applications. CAIS. **34**(1), 65 (2014)
35. Watson, H.J.: Update tutorial: big data analytics: concepts, technology, and applications. CAIS. **44**(1), 21 (2019)
36. Delen, D.: Real-World Data Mining: Applied Business Analytics and Decision Making. FT Press Analytics, Upper Saddle River (2014)
37. Pang, Z., Yuan, H., Zhang, Y.T., Packirisamy, M.: Guest editorial health engineering driven by the industry 4.0 for aging society. IEEE J. Biomed. Health. **22**(6), 170 (2018)
38. Riabacke, M., Danielson, M., Ekenberg, L.: State-of-the-art prescriptive criteria weight elicitation. Adv. Decision Sci. **2012**, 24 (2012)
39. Dekimpe, M.G.: Retailing and retailing research in the age of big data analytics. Int. J. Res. Mark. **37**(1), 3–14 (2020)
40. Liang, T.P., Liu, Y.H.: Research landscape of business intelligence and big data analytics: a bibliometrics study. Expert Syst. Appl. **111**, 2–10 (2018)
41. Niu, Y., Ying, L., Yang, J., Bao, M., Sivaparthipan, C.B.: Organizational business intelligence and decision making using big data analytics. Inform. Process. Manag. **58**(6), 102725 (2021)
42. Chan, J.O.: Digital transformation digital transformation in the era of big data and cloud computing. Int. J. Intell. Inf. Syst. **9**(3), 16–23 (2020)
43. Mayer-Schönberger, V., Cukier, K.: Big Data: a Revolution that Will Transform how we Live, Work, and Think. Houghton Mifflin Harcourt (2013)
44. Nicola, M.: The socio-economic implications of the coronavirus pandemic (COVID-19): a review. Int. J. Surg. **78**, 185–193 (2020)
45. Brandtner, P., Darbanian, F., Falatouri, T., Udokwu, C.: Impact of COVID-19 on the customer end of retail supply chains: a big data analysis of consumer satisfaction. Sustainability. **13**(3), 1464 (2021)
46. Ivanov, D., Dolgui, A., Das, A., Sokolov, B.: Digital supply chain twins: managing the ripple effect, resilience, and disruption risks by data-driven optimization, simulation, and visibility. In: Handbook of Ripple Effects in the Supply Chain, pp. 309–332. Springer, New York (2019)
47. Keshavarz, H., Mahdzir, A.M., Talebian, H., Jalaliyoon, N., Ohshima, N.: The value of big data analytics pillars in telecommunication industry. Sustainability. **13**(13), 7160 (2021)
48. Al-Alwan, M., Al-Nawafah, S., Al-Shorman, H., Khrisat, F., Alathamneh, F., Al-Hawary, S.: The effect of big data on decision quality: evidence from telecommunication industry. Int. J. Data & Net Sci. **6**(3), 693–702 (2022)
49. Ahmad, H., Mustafa, H.: The impact of artificial intelligence, big data analytics and business intelligence on transforming capability and digital transformation in Jordanian telecommunication firms. Int. J. Data & Net Sci. **6**(3), 727–732 (2022)
50. Wisesa, O., Andriansyah, A., Khalaf, O.I.: Prediction analysis for business to business (B2B) sales of telecommunication services using machine learning techniques. Majlesi J. Electr. Eng. **14**(4), 145–153 (2020)
51. Younus, A.M., Zaidan, M.N., Shakir Mahmood, D.: Effects of artificial intelligence, big data analytics, and business intelligence on digital transformation in UAE telecommunication firms. Acad. J. Dig. Eco. Stability. **18**, 16–26 (2022)
52. Abbas, H., Ze, Y., Ahmad, W.: Competitive Approaches of Strategic Alliance in the Big Data Environment, a Moderating Role of Big Data Predictive Analytics in the Case of Telecommunication Sector of Pakistan. Preprints (2021)
53. Diaz-Aviles, E., Pinelli, F., Lynch, K., Nabi, Z., Gkoufas, Y., Bouillet, E., Salzwedel, J.: Towards Real-time Customer Experience Prediction for Telecommunication Operators, pp. 1063–1072. 2015 IEEE International Conference on Big Data (Big Data) (2015)
54. Singh, S., Liu, Y., Ding, W., Li, Z.: Empirical evaluation of big data analytics using design of experiment: case studies on telecommunication data. Serv. Trans. Big Data. **3**(2), 1–20 (2016)
55. Moumtzidis, I., Kamariotou, M., Kitsios, F.: Digital transformation strategies enabled by internet of things and big data analytics: the use-case of telecommunication companies in Greece. Information. **13**(4), 196 (2022)

Index

M. Mora et al. (eds.), *Development Methodologies for Big Data Analytics Systems*, Transactions on Computational Science and Computational Intelligence, https://doi.org/10.1007/978-3-031-40956-1

GPSR Compliance
The European Union's (EU) General Product Safety Regulation (GPSR) is a set of rules that requires consumer products to be safe and our obligations to ensure this.

If you have any concerns about our products, you can contact us on

ProductSafety@springernature.com

In case Publisher is established outside the EU, the EU authorized representative is:

Springer Nature Customer Service Center GmbH
Europaplatz 3
69115 Heidelberg, Germany

www.ingramcontent.com/pod-product-compliance
Ingram Content Group UK Ltd.
Pitfield, Milton Keynes, MK11 3LW, UK
UKHW022213051026
15624UKWH00001B/136
* 9 7 8 3 0 3 1 4 0 9 5 5 4 *